Lecture Notes in Computer Science 11708

More information about this series at http://www.springer.com/series/7409

Carlos Ordonez · Il-Yeol Song ·
Gabriele Anderst-Kotsis ·
A Min Tjoa · Ismail Khalil (Eds.)

Big Data Analytics
and Knowledge Discovery

21st International Conference, DaWaK 2019
Linz, Austria, August 26–29, 2019
Proceedings

 Springer

Editors
Carlos Ordonez
University of Houston
Houston, TX, USA

Il-Yeol Song
Drexel University
Philadelphia, PA, USA

Gabriele Anderst-Kotsis
Johannes Kepler University of Linz
Linz, Austria

A Min Tjoa ⓘD
Software Competence Center Hagenberg
Hagenberg im Mühlkreis, Austria

Ismail Khalil
Johannes Kepler University of Linz
Linz, Austria

ISSN 0302-9743 ISSN 1611-3349 (electronic)
Lecture Notes in Computer Science
ISBN 978-3-030-27519-8 ISBN 978-3-030-27520-4 (eBook)
https://doi.org/10.1007/978-3-030-27520-4

LNCS Sublibrary: SL3 – Information Systems and Applications, incl. Internet/Web, and HCI

This Springer imprint is published by the registered company Springer Nature Switzerland AG
The registered company address is: Gewerbestrasse 11, 6330 Cham, Switzerland

Preface

Big Data Analytics and Knowledge Discovery remain hot research areas for both academia and the software industry, further fueled by advances in hardware and software. Important research topics associated to these major themes include data lakes (schema-free repositories), database design (ER modeling, prototyping), data integration (especially linking structured and semistructured data sources), big data management (mixing relational tables, text and any files), query languages (SQL and beyond), scalable analytic algorithms, parallel systems (cloud, parallel database systems, Spark, MapReduce, HDFS), theoretical foundations, and practical applications.

With a track record of 21 editions, the International Conference on Big Data Analytics and Knowledge Discovery (DaWaK) has established itself as a high-quality forum for researchers, practitioners, and developers in the field of Big Data Analytics. This year's conference (DaWaK 2019) builds on this tradition, facilitating the interdisciplinary exchange of ideas, theory, techniques, experiences, and future research directions. DaWaK 2019 aims to introduce innovative principles, methods, models, algorithms, industrial products, and experiences to solve challenging problems faced in the development of new generation data management and analytic systems in the Big Data era.

Our call for papers attracted 61 submissions, from which the Program Committee finally selected 22 papers, yielding an acceptance rate of 36%. Each paper was reviewed by an average of four reviewers and in some cases up to five. Accepted papers cover a number of broad research areas on both theoretical and practical aspects. Some trends found in accepted papers include new generations of data warehouses, data lakes, data pre-processing, data mining, cloud computing, query processing, sequences, graph analytics, privacy-preserving data mining, and parallel processing. On the other hand, the program featured interesting case-studies on social networks, Twitter sentiment analysis, understanding ground transportation modes, and E-commerce, among others.

Due to the history and reputation of DaWaK, editors of a well-known journal agreed to receive extended versions of best papers selected from our program. This year, we are pleased to have a special issue in: *Data and Knowledge Engineering* (DKE, Elsevier).

We would like to thank all authors for submitting their papers to DaWaK 2019 and we hope they submit again in the future. On the other hand, we express our gratitude to all the Program Committee members who provided high quality reviews. We appreciate the great efforts of Amin Anjomshoaa for helping extend the ConfDriver system with several innovations to improve paper reviews, to manage a conference-to-journal long-term review process. Finally, we would like to thank the DEXA conference organizers for the support and guidance. For conference attendants, we hope they enjoyed the technical program, informal meetings, and interaction with colleagues from all over the world. For the readers of these proceedings, we hope these papers are interesting and they give you ideas for future research.

August 2019   Carlos Ordonez
                                                               Il-Yeol Song

Organization

General Chair

Thomas Natschläger   Software Competence Center Hagenberg, Austria

Program Committee Chairs

Carlos Ordonez   University of Houston, USA
Il-Yeol Song   Drexel University, USA

Steering Committee

Gabriele Anderst-Kotsis   Johannes Kepler University Linz, Austria
Ismail Khalil   Johannes Kepler University Linz, Austria
A Min Tjoa   TU Wien, Austria

Program Committee and Reviewers

Alberto Abelló   Universitat Politecnica de Catalunya, Spain
Toshiyuki Amagasa   University of Tsukuba, Japan
Elena Baralis   Politecnico di Torino, Italy
Ladjel Bellatreche   ENSMA, France
Sadok. Ben Yahia   Faculty of Sciences of Tunis, Tunisia
Jorge Bernardino   ISEC, Polytechnic Institute of Coimbra, Portugal
Vasudha Bhatnagar   Delhi University, India
Omar Boussaid   University of Lyon/Lyon 2, France
Michael Brenner   University of Hannover, Germany
Stephane Bressan   National University of Singapore, Singapore
Wellington Cabrera   Teradata, USA
Joel Luìs Carbonera   Federal University of Rio Grande do Sul, Brazil
Sharma Chakravarthy   The University of Texas at Arlington, USA
Frans Coenen   The University of Liverpool, UK
Isabelle Comyn-Wattiau   ESSEC Business School Paris, France
Alfredo Cuzzocrea   University of Trieste, Italy
Laurent d'Orazio   University of Rennes 1, France
Soumyava Das   Teradata, USA
Karen Davis   Miami University, USA
Claudia Diamantini   Università Politecnica delle Marche, Italy
Josep Domingo-Ferrer   Universitat Rovira i Virgili, Spain
Dejing Dou   University of Oregon, USA
Markus Endres   University of Augsburg, Germany
Leonidas Fegaras   The University of Texas at Arlington, USA

Philippe Fournier-Viger   Harbin Institute of Technology, China
Filippo Furfaro   Universitat della Calabria, Italy
Pedro Furtado   Universidade de Coimbra, Portugal
Carlos Garcia-Alvarado   Autonomic LLC, USA
Kazuo Goda   The University of Tokyo, Japan
Matteo Golfarelli   DISI - University of Bologna, Italy
Sergio Greco   University of Calabria, Italy
Hyoil Han   Illinois State University, USA
Takahiro Hara   Osaka University, Japan
Frank Hoppner   Ostfalia University of Applied Sciences, Germany
Stéphane Jean   LIAS/ISAE-ENSMA, University of Poitiers, France
Selma Khouri   LCSI/ESI, Algeria, and LIAS/ISAE-ENSMA, France
Min-Soo Kim   DGIST, Republic of Korea
Uday Kiran Rage   University of Tokyo, Japan
Jens Lechtenboerger   Westfalische Wilhelms-Universitat Munster, Germany
Young-Koo Lee   Kyung Hee University, Republic of Korea
Jae-Gil Lee   KAIST, Republic of Korea
Carson Leung   University of Manitoba, Canada
Sebastian Link   The University of Auckland, New Zealand
Sofian Maabout   University of Bordeaux, France
Patrick Marcel   Universite Francois Rabelais Tours, France
Alejandro Mate   University of Alicante, Spain
Jun Miyazaki   Tokyo Institute of Technology, Japan
Anirban Mondal   Ashoka University, India
Yang-Sae Moon   Kangwon National University, Republic of Korea
Yasuhiko Morimoto   Hiroshima University, Japan
Makoto Onizuka   Osaka University, Japan
Alex Poulovassilis   Birkbeck, University of London, UK
Praveen Rao   University of Missouri-Kansas City, USA
Franck Ravat   IRIT, Université Toulouse I Capitole, France
Goce Ristanoski   Officeworks, Australia

Organizers

Institute for
Telecooperation

Contents

Big Data Systems

Graphs and Machine Learning

Databases

Applications

Detecting the Onset of Machine Failure Using Anomaly Detection Methods

Mohammad Riazi[1]([✉]), Osmar Zaiane[1]([✉]), Tomoharu Takeuchi[3],
Anthony Maltais[2], Johannes Günther[1], and Micheal Lipsett[2]

[1] Department of Computing Science, University of Alberta, Edmonton, AB, Canada
{riazi,zaiane}@ualberta.ca
[2] Department of Mechanical Engineering, University of Alberta,
Edmonton, AB, Canada
[3] Mitsubishi Electric Co., Information Technology R&D Center,
Kamakura, Kanagawa, Japan

Abstract. During the lifetime of any machine, components will at some
point break down and fail due to wear and tear. In this paper we propose a data-driven approach to anomaly detection for early detection of
faults for a condition-based maintenance. For the purpose of this study,
a belt-driven single degree of freedom robot arm is designed. The robot
arm is conditioned on the torque required to move the arm forward and
backward, simulating a door opening and closing operation. Typical failures for this system are identified and simulated. Several semi-supervised
algorithms are evaluated and compared in terms of their classification
performance. We furthermore compare the needed time to train and test
each model and their required memory usage. Our results show that the
majority of the tested algorithms can achieve a F1-score of more than
0.9. Successfully detecting failures as they begin to occur promises to
address key issues in maintenance like safety and cost effectiveness.

Keywords: Anomaly detection · Fault detection ·
Predictive maintenance · Machinery diagnostics · Onset of failure ·
Machine learning

1 Introduction

Numerous factors can contribute to the quality of products, some of these factors
are not under the manufacturers' control. One of the most common sources of
quality problems is *faulty equipment* that has not been properly maintained [10].
Hence, monitoring the condition of machines and components such as cooling
fans, bearings, gears, belts and maintaining a desirable working state is crucial.

Funding support is gratefully acknowledged from the Natural Sciences and Engineering
Research Council of Canada, the Alberta Machine Intelligence Institute and Mitsubishi
Electric Corporation, Japan.

© Springer Nature Switzerland AG 2019
C. Ordonez et al. (Eds.): DaWaK 2019, LNCS 11708, pp. 3–12, 2019.
https://doi.org/10.1007/978-3-030-27520-4_1

When a machine or a component fails, corrective maintenance is performed to identify the cause of failure and decide on repair procedures to maintain and re-initiate the machine to its normal working conditions. However, time is required for procuring and repairing. Therefore, a maintenance strategy needs to be considered to minimize the downtime of a service. As the time of failure is not known in advance, time-based maintenance strategies are predominantly used for maintaining the conditions of machines and equipment [11].

There are many advantages to using scheduled maintenance. First, due to regular checks, the operational reliability of the machine is extended and catastrophic failures are somewhat prevented. Second, the time required to maintain the equipment is drastically reduced as the necessary tools and resources are obtained in advanced. But, there still remains substantial disadvantages including, costs pertaining to tools and experts performing regular maintenance eventhough it might not be necessary. Another problem is unforeseen failure that might occur during maintenance intervals. Moreover, during a regular maintenance a part may end up being replaced regardless of its considerable Remaining Useful Life (RUL), which incurs additional costs. Last but not least, the risk of failure may increase during maintenance due to improper diagnosis and or because of intrusive measures taken during repairs.

Prognostics and Health Management (PHM) is a discipline that consists of technologies and methods to evaluate the reliability of a product in its actual life cycle conditions to detect the development of failure and mitigate system risk [3]. Sensor systems are needed for PHM to online monitor equipment. Such online monitoring allows to check the condition of equipment constantly throughout its life cycle to identify any potential failure. This strategy is referred to as Condition-Based Maintenance (CBM). PHM can be implemented in several ways: physics-based models, knowledge-based models and data-driven models.

During recent years, data-driven models also known as data mining methods or machine learning methods for machine health monitoring are becoming more attractive due to the substantial development of sensors and computing platforms. Data-driven methods use historical data to automatically learn a model of system behaviour. Features that encompass the health state of a machine are extracted using a series of procedures. These features are then used by a variety of machine learning methods to decide on the health state of the machine. Based on the availability of data supervised, semi-supervised, unsupervised or reinforcement learning could be used to achieve the desired result.

Unfortunately, there are no freely published datasets available on machine degradation through time. The purpose of this study is two fold: Design and development of a robot-arm platform that can generate data at different operational states; Apply and compare a number of semi-supervised anomaly detection techniques on the data collected for detecting the *onset* of failure.

2   Background

Fault detection is a major task across many applications [1] including quality control, finance, security and medical. In an industrial or manufacturing settings, an equipment or product fault is an abnormal behaviour that may lead to total failure of a system. In general, there is a time gap between the appearance of a fault and its ultimate failure. This time gap can be used by condition-based maintenance to reduce the probability of failure by monitoring the stages of failure and prevent fault from becoming a failure.

As mentioned previously, PHM can be implemented using different approaches: physics-based models based on mathematical models [7], knowledge-based models formalized using if-then rules, and data-driven-based models which mainly rely on anomaly detection from data.

2.1 Anomaly Detection Methods

There exists numerous anomaly detection methods. A rough grouping of some of these methods include, statistical algorithms, clustering-based, nearest neighbour-based, classification-based, spectral-based, space subsampling-based and deep learning methods.

Statistical-based methods assume that the data follows a specific distribution, so the model is created from a specified probability distribution. The simplest approach for detecting anomalies would be flagging data points that deviate from common statistical properties of a distribution. The advantage of these models is that they output a probability as a measure of outlierness.

In **clustering-based anomaly detection**, the assumption is that data points that are similar belong to a similar group. This is determined by the distance to the cluster centroid. The anomaly score is then calculated by setting a threshold for the distance to the cluster centroid; if the data point's distance to the centre of the cluster exceeds the set threshold it is flagged as anomalous.

The **nearest-neighbour-based anomaly detection** generally assumes that normal data samples appear in neighbourhoods that seem to be dense, while anomalies are far from their closest neighbours. Nearest neighbour methods can be generally grouped into distance-based and density-based methods. Both approaches require a similarity or a distance measure in order to make the decision on the degree of abnormality of a data instance. Some examples include, k-NN, Mahalanobis distance, and Local Outlier Factor (LOF).

Classification-based anomaly detection can be divided into one-class (normal labels only) and multi-class (multiple classes) classification depending on the availability of labels. Anomaly detection based on a classifier comprises of two steps [2]: During training, a classifier is learned using available labeled training data. In a second step, the test instances are classified as normal or abnormal using the learned classifier.

Spectral or subspace-based methods try to extract features that best describe the variability of the training data [2]. These methods assume that the normal data can be presented in a lower dimension subspace where normal data is distinguished from abnormal data.

Many outlier detection methods suffer from curse of dimensionality; as the dimensionality of a given dataset increases, distance based approaches fail because the relative distance between any pair of points become relatively the same [5]. To overcome this problem **sub-sampling-based methods** divide the high dimensional space into smaller sub-spaces and monitor the outlier ranking in different sub-spaces. Outliers are points that consistently rank high in the smaller sub-spaces. The T*-Framework [5], and Isolation forest [8] are such algorithms.

3   Experiment Design

A belt-driven, single degree of freedom, re-configurable robot manipulator has been customized for this investigation. The system was developed to reproducibly simulate faults that can occur during regular operation of the arm and to assess the effect of mass, friction and belt tension on motion.

3.1   The Robot Arm Platform

The robot arm is designed to move back and forth, following a predefined trajectory. Even though this seems like a simple machine, there exist many real life examples such as opening and closing gates in elevator doors that require fault monitoring due to constant use and importance of up-time for providing service.

The robot arm consists of many components that help with its operation, data acquisition and reproducibility. The robot is depicted in Fig. 1.

The system is controlled by a programmed teensy 3.2 microcontroller. It is actuated by a brushless stepper motor to drive the belt through a programmed range of motion with a Leadshine DM542 Microstep Drive. This range of motion is measured by two US Digital E2 optical encoders. The driving belt is adjustable in tension. To measure the instantaneous tension in the belt, there are two strain gauges. As the tension has an influence on the systems functioning, its value is needed to reset the belt tension back to its normal operating value. Another influence on the system, the ambient temperature is measured by a thermo couple. A third influence is the weights that are used to change the inertia. For most experiments, 5.5 kg weights are placed on the arm. Lastly, to minimize external vibrations the arm is connected to a 3-ton seismic mass.

3.2   Data Acquisition

The arm is equipped with four sensors: two full-bridge strain gauge configurations and two encoders. For the purpose of this experiment, only the encoder on the motor output shaft is used because we are interested in measuring the torque for moving the arm. The data acquisition system outputs seven values: Timestamp, cycle number, cycle mode (forward or backward), Motor shaft angle (in degrees), Motor output torque (Nm), Belt tension (N), and ambient temperature (degree cel.). A representative set of complete cycles of the torque under normal belt tension set at 150 N is shown in red in Fig. 2.

3.3   Failure Types

After establishing a normal torque profile, a number of typical faults were simulated. Multiple sensors on the arm guarantee reproducibility. These are shown in Figs. 1(a) and (b). In total, five different fault modes were simulated: The belt was loosened to two different levels of either (1) 120N or (2) 100N. For the third experiment the belt was tightened to (3) 168N. Sand was scattered on the surface for the fourth experiment. The fifth experiment simulated an increase of the ambient temperature to 40 °C.

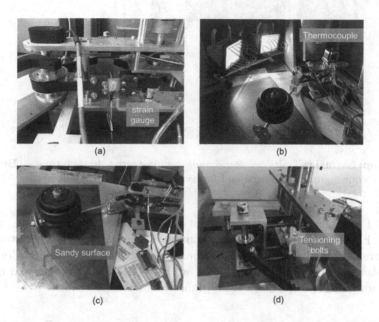

Fig. 1. Sensors installed on the belt and the arm and simulated faults. (a) Strain gauge for measuring the belt tension. (b) Thermocouple for measuring ambient temperature with halogen bulbs to generating heat to see the effect of temperature on the belt tension. (c) Friction using sandy surface. (d) Loose and tight belt tension faults using loosening and tightening of bolts and adjusting the height of idler pulley.

Table 1. Summary of data collected from the robot arm platform

	Temp (°C)	Tension (N.m)	No. samples
Normal	~23.5	~150	7193 (~22 h)
Loose L1	~23.5	~120	182
Loose L2	~23.5	~99–100	180
Tight	~23.5	~168	194
High temperature	~40–42	~166–169	210
Sand (friction)	~23.5	~150	183

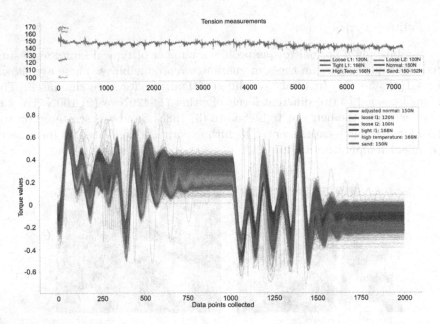

Fig. 2. Top: Normal vs faulty tension signals demonstrated through time — **Bottom:** Overlaid torque signals for normal and faulty data (Color figure online)

Table 1 summarizes the collected data. For a better understanding of the tensions, Fig. 2 shows the trend of tension over time for the normal state and the interval at which each fault falls considering the tension value. Moreover, to show how close and subtle the faults are in relation with the normal profile, overlaid torque signals for normal and fault conditions are shown.

3.4   Data Pre-processing

A majority of outlier detection algorithms use distance (Euclidean) as a measure of similarity and assume normality of input data. Hence feature normalization is required to level the ranges of the features and enforce the same effect in the computation of similarity [12]. For the purpose of this study, torque samples were rescaled so the features have the properties of a standard normal distribution with mean zero and a standard deviation of one.

Torque data is collected through a data acquisition system which is guaranteed to output reliable measurements. The torque is sampled at 200 Hz and each observation should contain 2000 sampled data points per cycle.

3.5   Feature-Based Representation of Torque Waveforms

Feature-based representation of time series have been used across many disciplines, e.g. medical or speech recordings [6]. A very simple example of representing a time series is using its mean and variance characteristics.

As a first attempt to construct features from the torque dataset (2000 data points per sample), we followed the work of [9] and used the first order statistics along other descriptive measures to extract a total of 11 features.

In a second attempt, additional features were extracted using the "Time Series Feature extraction based on scalable hypothesis tests" abbreviated as tsfresh python package [4]. The package contains numerous feature extraction methods and a robust feature selection algorithm. Using the tsfresh library, a total of 714 features were extracted, which was further pruned to 114.

To further prune the extracted features and select the most important and discriminating features, we used the gradient boosting machine from the Light-GBM library and the Pearson correlation to remove low importance and correlated features in both manually extracted feature sets and the tsfresh features.

The final set of features were reduced from 11 to 8, including mean, variance, skewness, kurtosis, mean, variance, skewness, kurtosis, and RMS for the manual features. From the automatic tsfresh features, 76 out of 114 were chosen.

3.6 Feature Learning and Extraction Using Auto-Encoders

We use a 1D segment from raw input data (torque data) as the input to our CNN auto-encoder because the torque signal is correlated only in the time domain, this is followed by 1D filters in the convolutional layers of the model.

, Evidently, a deep learning model with deeper architecture achieves a better performance compared to a shallow one. However, a deeper network results in a more complicated number of hyper-parameters and a variety of network architectures. This in fact makes the building of an appropriate and efficient model quite difficult [13]. For the purpose of this study we have particularly followed the work of Zhang et al. [14] to determine the 1D-CNNAE architecture.

The proposed 1D-CNNAE consists of a CNN encoder network and a mirrored decoder network with 5 layers of alternating convolution and pooling layers. The first convolutional layer consists of 64 filters (dimensionality of output space) and kernel size of 1, followed by a max-pooling layer. The subsequent layers are four alternating small (10×1) convolutional layers followed by max-pooling.

The learned features from the bottleneck layer of the CNNAE network is extracted and saved to use with the anomaly detection methods to further investigate the performance of the automatically extracted features.

4   Training Procedure

The normal dataset was split into train, validation and test sets with a ratio of 80%, 10%, and 10%, respectively. To prevent overfitting, the validation set was used for early stopping. The normal *test* dataset was set aside to be merged with the abnormal test datasets to obtain a more realistic test set. The algorithms were trained using the normal only dataset with the raw torque values (2000 features), the manually extracted features, tsfresh features and the CNNAE features. A contamination ratio which determines the amount of contamination of the data set, i.e., the proportion of outliers in the data set was predetermined (once assumed 90% of data are normal and once 95%) and used when fitting each model to define the threshold on the decision function. Once the detectors were fit, the outlier scores of the training set is determined using the corresponding decision function of a detector. The nearest rank method is then used to find the ordinal rank of the anomaly scores outputted by the detectors.

5   Results and Concluding Remarks

The trained models were used on the mixed normal and anomalous dataset to determine the anomaly scores. The anomaly scores were then categorized into anomalous and normal according to the set threshold (assuming 90% & 95% normal data). An anomaly score beyond the threshold is then flagged with label '1' as being anomalous and '0' if smaller than threshold.

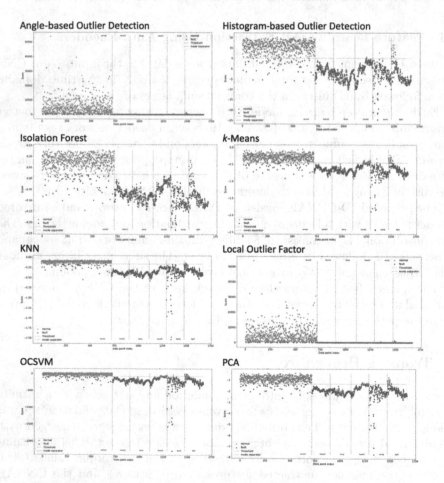

Fig. 3. Trained models applied to mixed data (normal + fault modes) with threshold set at 90%. Grey points mark normal, orange points mark abnormal data.

	k-NN	T*k-NN	LOF	T*LOF	ABOD	T*ABOD	iForest	T*iForest	HBOS	T*HBOS	OCSVM	T*OCSVM	PCA	T*PCA	CNNAE
Precision	0.854	0.852	0.847	0.832	0.853	0.802	0.855	0.854	0.857	0.858	0.856	0.852	0.861	0.852	0.846
Recall	1	1	1	1	1	0.796	1	1	1	1	1	1	1	0.988	0.941
F1-Score (Mean ± STD)	92.17 ±0.005	92 ±0.0071	91.71 ±0.01	90.8 ±0.0045	92.07 ±0.009	79.8 ±0.013	92.22 ±0.006	92.20 ±0.0071	92.35 ±0.006	92.40 ±0.0055	92.26 ±0.006	92.2 ±0.0084	92.54 ±0.006	91.6 ±0.0055	88.98 ±0.0035
g-means (Mean ± STD)	95.28 ±0.0039	95.2 ±0.0045	94.92 ±0.007	94.4 ±0.005	95.32 ±0.006	84.4 ±0.0055	95.65 ±0.0043	95.20 ±0.0045	95.14 ±0.0041	95.40 ±0.0055	95.54 ±0.004	95.2 ±0.0084	95.51 ±0.004	94.8 ±0.0045	90.1 ±0.005
Train time (sec)	0.19	13.7	0.25	2.55	33.43	1876	0.55	18.67	0.15	0.30	1.77	62.32	0.02	0.083	15.37
Test time (sec)	0.11		0.05		6.57		0.06		0.01		0.09		0		0.09
Mem. Footprint (KB)	705	9.66MB	2300	53.2MB	612	7.73MB	924	27.66MB	106	2.94MB	450	6.49MB	107	2.94MB	2.0MB

(a)

	k-NN	T*k-NN	LOF	T*LOF	ABOD	T*ABOD	iForest	T*iForest	HBOS	T*HBOS	OCSVM	T*OCSVM	PCA	T*PCA	CNNAE
Precision	0.927	0.852	0.915	0.832	0.924	0.802	0.926	0.854	0.923	0.858	0.923	0.852	0.925	0.852	0.915
Recall	1	1	0.976	1	1	0.796	0.996	1	0.968	1	0.978	1	1	0.988	0.903
F1-Score	96.23 ±0.011	92 ±0.0071	94.5 ±0.01	90.8 ±0.0045	96.07 ±0.011	79.8 ±0.013	96.03 ±0.0105	92.20 ±0.0071	94.55 ±0.008	92.40 ±0.0055	95 ±0.005	92.2 ±0.0084	96.13 ±0.006	91.6 ±0.0055	90.26 ±0.003
g-means	97.79 ±0.007	95.2 ±0.0045	96.68 ±0.006	94.4 ±0.005	96.07 ±0.006	84.4 ±0.0055	97.33 ±0.007	95.20 ±0.0045	93.14 ±0.005	95.40 ±0.0055	97.96 ±0.0031	95.2 ±0.0084	97.78 ±0.0037	94.8 ±0.0045	92.3 ±0.005
Train time (sec)	0.22	13.7	0.22	2.55	34.67	1881	0.59	17.6	0.18	0.056	1.85	62.37	0.02	0.083	15.32
Test time (sec)	0.13		0.04		7.77		0.06		0.0		0.11		0		0.09
Mem. Footprint (KB)	705	9.66MB	2300	53.2MB	612	7.73MB	924	27.66MB	106	2.94MB	450	6.49MB	107	2.94MB	2.0MB

(b)

Fig. 4. A side by side comparison of the anomaly detection algorithms applied to mixed data (normal + fault modes) with threshold set at (a) 90% and (b) at 95%.

Figure 4 shows a side-by-side comparison of each algorithm using the 8 manual features. The results show that the performance of all anomaly detectors using the hand crafted features is comparable to that of the raw torque measurements/features. Similarly, the 76 tsfresh features also produce comparable results to that of the 8 manually created and raw features, showing that the additional features have no or near to none performance increase in the models. We recommend the isolation forest anomaly detection if a single algorithm is to be choosen. Although based on the evaluation metrics, the algorithm is slightly slower in training time, some of its properties makes it the winner. These properties include: no need for calculating distance or density of samples; at the time of training a tree is constructed and at test time it passes sample points down the tree to calculate the average number of edges required to reach an external node. It is very robust in terms of choosing parameters; to configure it, one would need to (1) provide maximum samples to draw from training dataset to train each of the base estimators, (2) number of base estimators (the number of ensembles), (3) contamination, the ratio of outliers in the dataset (if used as a semi-supervised algorithm). Liu et al. [8], show that the choice of sample size is very robust and that the algorithm converges very fast with low sample size. We think the most important configuration parameter is the contamination ratio, which stands the same for all the other algorithms. However, according to Liu et al., the choice of contamination may be rectified by using a larger sampling size. In addition, the memory requirement for isolation forest is very low due to sub-sampling, making it a great practical choice for deployment in industry.

Figure 3 shows the anomalous scores of normal and faulty signals on a scatter plot; grey points identify as normal and orange points as anomalous for manually constructed features. The threshold (90%) is represented as a golden horizontal line, marking the cut-off at which a sample is flagged normal or anomalous.

6 Future Work

In a future study, a comparative assessment of the presented anomaly detection techniques will be made against the physics-based models of the normal and the different fault modes (loose belt, tight belt and friction).

References

1. Aggarwal, C.C.: Outlier analysis. In: Aggarwal, C.C. (ed.) Data Mining, pp. 237–263. Springer, Cham (2015). https://doi.org/10.1007/978-3-319-14142-8_8
2. Chandola, V., Banerjee, A., Kumar, V.: Anomaly detection: a survey. ACM Comput. Surv. 41(3), 15 (2009)
3. Cheng, S., Azarian, M.H., Pecht, M.G.: Sensor systems for prognostics and health management. Sensors 10(6), 5774–5797 (2010)
4. Christ, M., et al.: Time series feature extraction on basis of scalable hypothesis tests (tsfresh-a python package). Neurocomputing 307, 72–77 (2018)
5. Foss, A., Zaïane, O.R., Zilles, S.: Unsupervised class separation of multivariate data through cumulative variance-based ranking. In: International Conference on Data Mining (IEEE ICDM), pp. 139–148 (2009)
6. Fulcher, B.D., Little, M.A., Jones, N.S.: Highly comparative timeseries analysis: the empirical structure of time series and their methods. J R Soc Interface 10(83), 20130048 (2013)
7. Isermann, R.: Model-based fault-detection and diagnosis-status and applications. Annu Rev Control 29(1), 71–85 (2005)
8. Liu, F.T., Ting, K.M., Zhou, Z.-H.: Isolation forest. In: International Conference on Data Mining (IEEE ICDM), pp. 413–422 (2008)
9. Nanopoulos, A., Alcock, R., Manolopoulos, Y.: Feature-based classification of time-series data. Int. J. Comput. Res. 10(3), 49–61 (2001)
10. Reuters Editorial News: 4 Ways Predictive Maintenance Streamlines Manufacturing (2017). http://www.reuters.com/media-campaign/brandfeatures/intel/gated/TCM_1610337_intel_adlink_IOT_whitepaper_R6.pdf
11. Pecht, M.: Prognostics and health management of electronics. In: Encyclopedia of Structural Health Monitoring (2009)
12. Salama, M.A., Hassanien, A.E., Fahmy, A.A.: Reducing the influence of normalization on data classification. In: Proceedings of the International Conference on Computer Information Systems and Industrial Management Applications, pp. 609–613. IEEE (2010)
13. Snoek, J., Larochelle, H., Adams, R.P.: Practical Bayesian optimization of machine learning algorithms. In: Proceedings of the International Conference on Advances in Neural Information Processing Systems, pp. 2951–2959 (2012)
14. Zhang, W., et al.: A new deep learning model for fault diagnosis with good anti-noise and domain adaptation ability on raw vibration signals. Sensors 17(2), 425 (2017)

A Hybrid Architecture for Tactical and Strategic Precision Agriculture

Enrico Gallinucci, Matteo Golfarelli(✉), and Stefano Rizzi

DISI, University of Bologna, Viale Risorgimento 2, 40136 Bologna, Italy
{enrico.gallinucci,matteo.golfarelli,stefano.rizzi}@unibo.it

Abstract. In this paper we present a platform that implements a BI 2.0 architecture to support decision making in the precision agriculture domain. The platform, outcome of the Mo.Re.Farming project, couples traditional and big data technologies and integrates heterogeneous data from several owned and open data sources; its goal is to verify the feasibility and the usefulness of a data integration process that supports situ-specific and large-scale analyses made available by integrating information at different levels of detail.

Keywords: BI 2.0 · Precision agriculture · Data integration

1 Introduction

With the rise of precision agriculture, the world of agriculture has become a major producer and consumer of data. Indeed, recent technologies allow satellite images and sensor data to be generated with higher detail and frequency [1]; the set of available open data regularly increases in both quantity and quality, and new approaches to data collection, such as *crowd sensing* [2], are applied to the agriculture area as well. Properly handling such a mass of data requires emerging digital technologies, such as big data and IoT, to be adopted. The interest in adopting big data approaches for precision agriculture and precision farming is confirmed by increasing research activities in this area [3]. However, a careful analysis of the 34 research projects surveyed in [3] shows that most efforts are focused on applying machine learning techniques to ad hoc agricultural datasets, whereas data collection and integration systems have attracted less interest—which may give rise to the problem of *information silos* (i.e., information that can be hardly shared and reused).

Although the definition of a comprehensive and integrated architecture for precision agriculture has been poorly addressed in the scientific literature, both free and commercial data services are already available to obtain high-value data such as irrigation advices and vegetation indices. While most of these solutions are based on web applications to deliver data services, they strongly differ in

Partially supported by the Mo.Re.Farming Project (www.morefarming.it) funded by the POR FESR Program 2014–2020.

the way data are stored, processed, and made available, as well as in the type of data provided and in the professional figures and services they are oriented to.

In this paper we propose an innovative architecture, called *Mo.Re.Farming* (*MOnitoring and REmote system for a more sustainable FARMING*), for handling agricultural data in an integrated fashion. This architecture is oriented to data analysis and is inspired by Business Intelligence (BI) 2.0 approaches; it is hybrid in that it couples traditional and big data technologies to integrate heterogeneous data, at different levels of detail, from several owned and open data sources. Using the BI terminology, we distinguish between *tactical* and *strategic* services provided by Mo.Re.Farming. The former typically exploit data from a limited area and within a restricted time-span to provide detailed information to the users; the latter aggregate and analyze data from broader areas, spanning on longer time intervals. An example of tactical information is the current vegetation index of a specific field, which can be used to modulate the quantity of fertilizer to be spread on its surface. An example of strategic information is the time series of the average vegetation index for all the corn fields in the different provinces during the whole corn farming season, which enables irrigation forecast. Clearly, the production of these two kind of information involves a different quantity of raw data and a different level of detail. As agreed in the BI literature, these differences call for separated repositories and schemata to properly store information for the tactical and strategic levels.

A further feature of precision agriculture systems is the inherent presence of spatial information such as georeferenced satellite images, maps of fields, positions of sensors on the ground, and so on. To handle this feature, the adoption of a spatially enabled technology, typically a Geographic Information System (GIS), is required. In Mo.Re.Farming we exploit georeferencing as the basis to carry out a painless integration of the different data sources.

The proposed architecture has been developed in the context of the Mo.Re.Farming project, aimed at providing a Decision Support System for agricultural technicians in the Emilia-Romagna region (Italy) and to enable analyses related to the use of water and chemical resources, in terms of optimization and environmental impact. The Mo.Re.Farming project provides both the requirements for the architecture, and a case study to test it; thus, in this paper we describe the architecture from the technological and functional points of view with specific reference to its deployment within the project. The deployed architecture serves now as a hub for agricultural data in the Emilia-Romagna region; the integrated data are made available in open access mode and can be accessed through web interfaces and through a set of web services.

2   The Mo.Re.Farming Project

The main goal of the Mo.Re.Farming project is to verify the feasibility of a data integration approach to deliver precision agriculture services to a plethora of different stakeholders. The idea from which the project comes is that the higher the number of effectively integrated data sources, the higher the number of services to be delivered and stakeholders to be supported. In other words, *putting together information useful for single services enables and empowers further services*. The data sources that are currently ingested in Mo.Re.Farming are:

- **Satellite images:** images (about GB each) are taken from Sentinel-2 satellites, designed to support vegetation, land cover, and environmental monitoring. The Earth surface is subdivided into *granules*, and each image corresponds to a granule. The Emilia-Romagna region is covered by 7 granules; the current frequency of satellite passages is 3 a week. Images also contain quality indicators and metadata to enable cloud screening, georeferencing, and atmospheric corrections.
- **Field Sensors:** two sensors have been developed and installed on a set of sample fields: a smart pheromone trap and the waveguide-based spectrometry at the core of the humidity sensor. All field sensors are connected through a GPRS network to the Mo.Re.Farming server and data are downloaded only daily for energy saving purposes.
- **Crop Register:** yearly filled by farmers, it includes—for each rural land—a 49-valued classification of crops (e.g., tomato) and indicates whether or not the field is irrigated.
- **Administrative boundaries:** a vector layer including municipal, provincial, and regional boundaries. This layer is freely downloadable from the website of the Italian institute for statistics - ISTAT.
- **Rural Land Register:** a vector layer including municipal, field, and farm boundaries tagged with additional information such as field surface.
- **Weather data:** a vector layer with daily data about minimum, maximum, and average temperatures and rainfall on a regional grid of 858 sensors.

All the data listed above are inherently spatial, thus a natural way to analyze and visualize them is through a GIS system.

3   The Mo.Re.Farming Architecture

To meet both the tactical and strategic goals, in the Mo.Re.Farming project we adopted the three-tier architecture sketched in Fig. 1.

- The bottom level hosts a *data lake* [4], i.e., a storage repository that holds a vast amount of raw data in its native format, including structured, semi-structured, and unstructured data. The data lake adopts a multi-zone architecture, so as to logically separate the subsequent processing and enrichment activities, as well as to provide a safe environment for data scientists' ad-hoc analyses.
- The middle layer, called *Operational Data Store* (ODS), stores structured data at the finest level of detail for in-depth analysis and monitoring. While in the data lake no fixed schema is defined a priori, the ODS schema is defined at the design time.

– The top level of the architecture consists of a *spatial cube* to enable SOLAP (Spatial On-Line Analytical Processing). Multidimensional data are organized in cubes which can be analyzed using SOLAP operators such as *spatial slice* and *spatial drill*, which allow for aggregating measure values along hierarchies with SQL operators.

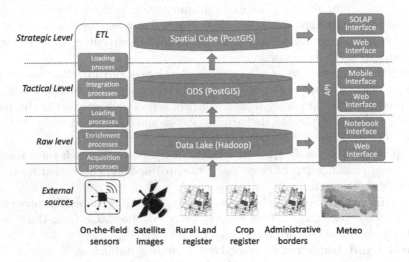

Fig. 1. The Mo.Re.Farming architecture

Data integration takes place at the tactical level and is mainly based on the spatial features of data. As described in more detail in Sect. 4, the relationships between heterogeneous data are found thanks to geopositioning even in absence of geographical references.

From the technological point of view the architecture is hybrid: while the data lake is Hadoop-based, the upper levels rely on two PostGIS DBMSs running on a centralized server. The reason for this lies in the current limitations of big data solutions in properly handling continuous field geographic data, that is, spatial phenomena that are perceived as having a value at each point in space and/or time [5]. Satellite images (i.e., rasters) need to be handled at the pixel level, thus requiring features that are typical of *array DBMSs* [6] such as Rasdaman and PostGIS; for instance, vegetation indices are computed through *map algebra* manipulations [7], i.e., algebraic operations on raster layers, ranging from arithmetical to statistical and trigonometric operations. At the time of project development, no big data GIS was able to execute this kind of operations (raster layers were merely used as a background to be overlapped with vector ones), which made the traditional GIS technology mandatory. Some considerations about the current state of spatial big data technologies are made in the conclusions.

4   Data Model and ETL Processes

In this section we discuss the data models adopted for data representation and storage in each architectural level, as well as the ETL processes that drive the flow of data between such levels.

4.1   Acquisition and Enrichment

At the data lake level, data are stored in files on the Apache Hadoop distributed file system (HDFS), which ensures system robustness and enables parallel processing. The processes that run in parallel are those concerning the acquisition and enrichment of satellite images, which are also the most computationally demanding ones. Specifically, the tasks (implemented in Python) are *Satellite images download*, *Atmospheric correction* (required to clean satellite images from reflections of solar light), *GeoTIFF creation*, and *Raster pyramid creation* (the GeoTIFF image is transformed into a pyramid, i.e., a multi-resolution hierarchy of tiled levels).

The acquisition and enrichment of data from the remaining external data sources is less complex;

- On-field sensors are connected to each other through a Robust Wireless Sensor Network and transmit their data via a GPRS connection. Enrichment of images (i.e., the recognition of the captured bugs and their counting) is done by running an ad-hoc image recognition software; the results are then stored on the ODS.
- Weather data are made available as open data on a daily basis. An ETL process directly puts the data on the ODS, as no enrichment is necessary.
- The rural land and crop registers and the administrative boundaries are all published as open data on a yearly basis. These data are manually downloaded and stored on the ODS; no enrichment is necessary here either.

4.2   Integration

Figure 2 shows the relational schema of the ODS, where integrated data are stored. Relations are grouped according to the corresponding data sources (shown as grey bounding boxes); links are shown in blue if they are determined by spatial join operations. The central role is clearly played by the Rural lands relation, which enables the spatial integration of the different data sources. In particular:

- Smart traps and humidity sensors (identified by spatial points) are associated to the rural land (represented by a multipolygon) they are contained in.
- Rural lands are associated to the municipality (represented by a multipolygon) they are contained in. The low precision of municipal boundaries results in an inexact overlapping with rural land boundaries, so each rural land is assigned to the municipality with whom the intersection area is the greatest. Rural lands are also associated to the three closest weather stations, to estimate weather conditions via a weighted triangulation.

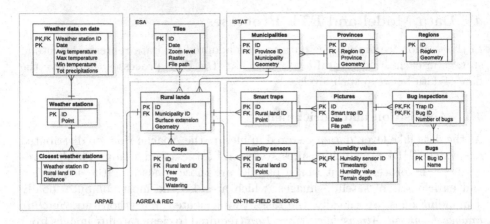

Fig. 2. Data schema of the ODS

– The integration between rural lands and tiles is done by trimming the tile's raster based on the shape of the rural land's multipolygon. Materializing the integration would require extensive storage, so it is computed on-the-fly when answering queries and when loading data to the spatial cube.

4.3   Loading

The functions offered at the strategic level are centered on a spatial cube, whose multidimensional schema (shown in Fig. 3 using the DFM notation [8]) was obtained with a data-driven approach and validated by experts in the field of agriculture. The Farming snapshot cube features four dimensions: Rural land, Date, Crop, and Watering regime. The Date dimension develops into a temporal hierarchy and contains the date of every satellite image downloaded. The Rural land dimension develops into a spatial hierarchy and provides the geometries of rural lands and administrative divisions. The Crop and Watering regime dimensions contain the list of crops and watering regimes, respectively. The events

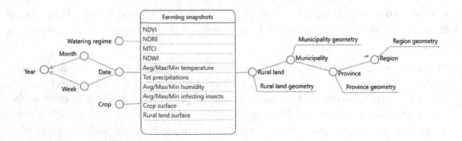

Fig. 3. Multidimensional schema of the spatial cube

represented in this cube consist of snapshots, one for each satellite image down-loaded, providing statistics for the crop of a given rural land in a given date. The available statistics, represented as measures in the cube, are:

- NDVI, NDRE (or red-edge NDVI), MTCI, NDWI: a set of vegetation indices, computed by analyzing—for each given date—the portion of satellite image that falls within the spatial boundaries of the rural land.
- Temperature, Precipitations, Humidity, Infesting insects: weather and sensor information computed by weighing the measurements at the closest stations.
- Crop surface, Rural land surface: the extent of the land for a crop and the overall extent of the rural land.

5   User Interface for Data Access

In Mo.Re.Farming the data collected and produced are mainly accessed via a web interface that guides the user experience through a set of dashboards, depending on the chosen architectural level.

The data lake is mainly used for back-end computations; it can be queried through notebook technology, a perfect tool for data scientists since it allows to couple advanced visualizations, data retrieval, programming, and documenting of research procedures. In Mo.Re.Farming we adopt Apache Zeppelin.

Fig. 4. Tactical-level web application showing statistics on a field in the municipality of Podenzano; the colors of the map are representative of the NDVI

The main access to the ODS level is through an ad-hoc dash-board, implemented in PHP and Javascript. The interface (available at semantic.csr.unibo.it/morefarming and shown in Fig. 4) is intended for both agri-cultural technicians and farmers, and allows several different information to be visualized. The option panel on the left allows to select: (i) the kind of image to be displayed; (ii) the reference date for the image; and (iii) one or more vector layers. The user can obtain statistics about any point or field by simply clicking on the map. The "P" icon on the map allows to obtain statistics on a custom polygon drawn on the map. Finally, the blue markers that locate smart traps can be clicked to visualize the photos in inverse chronological order, indicating the number and kinds of bugs recognized.

Finally, a SOLAP solution is implemented on the top layer through the Saiku software, which enables the execution of SOLAP queries on the spatial cube. The spatial features can be exploited by first drawing a polygon on the map through the tactical-level interface; then, the polygon is used as a spatial filter in the query to select only the rural lands that intersect that polygon (see Fig. 5).

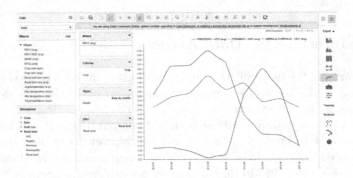

Fig. 5. Strategic-level web application, showing average monthly trends for NDVI on the crops available in a set of fields.

The web interface is complemented by a set of public web services that expose the available data by means of standard RESTful APIs (semantic.csr.unibo.it/morefarming/api).

6   Related Literature

The literature on big data approaches to precision farming and precision agriculture is growing quickly. Most efforts are focused on applying machine learning techniques to ad hoc agricultural datasets [3]. Another relevant part of the literature is devoted to sensors [1] and remote sensing [9]. All these works only marginally focus on the data management architecture; the only work in this direction is [10], which proposes an open and integrated *cyber-physical infrastructure*, i.e., a coordinated environment including several hardware components, software, and interactions. The applications proposed are exclusively at the tactical level, and no OLAP-like functionalities are discussed or implemented. In [11] the authors stress the advantages of making available to farmers heterogeneous but integrated data in right time. To this end they propose the PRIDE business model, but give no technical details on the architecture and data model.

Although the scientific community has paid little attention to the architectures and data integration, some free and commercial data services are available.

While most solutions are based on web applications to deliver data services, they differ in the way data are stored, processed, and made available, as well as in the type of data provided and in the professional figures and services they are oriented to. Among them, we mention *Global Land Cover* [12], *CropScape* [13], *Mundialis* [14], *Moses* [15], and *Earth Observation Data Services* [16].

Mo.Re.Farming presents several distinguishing features with reference to the previously described projects. In particular it integrates a high number of different layers while delivering detailed information in terms of both spatial and temporal dimensions (i.e., satellite images are captured daily at 10 m resolution). Furthermore, it provides support for both tactical and strategic precision agriculture by including a SOLAP interface.

7 Conclusion and Discussion

In this paper we described the architecture supporting the Mo.Re.Farming project, showing how the integration of weather, crops, and fields data boosts precision agriculture both at the strategic and tactical levels. The lessons we learnt from this experience are summarized below:

– Although the computational needs and the quantity of data to be stored pushes towards the adoption of big data solutions (e.g., Apache Hadoop), the level of development of spatio-temporal features on those platforms is still limited. In particular, no support to continuous field data is given in general-purpose big data solutions. At the time of writing, some open-source libraries for spatial big data operations (i.e., GeoSpark and GeoTrellis) have matured, and their combined functionalities cover the set of operations required for the management of satellite images. This would enable a migration of the project to a full big data solution, even though some effort would be necessary to properly integrate the available libraries.
– The precision agriculture landscape is currently characterized by several solutions, models, and services, each working on a subset of the available data and providing complementary but non-integrated information and services. Creating integrated hubs of information is mandatory to overcome these limitations and to deliver more effective information and services.
– Integration tasks are computationally demanding and can hardly be carried out on-the-fly. Specifically, while on-the-fly integration is still be feasible at the tactical level thanks to the low quantity of data involved, it can hardly be done at the strategic level the amount of data and processing is huge.
– The Mo.Re.Farming architecture represents an example of a data hub, conceived as a starting point for delivering integrated information and services. The advantages of carrying out integration at the physical level, i.e., by materializing the integrated data in the ODS, are (i) the possibility of late data reworking, (ii) a 360-degrees exploitation of data with no limitations due to different data owners, and (iii) the possibility of efficiently running complex analytics on heterogeneous data.

- Precision agriculture architectures should be as open as possible to enable complementary data exchange and fruition, so the need for a standard terminology is perceived. Though this issue has been addressed by some research papers (e.g., [17]), no complete and accepted solution is available yet.
- The number of potential data sources and data collection approaches will further increase in the future. We are working to build an intelligent and flexible data lake system, where new data in different formats can be ingested and processed based on the automatic recognition of the content.

References

1. Ojha, T., Misra, S., Raghuwanshi, N.S.: Wireless sensor networks for agriculture: the state-of-the-art in practice and future challenges. Comput. Electron. Agric. **118**, 66–84 (2015)
2. Minet, J., et al.: Crowdsourcing for agricultural applications: a review of uses and opportunities for a farmsourcing approach. Comput. Electron. Agric. **142**, 126–138 (2017)
3. Kamilaris, A., Kartakoullis, A., Prenafeta-Boldú, F.X.: A review on the practice of big data analysis in agriculture. Comput. Electron. Agric. **143**, 23–37 (2017)
4. Stein, B., Morrison, A.: The enterprise data lake: better integration and deeper analytics. PwC Technol. Forecast. Rethink. Integr. **1**, 1–9 (2014)
5. Vaisman, A.A., Zimányi, E.: A multidimensional model representing continuous fields in spatial data warehouses. In: Proceedings of ACM-GIS, Seattle, USA, pp. 168–177 (2009)
6. Baumann, P.: A database array algebra for spatio-temporal data and beyond. In: Proceedings of NGITS, pp. 76–93 (1999)
7. Tomlin, C., Berry, J.: Mathematical structure for cartographic modeling in environmental analysis. In: Proceedings of the American Congress on Surveying and Mapping Annual Meeting (1979)
8. Golfarelli, M., Rizzi, S.: Data Warehouse Design: Modern Principles and Methodologies. McGraw-Hill Inc, New York (2009)
9. Vibhute, A., Bodhe, S.: Applications of image processing in agriculture: a survey. Int. J. Comput. Appl. **52**(2), 34–40 (2012)
10. Chen, N., Zhang, X., Wang, C.: Integrated open geospatial web service enabled cyber-physical information infrastructure for precision agriculture monitoring. Comput. Electron. Agric. **111**, 78–91 (2015)
11. Sawant, M., Urkude, R., Jawale, S.: Organized data and information for efficacious agriculture using PRIDE model. Int. Food Agribus. Manage. Rev. **19**(A), 115–130 (2016)
12. Han, G., et al.: A web-based system for supporting global land cover data production. ISPRS **103**, 66–80 (2015)
13. Han, W., Yang, Z., Di, L., Mueller, R.: CropScape: a web service based application for exploring and disseminating US conterminous geospatial cropland data products for decision support. Comput. Electron. Agric. **84**, 111–123 (2012)
14. Neteler, M., Adams, T., Paulsen, H.: Combining GIS and remote sensing data using open source software to assess natural factors that influence poverty. In: Proceedings of World Bank Conference on Land and Poverty, Washington DC, USA (2017)

15. Di Felice, A., et al.: Climate services for irrigated agriculture: structure and results from the MOSES DSS. In: Proceedings of Symposium Società Italiana per le Scienze del Clima, Bologna, Italy, pp. 1–4 (2017)
16. Baumann, P., Dehmel, A., Furtado, P., Ritsch, R., Widmann, N.: The multidimensional database system RasDaMan. In: Proceedings of SIGMOD, pp. 575–577 (1998)
17. Song, G., Wang, M., Ying, X., Yang, R., Zhang, B.: Study on precision agriculture knowledge presentation with ontology. AASRI Proc. **3**, 732–738 (2012)

Urban Analytics of Big Transportation Data for Supporting Smart Cities

Carson K. Leung[1]([envelope]) [iD], Peter Braun[1], Calvin S. H. Hoi[1],
Joglas Souza[1], and Alfredo Cuzzocrea[2]

[1] University of Manitoba, Winnipeg, MB, Canada
kleung@cs.umanitoba.ca
[2] University of Trieste, Trieste, TS, Italy

Abstract. The advances in technologies and the popularity of the smart city concepts have led to an increasing amount of digital data available for urban research, which in turn has led to *urban analytics*. In urban research, researchers who conduct paper-based or telephone-based travel surveys often collect biased and inaccurate data about movements of their participants. Although the use of global positioning system (GPS) trackers in travel studies improves the accuracy of exact participant trip tracking, the challenge of labelling trip purpose and transportation mode still persists. The automation of such a task would be beneficial to travel studies and other applications that rely on contextual knowledge (e.g., current travel mode of a person). In DaWaK 2018, we made use of both the GPS and accelerometer data to classify ground transportation modes. In the current DaWaK 2019 paper, we explore additional parameters—namely, dwell time and dwell time history (DTH)—to further enhance the urban analytic capability. In particular, with these additional parameters, classification and predictive analytics of ground transportation modes becomes more accurate. This, in turn, helps the development of a smarter city.

Keywords: Data mining · Knowledge discovery · Big data · Classification ·
Ground transportation mode · Global positioning system (GPS) data ·
Geographic information system (GIS) data · Accelerometer data · Dwell time ·
Dwell time history (DTH)

1 Introduction

A *smart city* generally refers to an urban area that utilizes (i) various kinds of sensors and/or Internet of Things (IoT) devices to generate or collect big urban data—which may be of different veracity (e.g., uncertain and imprecise data [1–3])—at a high velocity and (ii) various kinds of data science solutions to analyze the generated or collected urban data for effective management of city assets and resources. Here, the data science solutions [4–7] for make use of various types of advanced methods, tools, and technologies from areas like big data analytics, knowledge discovery, databases, expert systems, artificial intelligence, mathematics, and statistics. As examples, these data science solutions for *urban data mining* or *urban analytics* [8–12] process, analyze, and mine these big urban data collected from citizens, devices, and assets to

© Springer Nature Switzerland AG 2019
C. Ordonez et al. (Eds.): DaWaK 2019, LNCS 11708, pp. 24–33, 2019.
https://doi.org/10.1007/978-3-030-27520-4_3

monitor and manage traffic and transportation systems. Other practical applications of the knowledge discovered from these big urban data include the monitoring and management of power plants, water supply networks, waste management, crime detection, information systems, schools, libraries, hospitals, and many other community services, as well as city planning. Many cities in the world are trying to establish themselves as smart cities by developing smart technologies. These technologies have been implemented in smart cities like Amsterdam, Barcelona, Columbus, Dublin, Madrid, New York, Stockholm, and Taipei. Recently, in Canada, the Impact and Innovation Unit (IIU) of the Privy Council Office (PCO)—together with Infrastructure Canada—started an Impact Canada Initiative (ICI)[1], which put up a Smart Cities Challenge[2] for finding solutions to achieve real and positive outcomes for smart cities. An objective is to enable citizens to move around the community and to reduce the average and maximum walking distance to transit. To achieve such an objective, it is important to classify the ground transportation modes (e.g., transit/bus) of the citizens. Understanding the transportation modes of the citizens in their trajectories help city planners in determining the appropriate supply of public transits (e.g., provide more frequent bus services at peak time, adjust bus routines to reduce citizens' walking distance to bus stops, increase bike lanes, build a rapid transit system such as metro or trains).

For urban research (specifically, travel studies in urban areas), traditional approaches usually use:

- paper-based and telephone-based *travel surveys* [13–15], which can be biased and contain inaccurate data about movements of their participants (e.g., participants tend to under-report short/irregular trips and car trips, but over-report public transit trips); and

- *commute diaries* [16, 17], which can be prone to errors in capturing data about people's daily commutes (e.g., participants tend to forget to record their commutes throughout the day, capture inaccurate information when recording trips at the end of the day, feel the mental burden of capturing commute diaries, and unwillingly/inaccurately record trips throughout the day [18, 19]).

To reduce the human workload and to utilize sensors for automatic processes, *travel surveys based on global navigational satellite systems (GNSS) or the Global Positioning System (GPS)* [20, 21] were used. These GNSS/GPS-based travel surveys rely on the GNSS trackers—which receive tracking information from systems like the GPS—to collect more objective and more accurate commute data from participants. However, the challenge of labeling trip purposes and classifying transportation modes persists. For instance, the manual segmentation of trajectories based on transportation mode can be labor intensive and is likely to be impracticable for big data [22].

The automation of such a task would obviously be beneficial to travel studies and other applications that rely on contextual knowledge (e.g., current travel mode of a

---

[1] https://impact.canada.ca/

[2] https://impact.canada.ca/en/challenges/smart-cities,   https://www.infrastructure.gc.ca/cities-villes/
index-eng.html

person). As an example for a contextual use for transportation, when the person is driving, a device like a smartphone could recognize that the person is driving in a car and give a notification about the current estimated time of arrival—assuming that the phone knows the destination based on previous user interaction or saved frequently visited locations. Another application for transportation mode classification is the automatic trip transportation mode labeling for trip history. This is similar to timeline in Google Maps (which keeps track of a user's location history and attempts to automatically classify trips with the major transportation mode). However, the resulting classification is observed to be not very accurate and needs corrections by the user. Moreover, it does not track when transportation modes were changed. Hence, a more accurate algorithm or system is needed.

In recent years, advances in technologies have enabled the use of some combinations of data from different sensors (e.g., GNSS/GPS, accelerometers) and other modern smartphone sensors (e.g., barometer, magnetometer, etc.). Some related works [23] use only GPS data, while some others [24, 25] use only accelerometer data. In addition, some related works [26] integrate both GPS and accelerometer data, while some others [27, 28] integrate GPS data with geographic information system (GIS) data. In DaWaK 2018, we [29] integrated GPS, GIS and accelerometer into a single system. To further enhance the accuracy of our classification, we explore the dwell time and its history in the current DaWaK 2019 paper.

In this paper, our *key contribution* is our enhanced classification system for ground transportation modes. We design new classification features based on both *dwell time* and *dwell time history* (*DTH*). Generally, in transportation, *dwell time* (also known as terminal dwell time) refers to the time a vehicle such as a public transit bus or train spends at a scheduled stop without moving. Dwell time usually measures the amount of time spent on boarding or disembarking passengers. Sometimes, it also measures the amount of time spent on waiting for traffic ahead to clear, merging into parallel traffic, or idling in order to get back on schedule. Different transportation modes are expected to exhibit different patterns in dwell time.

The remainder of this paper is organized as follows. The next section presents background and related works. Section 3 describes our enhanced transportation mode classification system. Evaluation results and conclusions are given in Sects. 4 and 5, respectively.

2 Background and Related Works

In DaWaK 2018, we [29] presented a ground transportation mode classification system that uses GPS data, accelerometer data, and GIS data. The end results of the system are segmented trips (or trip windows), where each segment is labelled with the ground transportation mode the person used for the period of the segment.

The system first collects trip traces (GPS locations) and trip accelerometer data, and stores them in a database (e.g., MongoDB). Here, the default GPS sampling rate was set to 1 Hz, and the default accelerometer sampling was set to 22 Hz. In addition, the system also collects the bus stop locations (GIS information) in a city—via its transit application programming interface (API)—when "bus" is one of the ground transportation modes for classification.

Once the trip data are collected, the system segments every trip (which is simply a collection of data points collected during a person's entire commute from origin to destination—say, from home to work) based on the transportation mode used in each segment. More specifically, data of a trip are divided into many small windows of equal time interval. When a transportation mode change occurs, data are assigned to different windows so that no two transportation modes are mixed within the same window. Segmenting the data into small windows led to a benefit that classification can be performed in real-time. For instance, as soon as sufficient amount of data has been collected to fill a new window, the window can be classified with a transportation mode. Once every window is classified with a transportation mode, the user simply concatenates the windows/trip segments (each of which is labelled with a transportation mode) and presents each label on a map with a color-scheme for different transportation modes. Once the trip is rendered on a map, the user can easily identify different legs of the trip by simply looking at the different colors of the trip.

Then, the system extracts three key types of features for transportation mode classification:

- GPS-based features (for capturing the geo-location and time information provided by GPS sensors), which include speed (e.g., average, maximum), average altitude, average location accuracy, travel distance, and a flag for indicating whether there is GPS signal or not;
- Accelerometer-based features (for capturing the measurement on acceleration of different transportation modes such as automobile), which include magnitude (e.g., minimum, 25th percentile, average, 75th percentile, maximum, standard deviation), lag-1 autocorrelation, pairwise correlations among the axes in 3-dimensional space, average roll, average pitch, and average yaw; as well as
- GIS-based features (for capturing the GIS information related to bus stop locations), which include the number of bus stops, the number of "stops" at bus stops, distance to closest bus stop, and a flag for indicating whether a commuter stops at bus stops.

Once these features are extracted, the system builds and trains a (random forest based) classification model, which can be used for predicting labels (i.e., ground transportation modes—such as "walk", "bike", "bus", and/or "car"—used by the user) for unseen data.

3 Our Enhanced Classification System for Urban Analytics

3.1 Dwell Time

To further enhance our ground transportation mode classification system described in Sect. 2, we consider *dwell time* (which is the time a person dwells near a bus stop). The idea is that, when people commute, they usually tend to have very long dwell times at their origin (e.g., home) and destination (e.g., workplace). To detect such long dwell times, the user needs to define an appropriate minimum threshold for their use cases.

For example, a threshold of 6 h would likely find trip legs that are temporally between staying at home and staying at work. A threshold of 1 h would likely capture trip legs between the times of staying home and staying at work (e.g., a grocery shopping trip on the way home, a drive to a restaurant during lunch break).

Moreover, if a person takes the bus during a commute then they usually spent some significant amount of time (e.g., minutes) at the departure, transfer, and arrival bus stops. While the bus is in motion, it stops at bus stops along the bus route, either to pick up new passengers or let passengers off the bus. Those kind of dwell times are usually shorter but more frequent. Those short dwelling times near a bus stop could also be used in the classifier training to better identify as the "bus" mode.

When measuring dwell time (in seconds), our enhanced system also fetches the nearest bus stop location based on a radius maximum threshold $D_{busStop}$, which default is set to 5 m. Note that $D_{busStop}$ cannot be too high or too low. On the one hand, if $D_{busStop} = 1$ m, then the system may not capture all dwell times because buses do not always stop exactly within 1 m of a bus stop. On the other hand, if $D_{busStop} = 50$ m, then the system may capture too many bus stops before the bus actually stops at the stop.

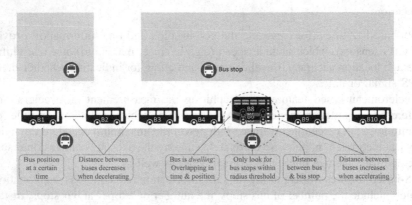

Fig. 1. Positions of a bus moving along part of its route and dwelling at a bus stop.

Figure 1 shows a bus moving along part of its route. The locations of the bus at each data window are represented by objects B1 to B10. At positions B5 to B8, the bus has stopped at the bus stop and has dwelled there for some time, which is represented by overlapping bus objects. When a bus dwells, we calculate the dwell time at a bus stop only if the bus is close enough to a bus stop, which is determined by the radius threshold $D_{busStop} = 5$ m. As there may nearby bus stop on the other side of the road for buses travelling in the opposite direction and some of these stops may locate across intersections, a radius maximum threshold $D_{busStop}$ helps fetch only bus stops that are on the correct side of the road.

3.2   Dwell Time History (DTH)

In addition to dwell time, our enhanced system also extracts the following features related to *dwell time history*:

- *Total long dwell time*, which is the total time of dwell times calculated from a set of consecutive windows with dwell times of at least 1 s, without a gap.
- *Number of long dwell time*, which is the total number of long dwell times encountered in all previous windows.

If the *total long dwell time* is high (as determined by the training of the classifier), then the current window is likely to be a "bus" mode window. When a person takes the bus, the person usually dwells for a relatively long time at the departure bus stop waiting for their bus to arrive. Those consecutive windows of dwelling are counted towards the long dwell time, which in turn are added to the total long dwell time. Once the person gets on the bus, the bus moves along its route encountering many bus stops and stopping at most (if not all) of them, then there will be many long dwell times for the whole trip. Thus, the total long dwell time would be a good contributor for transportation mode classification, and it is easy to keep track of in real time.

The number of long dwell time can be interpreted as the number of times a bus stops along its bus route. If this number is relatively high, then a person is likely to be commuting on a bus. Theoretically, a bus should stop more often at bus stops than a car

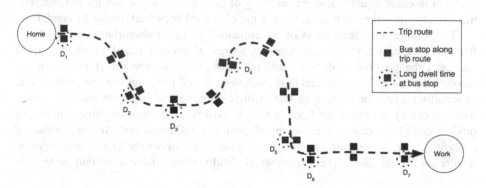

Fig. 2. A sample bus trip from home to work, on which the bus dwells at 7 bus stops.

following the same route. Thus, the number of long dwell times would be a good contributor for transportation mode classification.

Figure 2 shows a bus trip example going from home to workplace. Along the bus route, there are many bus stops on both sides of the street. D_1 to D_7 represent the bus stops where the bus stopped and dwelled for a longer time to let passenger on or off the bus. During the trip, we track the number of such long dwell times and the total long dwell time. Here, there are 7 long dwell times: $D_1 = 300$ s, $D_2 = 30$ s, $D_3 = 10$ s, $D_4 = 20$ s, $D_5 = 50$ s, $D_6 = 120$ s, and $D_7 = 10$ s. Then, the total long dwell time would be their sum, i.e., 540 s.

Observed that a *bus ride* tends to produce a history of different dwell times (depending the number of passengers getting on and off the bus) near the bus stops. On the other hand, a *metro ride* tends to produce a history of more regular dwell times at the platforms. In contrast, a *ride on taxi or a private vehicle* tends to produce fewer dwell times (mostly just at the origin and destination).

4 Evaluation

To evaluate our enhanced ground transportation mode classification system, we conducted experiments on a computer running Ubuntu 16.04 LTS as the main operating system. The CPU was an AMD Phenom II X6 1100T with 6 cores clocked at 3.3 GHz to 3.7 GHz. There are 16 GB of RAM and a solid state drive in the computer. We collected the GIS information (i.e., bus stop locations) from Winnipeg Transit Data Server by querying trip data via the Winnipeg Transit API. Users anonymously and securely uploaded their saved GPS- and accelerometer-based data via the mobile applications or dashboard. These trip information from users were then stored in a MongoDB, which supports basic geo-spatial query capabilities. The trip information was collected throughout a year, which contains trips with different weather and road conditions from summer to winter times. It captures the ground transportation mode (e.g., "walk", "car", "bus") used by the user at the time of commute. The trips for each transportation mode are shuffled and then split into two sets: (i) the training set and (ii) the testing/validation set. We used 70% of the data for stratified 10-fold cross-validation with a 50/50 partition split between the training and the testing data for each partition in order to determine the accuracy of the random forest based classifier. The module classifies unseen data and stores the classified trips back to the MongoDB.

To measure the effectiveness of our enhanced ground transportation mode classification system, we use the standard measures of precision and recall. *Precision* measures the positive predictive/classified value, i.e., the fraction of true positives among all positives (i.e., true and false positives). *Recall* measures the true positive rate or sensitivity, i.e., the fraction of true positives among true positives and false negatives. *Accuracy* measures the fraction of true positives and true negatives among all predications/classifications (i.e., among all positives and negatives). So, *our first set of experiments is to compute the precision, recall, and accuracy of the classification results on unseen data.* The experimental results show that our system accurately

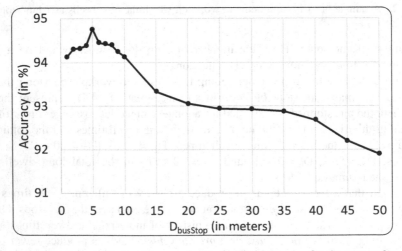

Fig. 3. Experimental results on radius maximum threshold $D_{busStop}$ for the nearest bus stops.

classified different segments of unseen data with highly accurate (e.g., above 88%) ground transportation modes.

Recall from Sect. 3.1 that the default radius maximum threshold $D_{busStop}$ for the nearest bus stop locations is set to 5 m. *Our second set of experiments is to vary this threshold to observe its effect on accuracy.* The experimental results shown in Fig. 3 demonstrate that $D_{busStop} = 5$ gave the most accurate classification. Note that, with very low $D_{busStop} = 1$ m, the accuracy is low because the system does not capture all dwell times as buses do not always stop within such a short distance of a bus stop. On the other hand, with very high $D_{busStop} = 50$ m, the accuracy is even low because the system captures too many bus stops.

Recall from Sect. 3.2 that a key contribution of this paper is our proposal of the concept of *dwell time history (DTH)*. Hence, *our third set of experiments is to measure the benefits (e.g., increase in classification accuracy) of using DTH.* The experimental results shown in Table 1 reveals that our enhanced system (which uses the DTH) led to higher classification accuracy. Specifically, when classifying ground transportation modes based on only GPS- and GIS-based data, the existing classification system from DaWaK 2018 gave an 88.96% accuracy. In contrast, our enhanced system in this DaWaK 2019 paper led to an additional +5.35% accuracy, for a 94.31% accuracy. Similarly, when classifying ground transportation modes based on accelerometer-, GPS- and GIS-based data, the existing DaWaK 2018 system gave a 96.97%. In contrast, our enhanced system in this DaWaK 2019 paper led to higher accuracy of 98.00%. These results demonstrate the benefits of using the DTH.

Table 1. Classification accuracy of not using vs. using the DTH.

Without DTH	Accuracy	With DTH	Accuracy
GPS + GIS data [29]	88.96%	GPS + GIS + DTH	94.31%
Acc + GPS + GIS data [29]	96.97%	Acc + GPS + GIS + DTH	98.00%

5   Conclusions

In this DaWaK 2019 paper, we extended and enhanced the ground transportation mode classification system presented in DaWaK 2018 by incorporating the *dwell time* and *dwell time history (DTH)*. The enhanced system combines GIS-, GPS- and accelerometer-based data, together with dwell time and DTH in classifying ground transportation mode used by commuters. Experimental results show that the use of DTH led to higher accuracy than their counterparts that do not use the DTH. The results demonstrate the benefits of incorporating dwell time and DTH—more accurate classification and prediction, which supports urban analytics of big transportation data and the development of smart cities.

Although we conducted urban analytics on Winnipeg bus data, the analytics and the presented system are expected to be applicable to other transportation modes (e.g., trams) and/or cities (e.g., Vancouver, Toronto). As ongoing and future work, we are exploring the use of *transfer learning* to migrate the knowledge learned from the

Winnipeg bus data to other transportation modes and/or other cities. Moreover, we are also conducting more exhaustive evaluation and exploiting other further enhancement to our system.

Acknowledgements. This project is partially supported by NSERC (Canada) and University of Manitoba.

References

1. Braun, P., Cuzzocrea, A., Jiang, F., Leung, C.K.-S., Pazdor, A.G.M.: MapReduce-based complex big data analytics over uncertain and imprecise social networks. In: Bellatreche, L., Chakravarthy, S. (eds.) DaWaK 2017. LNCS, vol. 10440, pp. 130–145. Springer, Cham (2017). https://doi.org/10.1007/978-3-319-64283-3_10
2. Chen, Y.C., Wang, E.T., Chen, A.L.P.: Mining user trajectories from smartphone data considering data uncertainty. In: Madria, S., Hara, T. (eds.) DaWaK 2016. LNCS, vol. 9829, pp. 51–67. Springer, Cham (2016). https://doi.org/10.1007/978-3-319-43946-4_4
3. Hoi, C.S.H., Leung, C.K., Tran, K., Cuzzocrea, A., Bochicchio, M., Simonetti, M.: Supporting social information discovery from big uncertain social key-value data via graph-like metaphors. In: Xiao, J., Mao, Z.-H., Suzumura, T., Zhang, L.-J. (eds.) ICCC 2018. LNCS, vol. 10971, pp. 102–116. Springer, Cham (2018). https://doi.org/10.1007/978-3-319-94307-7_8
4. Egho, E., Gay, D., Trinquart, R., Boullé, M., Voisine, N., Clérot, F.: MiSeRe-hadoop: a large-scale robust sequential classification rules mining framework. In: Bellatreche, L., Chakravarthy, S. (eds.) DaWaK 2017. LNCS, vol. 10440, pp. 105–119. Springer, Cham (2017). https://doi.org/10.1007/978-3-319-64283-3_8
5. Leung, C.K.: Big data analysis and mining. In: Advanced Methodologies and Technologies in Network Architecture, Mobile Computing, and Data Analytics, pp. 15–27 (2019). https://doi.org/10.4018/978-1-5225-7598-6.ch002
6. Leung, C.K., Jiang, F., Pazdor, A.G.M., Peddle, A.M.: Parallel social network mining for interesting 'following' patterns. Concurr. Comput. Pract. Exp. **28**(15), 3994–4012 (2016). https://doi.org/10.1002/cpe.3773
7. Fayyad, U.M., Piatetsky-Shapiro, G., Smyth, P.: From data mining to knowledge discovery: an overview. In: Advances in Knowledge Discovery and Data Mining, pp. 1–34 (1996)
8. Katakis, I. (ed.): Mining urban data (Part A). Inf. Syst. **54**, 113–114 (2015). https://doi.org/10.1016/j.is.2015.08.002
9. Andrienko, G., et al.: Mining urban data (Part B). Inf. Syst. **57**, 75–76 (2016). https://doi.org/10.1016/j.is.2016.01.001
10. Andrienko, G., et al.: Mining urban data (Part C). Inf. Syst. **64**, 219–220 (2017). https://doi.org/10.1016/j.is.2016.09.003
11. Behnisch, M., Ultsch, A.: Urban data mining using emergent SOM. In: Preisach, C., Burkhardt, H., Schmidt-Thieme, L., Decker, R. (eds.) Data Analysis, Machine Learning and Applications. Studies in Classification, Data Analysis, and Knowledge Organization, pp. 311–318. Springer, Heidelberg (2008). https://doi.org/10.1007/978-3-540-78246-9_37
12. Sokmenoglu, A., Cagdas, G., Sariyildiz, S.: Exploring the patterns and relationships of urban attributes by data mining. In: eCAADe 2010, pp. 873–881 (2010)
13. Ettema, D., Timmermans, H., van Veghel, L.: Effects of data collection methods in travel and activity research (1996)

14. Murakami, E., Wagner, D.P., Neumeister, D.M.: Using global positioning systems and personal digital assistants for personal travel surveys in the United States. In: Transport Surveys: Raising the Standard, pp. III-B:1–III-B:21 (2000)
15. Stopher, P.R..: Household travel surveys: cutting-edge concepts for the next century. In: Conference on Household Travel Surveys, pp. 11–23 (1995)
16. Maat, K., Timmermans, H.J.P., Molin, E.: A model of spatial structure, activity participation and travel behavior. In: WCTR 2004 (2004)
17. Stopher, P.R.: Use of an activity-based diary to collect household travel data. Transportation 19(2), 159–176 (1992). https://doi.org/10.1007/BF02132836
18. Arentze, T., et al.: New activity diary format: design and limited empirical evidence. TRR 1768, 79–88 (2001). https://doi.org/10.3141/1768-10
19. Schlich, R., Axhausen, K.W.: Habitual travel behaviour: evidence from a six-week travel diary. Transportation 30(1), 13–36 (2003). https://doi.org/10.1023/A:1021230507071
20. Forrest, T., Pearson, D.: Comparison of trip determination methods in household travel surveys enhanced by a global positioning system. TRR 1917, 63–71 (2005). https://doi.org/10.1177/0361198105191700108
21. Wolf, J., Guensler, R., Bachman, W.: Elimination of the travel diary: experiment to derive trip purpose from global positioning system travel data. TRR 1768, 125–134 (2001). https://doi.org/10.3141/1768-15
22. Biljecki, F., Ledoux, H., van Oosterom, P.: Transportation mode-based segmentation and classification of movement trajectories. IJGIS 27(2), 385–407 (2013). https://doi.org/10.1080/13658816.2012.692791
23. Zheng, Y., Chen, Y., Li, Q., Xie, X., Ma, W.: Understanding transportation modes based on GPS data for web applications. ACM TWeb 4(1), 1:1–1:36 (2010). https://doi.org/10.1145/1658373.1658374
24. Hemminki, S., Nurmi, P., Tarkoma, S.: Accelerometer-based transportation mode detection on smartphones. In: SenSys 2013, pp. 13:1–13:14 (2013). https://doi.org/10.1145/2517351.2517367
25. Shaque, M.A., Hato, E.: Use of acceleration data for transportation mode prediction. Transportation 42(1), 163–188 (2015). https://doi.org/10.1007/s11116-014-9541-6
26. Ellis, K., Godbole, S., Marshall, S., Lanckriet, G., Staudenmayer, J., Kerr, J.: Identifying active travel behaviors in challenging environments using GPS, accelerometers, and machine learning algorithms. Front. Public Health 2, 36:1–36:8 (2014). https://doi.org/10.3389/fpubh.2014.00036
27. Chung, E., Shalaby, A.: A trip reconstruction tool for GPS-based personal travel surveys. Transp. Plan. Technol. 28(5), 381–401 (2005). https://doi.org/10.1080/03081060500322599
28. Stenneth, L., Wolfson, O., Yu, P.S., Xu, B.: Transportation mode detection using mobile phones and GIS information. In: ACM SIGSPATIAL GIS 2011, pp. 54–63 (2011). https://doi.org/10.1145/2093973.2093982
29. Leung, C.K., Braun, P., Pazdor, A.G.M.: Effective classification of ground transportation modes for urban data mining in smart cities. In: Ordonez, C., Bellatreche, L. (eds.) DaWaK 2018. LNCS, vol. 11031, pp. 83–97. Springer, Cham (2018). https://doi.org/10.1007/978-3-319-98539-8_7

Patterns

Frequent Item Mining When Obtaining Support Is Costly

Joe Wing-Ho Lin[✉] and Raymond Chi-Wing Wong

The Hong Kong University of Science and Technology, Kowloon, Hong Kong
{whlinaa,raywong}@cse.ust.hk

Abstract. Suppose there are n users and m items, and the preference of each user for the items is revealed only upon probing, which takes time and is therefore costly. How can we quickly discover all the frequent items that are favored individually by at least a given number of users? This new problem not only has strong connections with several well-known problems, such as the frequent item mining problem, it also finds applications in fields such as sponsored search and marketing surveys. Unlike traditional frequent item mining, however, our problem assumes no prior knowledge of users' preferences, and thus obtaining the support of an item becomes costly. Although our problem can be settled naively by probing the preferences of all n users, the number of users is typically enormous, and each probing itself can also incur a prohibitive cost. We present a sampling algorithm that drastically reduces the number of users needed to probe to $O(\log m)$—regardless of the number of users—as long as slight inaccuracy in the output is permitted. For reasonably sized input, our algorithm needs to probe only 0.5% of the users, whereas the naive approach needs to probe all of them.

Keywords: Frequent item mining · Random sampling

1 Introduction

We propose a new data mining problem called HIDDEN FREQUENT ITEM MINING: Suppose we have a set of users $\mathcal{U}$ and a set of items $\mathcal{T}$, and that the preference of each user for the items is revealed only upon probing, which takes time and is therefore costly. Our aim is to quickly discover all the *frequent items* that are favored (i.e., the best choice) individually by at least a predefined number of users. This problem finds applications in areas such as sponsored search and marketing surveys.

Connections with Existing Problems. Our problem has strong connections with several well-recognized problems, including frequent item mining [5,12,13,20], heavy hitter finding [7,8,17], and estimation of population proportion problem in survey sampling [21,22]. As will be clear in Sect. 3, although our problem resembles the frequent item mining problem, one crucial difference

© Springer Nature Switzerland AG 2019
C. Ordonez et al. (Eds.): DaWaK 2019, LNCS 11708, pp. 37–56, 2019.
https://doi.org/10.1007/978-3-030-27520-4_4

sets it apart from its counterpart: in our problem, users' preferences are unknown in advance and therefore obtaining them is costly. In contrast, our counterpart assumes that the items purchased in each transaction (which may be viewed as the preferences of a user in our problem) are already known. We will also see that existing problems, such as heavy hitter finding, can be adapted to tackle a special case of our problem. Unfortunately, such an adaption results in an excessively high cost, thus rendering it impractical. Similarly, the estimation of population proportion problem in survey sampling can be employed to tackle a special case of our problem. However, existing solutions to this problem make extra assumptions in their results and are generally less efficient than our solution. Despite sharing similarities with existing studies, our proposed problem—to the best of our knowledge—has neither been proposed before, nor can it be efficiently solved by adapting solutions to existing problems.

Proposed Problem. Given a set $\mathcal{U}$ of n users and a set $\mathcal{T}$ of m items, we say that a user $u \in \mathcal{U}$ *favors* item $t \in \mathcal{T}$ if user u gives item t the highest score (or ranking), relative to other items in $\mathcal{T}$. We have at our disposal a *favorite-probing query* (or *query* for short) q, which probes—and subsequently returns— the favorite items $q(u)$ of a given user $u \in \mathcal{U}$. For simplicity, we sometimes say *"probe a user"* to mean that we issue a query to discover a user's favorite items. We assume that the preference of a user remains *unknown* until a favorite-probing query is issued to probe about it. In addition, we measure the *support* of an item by the number of users who favor the item. A user can favor multiple items simultaneously by giving the highest score to multiple items.

Problem Aim. The aim of our problem is to discover all the items whose individual support is at least pn, where $p \in (0, 1]$ is a parameter called *support proportion* and n is the number of users. In other words, each of these items must be favored by at least $p \times 100\%$ of all users. We call each of these items a *frequent item* (or sometimes *popular item*). Table 1 shows an example of our problem.

Table 1. Illustration. We have $n = |\mathcal{U}| = 6$, $m = |\mathcal{T}| = 4$, $\mathcal{U} = \{u_1, u_2, u_3, u_4, u_5, u_6\}$, and $\mathcal{T} = \{t_1, t_2, t_3, t_4\}$. If $p = 1/3$ so that $pn = (1/3) \cdot 6 = 2$, then both t_1 and t_2 are frequent items, since their individual support is at least $pn = 2$. However, neither t_3 nor t_4 is a frequent item, since their individual support is below 2. In this example, both users u_1 and u_5 favor two items each, whereas other users favor only one item each.

Item	Users favoring the item	Support
t_1	u_1, u_2, u_3	3
t_2	u_1, u_4, u_5, u_6	4
t_3	u_5	1
t_4	None	0

Worst-Case Lower Bound and Remedy. Readers might raise this question: Is it always possible to solve the proposed problem without probing the preferences of all users? The answer is no if an exact answer is always desired; for instance, if a particular item t is favored by $pn - 1$ users, then we must probe all n users until we can confirm whether item t is frequent. This is because leaving any user's preference unaccounted for opens up the possibility that the user favors item t. This in turn implies that *every* exact algorithm requires at least n queries in the worst case—meaning that our problem has a lower bound of n query. This result naturally motivates us to consider another question: would it be *fatal* to return an item whose support is only marginally lower than the threshold pn? We believe the answer is no, as supported by the applications to be discussed in Sect. 2.

Contributions. We are the first to propose the HIDDEN FREQUENT ITEM MINING problem, which finds applications in a variety of areas. Our problem also bears a strong relationship to numerous database problems. To efficiently solve our problem, we present the first sampling algorithm that outperforms the naive approach and other competing methods both in theory and in practice. Our analysis shows that to fulfil a strong probabilistic guarantee, our algorithm needs to issue only $s = \max\{\frac{8p}{\varepsilon^2} \ln \frac{m}{\delta}, \frac{12(p-\varepsilon)}{\varepsilon^2} \ln \frac{m}{\delta}\}^1$ queries, which equals $O(\log m)$ when all user parameters are fixed. One feature of s is its independence of n—suggesting that the number of users to probe is irrelevant to the number of users itself. To further extend the usability, we also describe how our algorithm can handle the scenario where users' preferences frequently require updates.

2   Applications

Application 1 (Marketing Survey). It is often a top priority for various industries to conduct marketing surveys to identify their most popular products for promotional purposes. For example, companies in the movie industry are eager to find out the popular movie actors to further promote those actors' movies. The same promotion strategy is also widely employed in various other industries, such as the music and retail industries. Companies in the music industry aim to identify the most popular musical artists to promote their albums, whereas those in the retail industries are keen to discover which brands of a certain type of product (e.g., red wine) are favored by customers.

Although such promotions could bring in substantial profits to the companies, the task of identifying popular products could also be costly due to the huge number of customers needed to probe. For example, the number of unique visitors to the well-known movie website IMDb (http://www.imdb.com) exceeds 250 million each month [1]. In addition, companies often have to offer financial incentives (e.g., coupons) to attract participation in a survey. Thus, the costs of

---

[1] ε is the error parameter and δ is the confidence parameter. They are fully discussed in Sect. 4.1.

conducting surveys include both manpower and funding. As a result, minimizing the number of customers needed to probe should be the goal of a company.

We can model the above marketing survey as our proposed problem. If we aim to discover the popular movie actors, then each customer is modeled as a user $u \in \mathcal{U}$ and each actor as an item $t \in \mathcal{T}$. The aim is therefore to identify all the actors that are favored individually by at least pn customers, while at the same time minimizing the total number of queries required.

Our problem can also discover *sleeping beauties* [11]—items that have potential to become popular but have yet to. For instance, although IMDB already provides a list of popular movies [2], the movies obtained from our problem may not appear in that list, because some movies of popular movie actors may not be popular or well-known. However, such movies possess high potential to go viral because they feature movie actors favored by current users.

Application 2 (Sponsored Search). Another major application of our problem is to help search engine companies uncover popular ads. Given a set of possible search terms—some of which may have never been entered by users before—and a set of ads, search engine companies would want to identify the ads that receive the highest *score* (or *Adrank*, as it is called in the literature) from many search terms. Those ads would therefore be shown frequently to users when they type in keywords appearing in the search term set. To model this application as our proposed problem, we let $\mathcal{U}$ be the set of search terms and $\mathcal{T}$ be the set of ads. The aim is therefore to identify a subset R of ads such that each ad in R receives the highest score from at least pn search terms. At the same time, however, we want to keep the number of score calculations to a minimum. It is worth noting that set R changes over time because any updates of advertisers' bid price for their ads could affect R. To tackle this problem, we will propose a fast update procedure in Sect. 5.4.

Discovering popular ads can benefit both search engine companies and advertisers. On the companies' side, they may want to raise the price that they charge for popular ads to further increase their revenue. On the advertisers' side, search engine companies can now inform the advertisers concerned that their ads are frequently shown to prospective customers. Given this assurance, those advertisers would more likely keep advertising in the search engines. This information also serves as a valuable indication to the advertisers that their current settings of bid prices and campaign budgets are reasonable. In fact, it is a well-known issue in the sponsored search community that advertisers often struggle to find out whether their current setting is effective in attracting users [6,24]. Therefore, our problem can alleviate this issue.

Application 3 (Crowdsourcing). Our problem can also help crowdsourcing platforms discover the most popular task types among their users. A crowdsourcing platform allows task requesters to put tasks of various types (e.g., image tagging and audio transcription) onto the website and allows workers to work on the tasks offered by the requesters [10]. There are several well-established crowdsourcing platforms for both task requesters and workers, such as Amazon Mechanical Turk (AMT) (http://www.mturk.com/).

Suppose that a new worker comes to a crowdsourcing platform to look for tasks to work on. However, since the platform does not possess any personal information about the new worker, what type of task can the website recommend to him? A natural step for the website to take is to recommend a set of popular task types that are the favorites of its current workers to the new visitor. In order to identify this set, we can model the scenario as our problem, with the user set $\mathcal{U}$ representing the current workers and the item set $\mathcal{T}$ the task types. Our goal is therefore to discover the task types that are favored by many workers. As with Application 1, because a query requires both manpower and financial incentives, a crowdsourcing company should aim to identify the set of popular task types by probing as few workers as possible.

3 Related Work

3.1 Frequent Item Mining

The *frequent item mining* problem [5,12,13] is related to our problem and can be stated as follows: Given a set of items and a set of transactions—each of which is a subset of the set of items—we identify all the itemsets, which are subsets of the item set, such that each of the itemsets appears in at least a given number of the transactions.

For the sake of comparison, we can view the favorite items of a particular user in our problem as a transaction in the frequent item mining problem. There are several crucial differences between the two problems. (1) In our problem, because a user in fact does not make a "transaction" (i.e., the favorite items), it makes sense to assume that the "transaction" is hidden. Therefore, the favorite items of a user remain unknown until we issue a favorite-probing query to identify them. In the frequent item mining problem, by contrast, a transaction is indeed made by a customer and thus it makes sense instead to assume that it is known. Thus, there is no need to issue a favorite-probing query to find the "favorite items" of a user in the frequent item mining problem. In short, while our counterpart assumes that user preferences are known, our problem does not. (2) In our problem, a transaction represents the favorite items of a user. However, in the frequent item mining problem, a transaction generally does not correspond to the favorite items of a user, because, for instance, the items in a transaction might be bought for other people.

3.2 Heavy Hitter Finding

The *heavy hitter finding problem* [7,8] is similar to the frequent item mining problem, in that a heavy hitter corresponds to a frequent item in frequent item mining. The difference is that in heavy hitter finding, the transactions come as a *data stream*, whereas in traditional frequent item mining, the transactions are fixed and remain constant throughout. The heavy hitter finding problem also assumes that the items in each transaction are known and is therefore different from our problem.

3.3 Survey Sampling

Our problem bears similarities to problems in survey sampling. A particularly relevant one is the *estimation of population proportion* problem [21,22]. This problem can be stated as follows: given a set of user and a set of categories (or items in the language of our problem), and each user belongs to exactly one category, the aim is to estimate the proportion of users belonging to each category. There are two major differences that distinguish our problem from the problem in survey sampling. (1) In our problem, a user is allowed to favor multiple items, whereas under the setting of the estimation of population proportion problem, each user can favor *only one* item. Therefore, our problem is far more general than its counterpart. (2) In the estimation of population proportion literature, results are commonly derived based on the following two assumptions [21,22]: (1) the data is normally distributed, or at least the normal approximation can be applied to the data. (2) The *finite population correction factor* can be ignored. In this paper, we do not make either of these assumptions. Despite the differences above, we compare the performance of the algorithm proposed in [21,22] with ours in Sect. 5.1.

4 Problem Definition

Let $\mathcal{U}$ and $\mathcal{T}$ be two given sets, where $\mathcal{U}$ is the *user set* containing $n = |\mathcal{U}|$ users and $\mathcal{T}$ the *item set* containing $m = |\mathcal{T}|$ items. Each user $u \in \mathcal{U}$ is associated with a unique score function that takes as input the item set $\mathcal{T}$ and returns as output the item(s) $t^* \in \mathcal{T}$ that receives the highest score (or highest ranking) from user u, relative to other items in $\mathcal{T}$. We say that a user u favors (or prefers/endorses) the item $t^* \in \mathcal{T}$ if and only if user u gives item t the highest score.

We have at our disposal a *favorite-probing query* q that takes as input a user $u \in \mathcal{U}$ and returns as output the favorite item(s) $q(u) \in \mathcal{T}$ of that user. We measure the *support* of an item $t \in \mathcal{T}$, denoted by support(t), by the number of users who favor that item. We say that an item $t \in \mathcal{T}$ is *frequent* (or sometimes *popular*) if support(t) $\geq pn$, where $p \in (0, 1]$ is a parameter called the *support proportion*.

Aim. Given the user set $\mathcal{U}$ and item set $\mathcal{T}$, the HIDDEN FREQUENT ITEM MINING problem is to find a subset $R \subseteq \mathcal{T}$ of the item set such that R contains all the frequent items and no items that are not frequent. Formally, $R = \{\, t \in \mathcal{T} \mid$ support(t) $\geq pn \,\}$.

4.1 ε-approximation

We present a relaxed version of the HIDDEN FREQUENT ITEM MINING problem. We call it the ε-approximation of the HIDDEN FREQUENT ITEM MINING problem. We first classify items into different types according to their support. Table 2 provides a summary of the classification.

Definition 1 (Classification of items). *Let $p, \varepsilon \in (0, 1]$ be two parameters. We say that an item $t \in \mathcal{T}$ is **frequent** if $support(t) \geq pn$; we say that t is **potentially-frequent** if $(p - \varepsilon)n \leq support(t) < pn$. Otherwise, we say that t is **infrequent**.*

Problem Formulation. An algorithm for the ε-approximation of the HIDDEN FREQUENT ITEM MINING problem accepts three user-specified parameters: (i) support proportion $p \in (0, 1]$, (ii) error parameter $\varepsilon \in (0, 1]$ such that $\varepsilon < p$, and (iii) failure parameter $\delta \in (0, 1]$. Given these parameters, the algorithm provides the following performance guarantee.

Definition 2 (ε-approximation guarantee). *An algorithm is said to achieve the ε-approximation guarantee (or guarantee for short) for the HIDDEN FREQUENT ITEM MINING problem if, with probability at least $1 - \delta$, the algorithm satisfies the following properties simultaneously:*

P1 *All frequent items (i.e., items whose individual support is at least pn) in item set $\mathcal{T}$ are returned. In other words, the algorithm will produce no **false negatives** (i.e., recall $= 100\%$).*

P2 *No infrequent items (i.e., items whose individual support is less than $(p-\varepsilon)n$) in item set $\mathcal{T}$ are returned. In other words, the algorithm will not return an item that has a support lower than the minimum tolerable value.*

Remark 1 (Treatment of potentially-frequent items). Notice that potentially-frequent items may or may not be returned. If such an item happens to be returned, then our algorithm is said to have returned a *false positive*—the only scenario in which our algorithm errs. Still, all such false positives possess a desirable property that their support is fairly high: they are only marginally lower than the threshold value sought by the user. In most cases, therefore, returning such a high-support false positive should not be fatal and should be tolerable to the user. As we see in Sect. 5.3, our proposed algorithm also possesses another desirable property that for every potentially-frequent item, the lower its support, the lower the chance of it being returned.

Table 2. Summary of Definitions 1 and 2.

Scenario	Classification (Definition 1)	Decision (Definition 2)
$support(t) \geq pn$	t is a frequent item	Return t
$(p - \varepsilon)n \leq support(t) < pn$	t is a potentially-frequent item	May or may not return t (i.e., inconclusive)
$support(t) < (p - \varepsilon)n$	t is an infrequent item	Do not return t

Example 1. Suppose that $n = 10^5$, $m = 10^3$, $p = 10\%$, $\varepsilon = 1\%$, $\delta = 1\%$. Then, with probability at least $1 - \delta = 99\%$, an algorithm that achieves the ε-approximation guarantee returns all items with an individual support at least $pn = 0.1 \cdot 10^5 = 10,000$ (Property P1). Moreover, the algorithm does not return any items with a support less than $(p - \varepsilon)n = (0.1 - 0.01)10^5 = 9,000$ (Property P2). This leaves those items with a support between 9,000 and 10,000; these items may or may not be returned (Remark 1).

5   Algorithm

Notation: Discover(*t*). For every item t, we denote by discover(t) the support that is *discovered* by our algorithm for item t. This is different from support(t), which means the *true* support of item t.

5.1   Support-Sampling (SS)

We propose a fast algorithm that provides the ε-approximation guarantee. Our algorithm, termed SUPPORT-SAMPLING (SS), works as follows: It first randomly selects a user, and then it issues a favorite-probing query to identify the user's favorite item(s). It then increments the discovered support of the item(s) accordingly. Our algorithm will repeat the above process for $s = \max\{\frac{8p}{\varepsilon^2} \ln \frac{m}{\delta}, \frac{12(p-\varepsilon)}{\varepsilon^2} \ln \frac{m}{\delta}\}$ times. In other words, it in total selects (without replacement) s users and finds their respective favorite item(s). After sampling all these s users, SS returns all items whose discovered support is at least $(p - \varepsilon/2)s$, along with the probability ($= 1 - \delta$ if not early terminated) of successfully achieving the ε-approximation guarantee. Notice that the incremental nature of SS makes early termination with a performance guarantee possible. Specifically, at any point of execution, the user can ask SS to return a set of items that (potentially) satisfy the ε-approximation guarantee, together with the probability of success. We call our algorithm *support sampling* because each sample (i.e., a selected user) can be regarded as a piece of supporting evidence that an item is frequent. The pseudocode of SS is given in Algorithm 1.

Comparison with Survey Sampling. Recall from Sect. 3.3 that our proposed problem shares similarities with the estimation of population proportion problem in survey sampling [21,22]. In particular, the method proposed in [21,22] can be adapted to solve a special case of our problem—when each user favors exactly one item. However, the method becomes *invalid* if any user favors more than one item [21,22]. The method requires a sample size of $\max_{1 \leq i \leq m} z^2(1/i)(1-1/i)/\varepsilon^2$, where z is the upper $(\delta/2i) \times 100$th percentile of the standard normal distribution. This sample size is significantly larger than ours when the error tolerance ε is small, which is often the case in practice. For example, our sample size is only half that of our counterpart when $p = 1\%$, $\varepsilon = 0.1\%$, $\delta = 1\%$, $m = 1000$ and for any arbitrary value of n (because both sample sizes are independent of n).

Algorithm 1. SUPPORT-SAMPLING (SS)

Input: $\mathcal{U}, \mathcal{T}, p, \varepsilon, \delta$

Output: A set R of items, the probability of success.

1: $s = \max\left\{\frac{8p}{\varepsilon^2}\ln\frac{m}{\delta}, \frac{12(p-\varepsilon)}{\varepsilon^2}\ln\frac{m}{\delta}\right\}$ //Sample size

2: **for each** item $t \in \mathcal{T}$ **do**

3:     discover$(t) = 0$ //Initialization

4: **for** $i = 1$ **to** s **do**

5:     select a user u uniformly at random and independently from $\mathcal{U}$

6:     issue a favorite-probing query $q(u)$ to identify the favorite item(s) t^* of user u

7:     increment (all) discover(t^*)

8:     **if** user chooses early termination **then**

      //See Lemma 3 for derivation of failure probability

9:         failure $= \max\left\{m\exp\left(-\frac{i\varepsilon^2}{8p}\right), m\exp\left(-\frac{i\varepsilon^2}{12(p-\varepsilon)}\right)\right\}$

      //R is the set of returned items

10:       $R = \{t \in \mathcal{T} \mid \text{discover}(t) \geq (p - \varepsilon/2)i\}$

11:       **return** $(R, 1 - \text{failure})$ //Early termination

12: $R = \{t \in \mathcal{T} \mid \text{discover}(t) \geq (p - \varepsilon/2)s\}$

13: **return** $(R, 1 - \delta)$

The method in [21,22] has several disadvantages. As detailed in Sect. 3.3, their results assume that data is normally distributed and the finite population correction factor can be ignored. Our derivation, however, makes no such assumptions. Furthermore, to evaluate their sample size, we need to solve a maximization problem, which is inconvenient in practice. In contrast, our derivation gives rise to a simple closed-form formula that can be readily evaluated.

5.2 Analysis of Support-Sampling

We show that SS provides the ε-approximation guarantee. In the proofs, we allow each user to favor multiple items. Proofs are presented in the Appendix.

Proof Strategy. We first consider individually the probability that SS fails to achieve Properties P1 and P2. We then use the *union bound* to find an upper bound for the probability that our algorithm *fails*, whereby deriving the required sample size to achieve the ε-approximation guarantee. We start by proving three lemmas that are useful in showing the ε-approximation guarantee of SS. All proofs can be found in the Appendix.

Lemma 1. SUPPORT-SAMPLING *fails to achieve Property P1 of ε-approximation guarantee with probability at most* $ke^{-\frac{s\varepsilon^2}{8p}}$, *where k is the number of frequent items.*

Lemma 2. SUPPORT-SAMPLING *fails to achieve Property P2 of ε-approximation guarantee with probability at most* $(m - k)e^{-\frac{s\varepsilon^2}{12(p-\varepsilon)}}$.

Lemma 3. SUPPORT-SAMPLING *fails to achieve the ε-approximation guarantee with probability at most* $\max\{me^{-\frac{s\varepsilon^2}{8p}}, me^{-\frac{s\varepsilon^2}{12(p-\varepsilon)}}\}$.

Theorem 1. SUPPORT-SAMPLING *achieves the ε-approximation guarantee by sampling s users, where* $s = \max\{\frac{8p}{\varepsilon^2}\ln\frac{m}{\delta}, \frac{12(p-\varepsilon)}{\varepsilon^2}\ln\frac{m}{\delta}\}$.

Proof. See Appendix.

Remark 2. Theorem 1 implies that if all parameters are fixed, the sample size s becomes $O(\log m)$.

Example 2 (Performance Comparison). Table 3 shows the number of queries needed by SS under a variety of input settings. SS consistently requires only a tiny number of queries—typically less than 0.5% of that needed by the naive one.

Table 3. Sample size needed by SS under various input settings. Input in the first row is the default; numbers in bold are values different from the default.

Input					Sample size	Ratio
p	ε	δ	n	m	s	s/n
.1	.02	.05	10^7	10^7	$4.6 \cdot 10^4$	0.46%
.2	**.04**	.05	10^7	10^7	$2.3 \cdot 10^4$	0.23%
.2	**.06**	.05	10^7	10^7	$0.9 \cdot 10^4$	0.09%
.1	.02	**.01**	10^7	10^7	$5.0 \cdot 10^4$	0.50%
.1	.02	.05	$\mathbf{10^8}$	10^7	$4.6 \cdot 10^4$	0.05%
.1	.02	.05	$\mathbf{10^9}$	10^7	$4.6 \cdot 10^4$	0.005%
.1	.02	.05	10^7	$\mathbf{10^8}$	$5.1 \cdot 10^4$	0.51%
.1	.02	.05	10^7	$\mathbf{10^9}$	$5.7 \cdot 10^4$	0.57%

5.3 Analysis of Potentially-Frequent Items

In this section, we give further insight into how often SS returns potentially-frequent items—those with a support marginally lower than pn. Our result shows that if a potentially-frequent item t has a support between $(p - \varepsilon/2)n$ and pn, then it is *likely* that it will be returned. On the contrary, if its support is between $(p-\varepsilon)n$ and $(p-\varepsilon/2)n$, then it is *unlikely* that it will be returned. More generally, our results signify that for every potentially-frequent item, the lower its support, the less likely our algorithm will return it. Therefore, SS possesses another desirable property that if a potentially-frequent item happens to be returned, then the chance is higher that it has a support close to pn rather than to $(p - \varepsilon)n$. Our discussion is supported by the results below. Proofs are presented in the Appendix.

Lemma 4. *Let random variables X and Y be the discovered support of items t_h and t_ℓ respectively, such that $X \sim B(s, p_h)$ and $Y \sim B(s, p_\ell)$, where $p_h = support(t_h)/n$ and $p_\ell = support(t_\ell)/n$. If $support(t_\ell) \leq support(t_h)$, or equivalently $p_\ell \leq p_h$, then*

$$Pr[Y \geq c] \leq Pr[X \geq c], \text{ where } c = (p - \varepsilon/2)s.$$

Lemma 4 suggests that as the support of a potentially-frequent item decreases, so does the chance of SS returning it.

We now derive tight bounds for the probability that a potentially-frequent item is returned.

Proposition 1. *Let the support of a given potentially-frequent item t be $pn - r$, where $\varepsilon n/2 < r \leq \varepsilon n$. The probability that item t is returned by SS is **at most** $e^{-2s\left(\frac{r}{n} - \frac{\varepsilon}{2}\right)^2}$.*

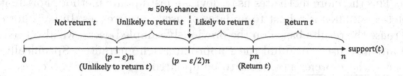

Fig. 1. Decisions by SUPPORT-SAMPLING for any item t.

Example 3. It is easy to see that the bound decreases exponentially as r increases from $\varepsilon n/2$ to εn. To examine how rapidly the bound falls, we adopt the default input in Table 3. Now, if $r = \frac{\varepsilon n}{1.5}$, then the probability that an item with a support $= pn - \frac{\varepsilon n}{1.5} = (p - \frac{\varepsilon}{1.5})n$ is returned is at most 1.6×10^{-3}. As r increases from $\frac{\varepsilon n}{1.5}$ to εn, the probability decreases exponentially from 1.6×10^{-3} to 6.3×10^{-26}.

Our next result shows that if an item has support between $(p - \varepsilon/2)n$ and pn, it is *likely* to be returned.

Proposition 2. *Let the support of a given potentially-frequent item t be $pn - r$, where $0 < r < \varepsilon n/2$. The probability that item t is returned by SS is **at least** $1 - e^{-2s\left(\frac{r}{n} - \frac{\varepsilon}{2}\right)^2}$.*

Special Case: $r = \varepsilon n/2$. A glance at Propositions 1 and 2 suggests that both are *undefined* for $r = \varepsilon n/2$. The reason is that for an item t with support$(t) = (p - \varepsilon/2)n$, its *expected* discovered support equals $(p - \varepsilon/2)s$, and our algorithm returns item t only if its discovered support is at least $(p - \varepsilon/2)s$—which is exactly the expected discovered support. Unfortunately, in general, a tail inequality—including Hoeffding's inequalities—is only capable of bounding the probability that the value assumed by a random variable *exceeds* or *falls behind* the expected value. Thus, we cannot use a tail inequality to bound the probability that item t is returned.

Even so, we argue that the probability of returning item t is approximately 0.5 by appealing to the *Central Limit Theorem* [23]. Notice that the discovered support of item t follows the binomial distribution $B(s, p - \varepsilon/2)$. Since s is typically larger than 30, which is a rule of thumb for applying the Center Limit Theorem [23], the Central Limit Theorem tells us that the binomial distribution is approximately a normal distribution with mean $\mu = (p - \varepsilon/2)s$ [23]. Now, because the probability for a normally distributed random variable to be at least as large as its mean is 0.5, it follows that the probability of item t being returned is approximately 0.5. Figure 1 summarizes our discussion in this section.

5.4 Updates of Users' Preferences

In some applications, a user's preference can be ever-changing. In the sponsored search application, for example, the score that a search term assigns to an ad will change accordingly whenever the advertiser concerned adjusts the bid price for that ad. This therefore motivates us to devise a fast update method. Fortunately, SUPPORT-SAMPLING is robust to updates of users' preferences. In fact, not only can we reuse the results based on the previously sampled users, we also need not sample any *new* users to fulfill the ε-approximation guarantee. Specifically, for each user u whose preference needs to be updated, we consider two cases.

Case 1: u was sampled. We issue an additional query to identify the new favorite item(s) t of user u and increment discovered(t). we also decrement the discovered support of the item(s) that was u's favorite.

Case 2: u was not sampled. We do not need to issue any additional query, nor do we need to update the result of our algorithm.

Table 4. Input values (defaults shown in bold).

Input	Values
n	1m, 2m, 4m, **8m**, 10m
m	40k, 80k, 160k, **320k**, 640k
p	$0.05, \mathbf{0.10}, 0.15, 0.20, 0.25, 0.30$
ε	$\mathbf{0.1p}, 0.15p, 0.2p, 0.25p, 0.3p$

Correctness. After the above procedure, SS still maintains the ε-approximation guarantee. For Case 1, the score function update of a sampled user u will not affect the selection probability of any user, because the selection probability depends only on the number of users, rather than on the users' score functions. For Case 2, we note that user u's preference will not be considered with or without a score function update, because (1) u was not sampled and (2) the update will not affect the selection probability, as in Case 1.

Example 4. Suppose we update the preferences of k $(1 \leq k \leq n)$ users, each of whom is randomly chosen without replacement. While the naive one needs to issue exactly k queries, our algorithm needs to issue a query only if the user was sampled previously. So, the number of queries needed by our algorithm for the update follows a *hypergeometric* distribution [23]. Thus, the expected number of queries needed is $k \times (s/n)$. Using the default setting shown in Table 3, we have $s/n \approx 0.46\%$. Therefore, the expected number of queries is just 0.46% of the number required by the naive algorithm.

6 Experiments

We experimentally evaluate our proposed algorithms using both real and synthetic datasets.

Setup. All experiments were run on a machine with a 3.4 GHz CPU and 32 GB memory. The OS is Linux (CentOS 6). All algorithms were coded in C++ and compiled using g++.

We compare our proposed algorithms, SUPPORT-SAMPLING (SS), with the Naive Solution (NS) as well as three adapted existing algorithms: Survey Sampling [21,22] (Sect. 3.3), Top-k query [15,19] (a well-known problem in the database literature) and Sticky Sampling [17], which is a classic algorithm for the heavy hitter problem (Sect. 3.2). The adaption of Top-k query and Sticky Sampling is made possible by requiring the algorithms to issue a favorite-probing query for each user to identify the user's favorite items as a first step. In particular, because Sticky Sampling is a streaming algorithm, we implement it in such a way that a new user arrives only after we finish processing the user immediately preceding him/her, so as to allow enough time for it to discover the favorite items of each user. The performance metrics are the execution time and accuracy. For randomized algorithms such as SS and Sticky Sampling, we ran them 100 times for each experiment setting to report its average execution time and average accuracy. The default values of the parameters p and ε are 10% and 1% respectively.

Real Datasets. We use two real datasets to simulate the marketing survey application (Application 1): Yahoo! musical artist ratings dataset [4] (Yahoo! dataset) and Netflix movie ratings dataset [3] (Netflix dataset). Yahoo! dataset contains over 100 million user ratings of 98,211 musical artists ($m = 98,211$) by 1,948,882 users ($n = 1,948,882$), whereas Netflix dataset consists of more than 100 million users ratings of 17,770 movies ($m = 17,770$) by 480,189 users ($n = 480,189$). In real datasets, a query $q(u)$ on user u corresponds to finding the artists (or movies for the Netflix dataset) that receive the highest rating from user u, relative to other artists/movies. We assume those artists/movies that did not receive a rating from user u were not favored by u. We must also emphasize that although the ratings given by each user are *known* before we issue a query, in practice this information is *unknown* beforehand. To account for the need to obtain user ratings in practice, we multiply the execution time of each algorithm by 10,000.

Synthetic Datasets. We use synthetic datasets to simulate the sponsor search application (Application 2). Following related studies (e.g., [18,25]), we view each item $t \in \mathcal{T}$ as a point in a space, and we set the dimension (i.e., number of attributes) d to 5. To exploit the generality of our problem, each user is randomly associated with either a linear $\sum_{i=1}^{d} w_i x_i$ or quadratic $\sum_{i=1}^{d} w_i x_i^2$ score function, where w_i is the attribute weight of the score function for the ith attribute of an item, and x_i is the ith attribute value of an item. Note that although linear score functions are often employed in the literature (e.g., [18,25]) for its simplicity, non-linear score functions, such as quadratic ones used in our experiments, can better model users' preferences in many cases [16]. We independently generate each attribute value of an item from the uniform distribution with support (0,1), and so is each attribute weight w_i of a user's score function.

In synthetic datasets, a query $q(u)$ on user u corresponds to computing $\text{argmax}_{t \in \mathcal{T}} \text{score}(u,t)$, where $\text{score}(u,t)$ is the score that user u gives to item t. Experiments on synthetic datasets were conducted using various input settings shown in Table 4. The settings are similar to existing studies (e.g., [18,25]), except that dataset sizes are proportionally increased. As in [25], we set m (number of items) to be smaller than n (number of users). This is because in practice—and also in our real datasets—the item set size is often substantially smaller than the user set size. In all experiments, the failure parameter δ is set to 5%, due to its negligible impact on the execution time.

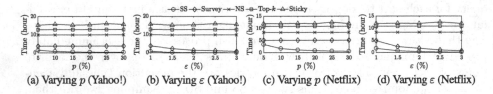

(a) Varying p (Yahoo!)   (b) Varying ε (Yahoo!)   (c) Varying p (Netflix)   (d) Varying ε (Netflix)

Fig. 2. Performance of SS versus other competitors on real datasets.

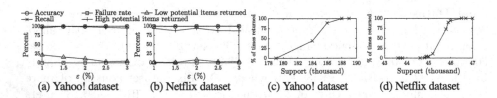

(a) Yahoo! dataset   (b) Netflix dataset   (c) Yahoo! dataset   (d) Netflix dataset

Fig. 3. Statistics of SS on real datasets.

6.1   Results on Real Datasets

We compare the execution time of various algorithms and then discuss the accuracy of SS on real datasets.

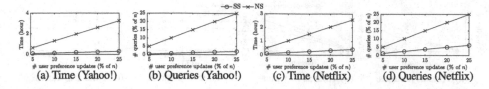

Fig. 4. SS vs. NS on updating users' preferences on real datasets.

Comparison on Efficiency. As Fig. 2 shows, SS consistently requires far less time than NS and other competitors, and their performance gap widens as p or ε increases. In particular, when both p and ε are at their default values, SS takes only about 8% of the time by NS to process the Yahoo! dataset, and about 20% of the time by the survey method, which is the second best. Notice that Top-k and Sticky Sampling are even more inefficient than NS because in order to adopt them, it is necessary to first issue a favorite-probing query for each user to identify the user's favorite items, which is exactly the strategy used by NS.

Accuracy of SS. We start by introducing some terms. We call a potentially-frequent item t a *high potential item* if $(p - \varepsilon/2)n \leq \mathrm{support}(t) < pn$; otherwise, we call it a *low potential item*.

Figure 3 shows the statistics of SS on real datasets. The *Failure rate* in Figs. 3(a) and (b) refers to the proportion of times SS fails to achieve the ε-approximation guarantee. Observe that although the failure parameter δ is set to 5%, SS never failed to achieve the ε-approximation guarantee in all the experiments, and thus recall is always 100%. This is because our theoretical analysis for SS is conservative, and so *in practice* the failure probability is generally much lower than the failure parameter δ. The accuracy of SS is consistently higher than 99.7% in all experiments. The very slight inaccuracy stems from the fact that some potentially-frequent items are returned. However, notice that those returned potentially-frequent items are often high potential items, rather than low potential ones. This also verifies our findings in Sect. 5.3: high potential items are likely to be returned, whereas low potential items are unlikely to. Figures 3(c) and (d) show the fraction of times over the 100 runs that each potentially-frequent item is returned, when parameters p and ε are at their default values.

Comparison on Updating Users' Preferences We compare the performance of SS and NS on updating users' preferences. Recall from Sect. 5.4 that SS can handle updates of users' preferences efficiently, whereas NS needs to issue an additional query for each user whose preference requires an update. Figures 4(a) and (b) show that for the Yahoo! dataset, both the execution time and number of queries needed by NS for updating users' preferences grow far more rapidly than SS. Similar trends can also be observed from the Netflix dataset (Figs. 4(c) and (d)).

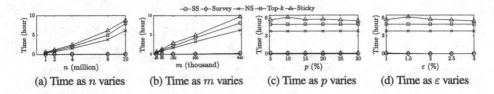

| (a) Time as n varies | (b) Time as m varies | (c) Time as p varies | (d) Time as ε varies |

Fig. 5. Performance of SS versus other competitors on synthetic datasets.

6.2 Results on Synthetic Datasets

We compare SS with other algorithms on synthetic datasets with sizes specified in Table 4.

Scalability with n. The execution time of NS, Top-k query and Sticky Sampling grow linearly with n (Fig. 5(a)). The linear relationship makes them impractically slow when n is large. For instance, NS requires more than 6 h to complete when n reaches 10M. In stark contrast to its competitors, SS is *insensitive* to the increase of n, consistently taking less than 4 min to execute regardless of the size of n.

Scalability with m. Figure 5(b) shows that the execution time of our competitors grows linearly with m. For example, NS takes almost 6.2 h to process 640K items. By contrast, the execution time taken by SS grow extremely slowly with m, and it requires less than 8 min to process 640K items. The efficiency of SS can be explained by referring to the sample size formula $s = \max\{\frac{8p}{\varepsilon^2} \ln \frac{m}{\delta}, \frac{12(p-\varepsilon)}{\varepsilon^2} \ln \frac{m}{\delta}\}$, where s grows only logarithmically with m.

Influence of Support Proportion p. As Fig. 5(c) show, SS requires less time as p increases. The cost of our competitors remain constant because of their independence of p.

Influence of Error Rate ε. Similar to the situation when p increases, the cost of SS falls as ε increases (Fig. 5(d)). In particular, SS requires only about 1.5% of the execution time and queries by NS when $\varepsilon = 1\%$.

7 Conclusion

We proposed an interesting problem called HIDDEN FREQUENT ITEM MINING, which has important applications in various fields. We devise a sampling algorithm that sacrifices slight accuracy in exchange for substantial improvement in efficiency.

Acknowledgement. The research is supported by HKRGC GRF 14205117.

A   Appendix: Probabilistic Inequalities

We outline several probabilistic inequalities that will be employed in our proofs.

Proposition 3 (Chernoff's bound [9] and Hoeffding's inequality [14]).
*Let $X_1, X_2, \ldots, X_s$ be independent Bernoulli variables and $Pr[X_i = 1] = p$,
where $0 < p < 1$, for $i = 1, 2, \ldots, s$. Let $X = \sum_{i=1}^{s} X_i$ and $E[X] = \mu$ such
that $\mu_\ell \le \mu \le \mu_h$, where $\mu_\ell, \mu_h \in \mathbb{R}$. Then, for any $\varepsilon < 1$,*

$$Pr[X \ge (1+\varepsilon)\mu_h] \le e^{\frac{-\mu_h \varepsilon^2}{3}}, \quad Pr[X \le (1-\varepsilon)\mu_\ell] \le e^{\frac{-\mu_\ell \varepsilon^2}{2}}, \tag{1}$$

$$Pr[X \ge \mu + \varepsilon] \le e^{-\frac{2\varepsilon^2}{s}}, \quad Pr[X \le \mu - \varepsilon] \le e^{-\frac{2\varepsilon^2}{s}}. \tag{2}$$

B   Appendix: Proofs

Proof (Lemma 1). Let X be the discovered support of a given frequent item t
(i.e., an item whose support is at least pn). For $i = 1, 2, \ldots, s$, let X_i be the
indicator random variable such that $X_i = 1$ if the ith sampled user favors item
t, and $X_i = 0$ otherwise. We first note that $X = \sum_{i=1}^{s} X_i$. In addition, since
our algorithm samples users independently and uniformly at random, we have
$E[X_i] = Pr[X_i = 1] \ge \frac{pn}{n} = p$, for all $i = 1, 2, \ldots, s$. So, we conclude that
$E[X] = E[\sum_{i=1}^{s} X_i] = \sum_{i=1}^{s} E[X_i] \ge ps$.

Now, we bound the probability that item t is not returned. According to our
algorithm, this event occurs if and only if the discovered support of item t is less
than $(p - \varepsilon/2)s$:

$$\mathbf{Pr}\left[X < \left(p - \frac{\varepsilon}{2}\right)s\right] = \mathbf{Pr}\left[X < \left(1 - \frac{\varepsilon}{2p}\right)ps\right] \le e^{-\frac{ps\left(\frac{\varepsilon}{2p}\right)^2}{2}} = e^{-\frac{s\varepsilon^2}{8p}}.$$

The inequality above is established by Chernoff's bound (Proposition 3).
Now, without loss of generality, suppose that there are k frequent items in the
item set $\mathcal{T}$, where $1 \le k \le m$ is unknown. As will be clear shortly, the value of
k is irrelevant to the analysis of our algorithm. By using the union bound (a.k.a
Boole's inequality), therefore, the probability that at least one of the k frequent
items is not returned by our algorithm is bounded above by $ke^{-\frac{s\varepsilon^2}{8p}}$.   □

Proof (Lemma 2). Now, let Y be the discovered support of a given infrequent
item (i.e., an item whose support is less than $(p-\varepsilon)n$). By using the same line of
argument for random variable X in Lemma 1, we can show that $E[Y] < (p-\varepsilon)s$.

Now, we bound the probability that the given infrequent item is returned.
By the design of our algorithm, this event occurs if and only if the discovered
support of that item is at least $(p - \varepsilon/2)s$:

$$\mathbf{Pr}\left[Y \ge \left(p - \frac{\varepsilon}{2}\right)s\right] = \mathbf{Pr}\left[Y \ge \left(1 + \frac{\varepsilon}{2(p - \varepsilon)}\right)(p - \varepsilon)s\right] \le e^{-\frac{(p-\varepsilon)s\left[\frac{\varepsilon}{2(p-\varepsilon)}\right]^2}{3}}$$

$$= e^{-\frac{s\varepsilon^2}{12(p-\varepsilon)}}.$$

The inequality is established by Chernoff's bound (Proposition 3). Since there are k items whose support is at least pn, it follows that there are at most $m - k$ items whose support is less than $(p - \varepsilon)n$ (because there are m items). By the union bound, the probability that our algorithm returns at least one of the infrequent items is at most $(m - k)e^{-\frac{s\varepsilon^2}{12(p-\varepsilon)}}$. $\square$

Proof (Lemma 3). From Lemmas 1 and 2, we know that the probability that our algorithm fails to achieve Property P1 (resp. Property P2) of the ε-approximation guarantee is at most $ke^{-\frac{s\varepsilon^2}{8p}}$ (resp. $(m-k)e^{-\frac{s\varepsilon^2}{12(p-\varepsilon)}}$). By appealing to the union bound again, the probability that our algorithm fails to achieve the ε-approximation guarantee is at most $ke^{-\frac{s\varepsilon^2}{8p}} + (m-k)e^{-\frac{s\varepsilon^2}{12(p-\varepsilon)}}$. We simplify the expression to identify the sample size s:

$$e^{-\frac{s\varepsilon^2}{8p}} \le e^{-\frac{s\varepsilon^2}{12(p-\varepsilon)}} \iff 12(p - \varepsilon) \ge 8p \iff p \ge 3\varepsilon . \tag{3}$$

$\boxed{\text{Case 1: } p < 3\varepsilon.}$ By referring to Eq. (3), we know that $e^{-\frac{s\varepsilon^2}{8p}} > e^{-\frac{s\varepsilon^2}{12(p-\varepsilon)}}$. Hence

$$ke^{-\frac{s\varepsilon^2}{8p}} + (m - k)e^{-\frac{s\varepsilon^2}{12(p-\varepsilon)}} < ke^{-\frac{s\varepsilon^2}{8p}} + (m - k)e^{-\frac{s\varepsilon^2}{8p}} = me^{-\frac{s\varepsilon^2}{8p}} .$$

$\boxed{\text{Case 2: } p \ge 3\varepsilon.}$ Again by Eq. (3), we have

$$ke^{-\frac{s\varepsilon^2}{8p}} + (m - k)e^{-\frac{s\varepsilon^2}{12(p-\varepsilon)}} \le ke^{-\frac{s\varepsilon^2}{12(p-\varepsilon)}} + (m - k)e^{-\frac{s\varepsilon^2}{12(p-\varepsilon)}} = me^{-\frac{s\varepsilon^2}{12(p-\varepsilon)}} .$$

Taking the maximum of both cases, we finish the proof. $\square$

Proof (Theorem 1). We use the result of Lemma 3.

$\boxed{\text{Case 1: } p < 3\varepsilon.}$ In this case, the probability that SUPPORT-SAMPLING fails is at most $me^{-\frac{s\varepsilon^2}{8p}}$. By setting this quantity to δ and then solve for s, we have $s = \frac{8p}{\varepsilon^2} \ln \frac{m}{\delta}$.

$\boxed{\text{Case 2: } p \ge 3\varepsilon.}$ Similarly, the probability that SUPPORT-SAMPLING fails in this case is at most $me^{-\frac{s\varepsilon^2}{12(p-\varepsilon)}}$. By setting this quantity to δ and then solve for s, we have $s = \frac{12(p-\varepsilon)}{\varepsilon^2} \ln \frac{m}{\delta}$. Taking the maximum of both cases, we finish the proof. $\square$

Proof (Lemma 4). For $i = 1, 2, \ldots, s$, let X_i and Z_i be Bernouilli distributions with mean p_h and p_ℓ/p_h, respectively. Let $Y_i := X_i Z_i$ so that $Y_i \le X_i$ with probability 1, and that Y_i is a Bernouilli distribution with mean p_ℓ. Now, notice that $X = \sum_{i=1}^{n} X_i$ and $Y = \sum_{i=1}^{n} Y_i$. Also, since $Y_i \le X_i$, it follows that $Y \le X$ with probability 1. Therefore, the event $Y \ge c$ implies the event $X \ge c$. Consequently, we have $\mathbf{Pr}\,[Y \ge c] \le \mathbf{Pr}\,[X \ge c]$. $\square$

Proof (Proposition 1). Let Y be the discovered support of a given potentially-frequent item t with $\text{support}(t) = pn - r$, where $\varepsilon n/2 < r \le \varepsilon n$. Then, using the same argument in Lemma 1, we can show that $\mathbf{E}[Y] = \left(\frac{pn-r}{n}\right)s = \left(p - \frac{r}{n}\right)s$. Hence, the probability that item t is returned by SUPPORT-SAMPLING is given by $\mathbf{Pr}\left[Y \geq \left(p - \frac{\varepsilon}{2}\right)s\right] = \mathbf{Pr}\left[Y \geq \left(p - \frac{\varepsilon}{2} + \frac{r}{n} - \frac{r}{n}\right)s\right] = \mathbf{Pr}\left[Y \geq \left(p - \frac{r}{n}\right)s + \left(\frac{r}{n} - \frac{\varepsilon}{2}\right)s\right] = \mathbf{Pr}\left[Y \geq \mathbf{E}[Y] + \left(\frac{r}{n} - \frac{\varepsilon}{2}\right)s\right]$. Now, since $r > \frac{\varepsilon n}{2}$, it follows that $\left(\frac{r}{n} - \frac{\varepsilon}{2}\right)s > 0$. Hence, we can apply Hoeffding's inequality (Proposition 3) to bound the equation: $\mathbf{Pr}\left[Y \geq \mathbf{E}[Y] + \left(\frac{r}{n} - \frac{\varepsilon}{2}\right)s\right] \leq e^{-\frac{2\left(\frac{r}{n} - \frac{\varepsilon}{2}\right)^2 s^2}{s}} = e^{-2s\left(\frac{r}{n} - \frac{\varepsilon}{2}\right)^2}.$ $\qquad\square$

Proof (Proposition 2). The proof is similar to that of Proposition 1 and is omitted due to the space constraint. $\qquad\square$

References

1. About IMDb (2019). http://www.imdb.com/pressroom/about/
2. IMDb Charts (2019). http://www.imdb.com/chart/top
3. Netflix datasets (2019). www.netflixprize.com
4. Yahoo! datasets (2019). https://webscope.sandbox.yahoo.com/
5. Agrawal, R., Srikant, R., et al.: Fast algorithms for mining association rules. In: Proceedings of 20th International Conference on Very Large Data Bases, VLDB, vol. 1215, pp. 487–499 (1994)
6. Chakrabarty, D., Zhou, Y., Lukose, R.: Budget constrained bidding in keyword auctions and online knapsack problems. In: Workshop on Sponsored Search Auctions, WWW 2007 (2007)
7. Cormode, G., Hadjieleftheriou, M.: Finding frequent items in data streams. Proc. VLDB Endow. **1**(2), 1530–1541 (2008)
8. Dimitropoulos, X., Hurley, P., Kind, A.: Probabilistic lossy counting: an efficient algorithm for finding heavy hitters. ACM SIGCOMM Comput. Commun. Rev. **38**(1), 5 (2008)
9. Dubhashi, D.P., Panconesi, A.: Concentration of Measure for the Analysis of Randomized Algorithms. Cambridge University Press, New York (2009)
10. Franklin, M.J., Kossmann, D., Kraska, T., Ramesh, S., Xin, R.: CrowdDB: answering queries with crowd sourcing. In: Proceedings of the 2011 ACM SIGMOD International Conference on Management of Data, pp. 61–72. ACM (2011)
11. Garfield, E.: Premature discovery or delayed recognition-why? Current Contents **21**, 5–10 (1980)
12. Han, J., Pei, J., Kamber, M.: Data Mining: Concepts and Techniques. Elsevier, New York (2011)
13. Han, J., Pei, J., Yin, Y.: Mining frequent patterns without candidate generation. ACM SIGMOD Rec. **29**, 1–12 (2000)
14. Hoeffding, W.: Probability inequalities for sums of bounded random variables. J. Am. Stat. Assoc. **58**(301), 13–30 (1963)
15. Ilyas, I.F., Beskales, G., Soliman, M.A.: A survey of top-k query processing techniques in relational database systems. ACM Comput. Surv. (CSUR) **40**(4), 11 (2008)
16. Kessler Faulkner, T., Brackenbury, W., Lall, A.: k-regret queries with nonlinear utilities. Proc. VLDB Endow. **8**(13), 2098–2109 (2015)

17. Manku, G.S., Motwani, R.: Approximate frequency counts over data streams. In: Proceedings of the 28th International Conference on Very Large Data Bases, pp. 346–357. VLDB Endowment (2002)
18. Peng, P., Wong, R.C.-W.: k-hit query: top-k query with probabilistic utility function. In: Proceedings of the 2015 ACM SIGMOD International Conference on Management of Data (2015)
19. Shanbhag, A., Pirk, H., Madden, S.: Efficient top-k query processing on massively parallel hardware. In: Proceedings of the 2018 International Conference on Management of Data, pp. 1557–1570. ACM (2018)
20. Solanki, S.K., Patel, J.T.: A survey on association rule mining. In: 2015 Fifth International Conference on Advanced Computing & Communication Technologies, pp. 212–216. IEEE (2015)
21. Thompson, S.K.: Sample size for estimating multinomial proportions. Am. Stat. 41(1), 42–46 (1987)
22. Thompson, S.K.: Sampling. Wiley, Hoboken (2012)
23. Walpole, R.E., Myers, R.H., Myers, S.L., Ye, K.: Probability and Statistics for Engineers and Scientists. Macmillan, New York (2011)
24. Zhang, W., Zhang, Y., Gao, B., Yu, Y., Yuan, X., Liu, T.-Y.: Joint optimization of bid and budget allocation in sponsored search. In: Proceedings of the 18th ACM SIGKDD International Conference on Knowledge Discovery and Data Mining, pp. 1177–1185. ACM (2012)
25. Zhang, Z., Jin, C., Kang, Q.: Reverse k-ranks query. Proc. VLDB Endow. 7, 785–796 (2014)

Mining Sequential Patterns of Historical Purchases for E-commerce Recommendation

Raj Bhatta, C. I. Ezeife[✉], and Mahreen Nasir Butt

School of Computer Science, University of Windsor,
401 Sunset Avenue, Windsor, ON N9B3P4, Canada
{bhatt11y,cezeife,nasir11}@uwindsor.ca

Abstract. In E-commerce Recommendation system, accuracy will be improved if more complex sequential patterns of user purchase behavior are learned and included in its user-item matrix input, to make it more informative before collaborative filtering. Existing recommendation systems that attempt to use mining and some sequences are those referred to as LiuRec09, ChoiRec12, SuChenRec15, and HPCRec18. These systems use mining based techniques of clustering, frequent pattern mining with similarity measures of purchases and clicks to predict the probability of purchases by users as their ratings before running collaborative filtering algorithm. HPCRec18 improved on the user-item matrix both quantitatively (finding values where there were 0 ratings) and qualitatively (finding specific interest values where there were 1 ratings). None of these algorithms explored enriching the user-item matrix with sequential pattern of customer clicks and purchases to capture better customer behavior. This paper proposes an algorithm called HSPRec (Historical Sequential Pattern Recommendation System), which mines frequent sequential click and purchase patterns for enriching the (i) user-item matrix quantitatively, and (ii) qualitatively. Then, finally, the improved matrix is used for collaborative filtering for better recommendations. Experimental results with mean absolute error, precision and recall show that the proposed sequential pattern mining based recommendation system, HSPRec provides more accurate recommendations than the tested existing systems.

Keywords: E-commerce recommendation ·
Sequential pattern mining · Collaborative filtering ·
User-item matrix quality · Consequential bond

1 Introduction

The main goal of a recommender system is to generate meaningful recommendations to a user for items that might interest them given the user-item

This research was supported by the Natural Science and Engineering Research Council (NSERC) of Canada under an Operating grant (OGP-0194134) and a University of Windsor grant.

© Springer Nature Switzerland AG 2019
C. Ordonez et al. (Eds.): DaWaK 2019, LNCS 11708, pp. 57–72, 2019.
https://doi.org/10.1007/978-3-030-27520-4_5

rating matrix, which indicates the rating of each item by each user. An important application of recommendation system is in the E-commerce domain [9]. In E-commerce environment, the implicit ratings derived from historical purchase and/or clickstream data are used as rating of items in the user-item rating matrix [6,7]. Many users may not be ready to provide explicit ratings for many items and there is a large set of continuously growing items (products), a very small percentage of which, each user may have purchased. In addition, users' purchase behavior change with time so that sequences of clicks and purchases play important role in capturing more realistic users' purchase behavior. Thus, one of the main challenges in the field of recommendation is how to integrate sequential pattern of purchases generated from historical and/or clickstream E-commerce data to capture more complex purchase behavior.

Collaborative filtering is one of the most widely used recommendation techniques. It accepts a user-item rating matrix (R), having ratings of each item (i) by user (u_i) denoted as r_{u_i}. The goal of collaborative filtering is to predict r_{u_i}, a rating of user u on item i which may be unknown, by going through the following four major steps [1]:

1. Compute the mean rating for each user u_j using all of their rated items.
2. Calculate the similarity between a target user v and all other users u_j. Similarity can be computed with Cosine Similarity (v, u_j) or Pearson Correlation coefficient [1] function.
3. Find similar users of target user v as their Top-N users.
4. Predict rating for target user v for item i using only ratings of v's Top-N peer group.

Collaborative filtering considers the searching of closest neighbors to generate matching recommendations. However, what people want from recommender systems is not whether the system can predict rating values accurately, but recommendations that match their interests according to time span. Thus, E-commerce recommendation system accuracy will be improved if more complex sequential patterns of user historical purchase behavior are learned and included in the user-item matrix to make it quantitatively and qualitatively rich before applying collaborative filtering.

1.1 Sequential Pattern Mining

Sequential pattern mining algorithms (for example GSP [10]) discover repeating sequential patterns (known as frequent sequences) from input historical sequential databases that can be used later to analyze user purchase behavior by finding the association between sequences of itemsets. In other words, it is the process of extracting sequential patterns whose support exceeds or is equal to a pre-defined minimum support threshold. Formally, given (a) a set of sequential records (called sequences) representing a sequential database D, (b) a minimum support threshold minsup, (c) a set of k unique candidate one length events (1-items), $C_1 = i_1, i_2, \ldots, i_n$, the problem of mining sequential patterns is that of finding the set of all frequent sequences FS in the given sequence database D of items I at the given minimum support.

Sequence database is composed of a collection of sequences $\{s_1, s_2, .., s_n\}$ that are arranged with respect to time [5]. A sequence database can be represented as a tuple ⟨SID, sequential pattern⟩, where SID: represents the sequence identifier and sequential pattern contains list of item sets with each item set containing 1 or more items. Table 2 shows an example of daily purchase sequential database. Sequential pattern mining algorithms available to find the frequent sequences from sequential databases include the GSP (Generalized Sequential Pattern Mining) [10].

Table 1. Daily click sequential database

SID	Click sequence
1	<(1,2,3), (7,5,3), (1,6), (6), (1,5)>
2	<(1,4), (6,3), (1,2), (1,2,5,6)>
3	<(1,5), (6,5,2), (6), (5)>
4	<(2,7), (6,6,7)>
5	<(1,5)>

Table 2. Daily purchase sequential database

SID	Purchase sequence
1	<(1,2), (3), (6), (7), (5)>
2	<(1,4), (3), (2), (1,2,5,6)>
3	<(1), (2), (6), (5)>
4	<(2), (6,7)>

An Example Sequential Pattern Mining with the GSP: Let us consider daily purchase sequential database (Table 2) as input, minimum support = 2, 1-candidate set $(C_1) = \{1, 2, 3, 4, 5, 6, 7\}$.

1. Find 1-frequent sequence (L_1) satisfying minimum support: Check the minimum support threshold of each singleton item and remove items that do not have occurrence count in the database greater than or equal to the minimum support threshold, to generate large or frequent 1-sequences L_1. In our case, $L_1 = \{<(1)>, <(2)>, <(3)>, <(5)>, <(6)>, <(7)>\}$.

2. Generate candidate sequence (C_2) using $L_1 \bowtie_{GSPjoin} L_1$: In this step, to generate larger 2-candidate sequence set, use 1-frequent sequence found in previous step to join itself GSPjoin way. The GSPjoin of L_{k-1} with L_{k-1} generates larger k-candidate set, by having every sequence (W_i) found in first L_{k-1} join with the other sequence (W_k) in the second L_{k-1} if sub-sequences obtained by removal of the first element of W_i and the last element of W_k are the same.

3. Find 2-frequent sequences (L_2) by scanning the database and counting the support of each C_2 sequence to retain only those sequences with support greater than or equal to the minimum support threshold.

4. Repeat the processes of candidate sequence generate (C_k) and frequent sequence count (L_k) steps until the result of either a candidate sequence generate step or a frequent sequence count step yields an empty set.

5. Output: Output frequent sequences as the union of all L's or L = $\{L_1 \cup L_2 \cup \ldots \cup L_m\}$ if the last found L is L_m.

1.2　Historical Data

E-commerce historical data consist of a list of items clicked and/or purchased by a user over a specific period of time. A fragment of E-commerce historical database data is presented in Table 3 with schema {Uid, Click, Clickstart, Clickend, Purchase, Purchasetime} where Uid represents User identity, Click represents list of items clicked by the user. Clickstart and Clickend represent the timestamps when the user started clicking items and when the click is terminated. Furthermore, purchase contains list of items purchased by the user and Purchasetime represents timestamp when the purchase happened.

Table 3. Historical e-commerce data showing clicks and purchases

Uid	Click	Clickstart	Clickend	Purchase	Purchasetime
1	1, 2, 3	2014-04-04 11:25:14	2014-04-04 11:45:19	1, 2	2014-04-04 11:30:11
1	7, 5, 3	2014-04-05 15:30:07	2014-04-05 15:59:36	3	2014-04-05 15:56:32
1	1, 6	2014-04-06 4:10:01	2014-04-06 04:30:29	6	2014-04-06 4:18:26
1	6	2014-04-07 8:50:29	2014-04-07 9:50:07	7	2014-04-07 8:59:21
1	1, 5	2014-04-08 14:10:24	2014-04-08 14:25:18	5	2014-04-08 14:19:55
2	1, 4	2014-04-13 4:01:11	2014-04-13 4:30:15	1, 4	2014-04-13 04:04:34
2	6, 3	2014-04-15 9:30:34	2014-04-15 9:40:11	3	2014-04-15 09:34:37
2	1, 2	2014-04-17 13:40:11	2014-04-17 13:59:11	2	2014-04-17 13:54:48
2	1, 2, 5, 6	2014-04-17 11:30:18	2014-04-17 11:50:19	1, 2, 5, 6	2014-04-17 11:44:55
3	1, 5	2014-04-20 09:40:45	2014-04-20 10:10:15	1	2014-04-20 10:02:53
3	6, 5, 2	2014-04-21 11:59:59	2014-04-21 12:10:39	2	2014-04-21 12:07:15
3	6	2014-04-22 17:05:19	2014-04-22 17:30:06	6	2014-04-22 17:10:28
3	5	2014-04-23 11:00:05	2014-04-23 11:20:15	5	2014-04-23 11:06:37
4	2, 7	2014-04-23 12:00:11	2014-04-23 12:30:10	2	2014-04-23 12:06:37
4	6, 6, 7	2014-04-26 9:45:11	2014-04-26 10:20:13	6, 7	2014-04-26 10:06:37
5	1, 5	2014-04-27 16:30:25	2014-04-27 16:45:45	?	

1.3　Consequential Bond (CB)

E-commerce data contains information showing some relationship between user clicks and purchases of products, used to derive a consequential bond between clicks and purchases as introduced by Xiao and Ezeife [12] in their HPCRec18 system. The term consequential bond originated from the concept that a customer who clicks on some items will ultimately purchase an item from a list of clicks in most of the cases. For example, the historical data in Table 3 shows that user 1 clicked items {1, 2, 3} and purchased {1, 2}. Thus, there is a relationship between clicks and purchases used to derive the consequential bond between clicks and purchases.

1.4   Problem Definition

Given E-commerce historical click and purchase data over a certain period of time as input, the problem being addressed by this paper is to find the frequent periodic (daily, weekly, monthly) sequential purchase and click patterns in the first stage. Then, these sequential purchase and click patterns can be used to make user-item matrix qualitatively (specifying level of interest or value for already rated items) and quantitatively (finding possible rating for previously unknown ratings) rich before applying collaborative filtering (CF) to improve the overall accuracy of recommendation.

1.5   Contributions

A main limitation of existing related systems such as (HPCRec18 [12]) is that they treated the entire clicks and purchases of items equally and did not integrate frequent sequential patterns to capture more real life customer sequence patterns of purchase behavior inside consequential bond. Thus, in this paper, we propose a system called Historical sequential pattern recommendation (HSPRec) to discover frequent historical sequential pattern from clicks and purchases so that discovered frequent sequential patterns are first used to improve consequential bond and user-item frequency matrix to further improve recommendations.

HSPRec Feature Contributions

1. Using sequential patterns to enhance consequential bonds between clicks and purchases.
2. Using sequential patterns to enhance user-item rating matrix quantitatively by filling missing ratings.
3. Improving accuracy of recommendation qualitatively and quantitatively with consequential bond and sequential patterns respectively to generate a rich user-item matrix for collaborative filtering to improve recommendations.

Procedural Contribution

To make the specified feature contributions, this paper proposes the HSPRec system (Algorithm 1), which first creates sequential databases of purchases and clicks, finds sequential patterns of purchases and clicks with the GSP algorithm [10]. Then, it uses sequential pattern rules to find next possible purchases and consequential bonds between clicks and purchases before collaborative filtering.

2   Related Work

1. Segmentation Based Approach-LiuRec09 ([7]): This approach is based on forming a segmentation of user on the basis of Recency, Frequency, Monetary (RFM) using K-means clustering method, where Recency is the period since the last purchase, Frequency is the number of purchases and Monetary is the amount of money spent. Once the RFM segmentation is created, users are further segmented using transaction matrix. The transactions matrix captures the list of items purchased or not purchased by users over a monthly period in a given product list. From the transaction matrix, users' purchases are further segmented into T-2, T-1, and T, where T represents the current purchases, T-1 and T-2 represents two previous purchases. Finally, association rule mining is used to match and select Top-N neighbors from the cluster to which a target user belongs using binary choice to derive the prediction score of an item not yet purchased by the target user based on the frequency count of the item, scanning the purchase data of k-neighbors. The major drawbacks of LiuRec09 [7] are: (a) It does not learn sequential purchases for user-item matrix creation (b) The utility of item such as price are ignored during the recommendation generation.

2. User Transactions Based Preference Approach-ChoiRec12 ([3]): Users are not always willing to provide a rating or they may provide a false rating. Thus, ChoiRec12 developed a system that derives preference ratings from a transaction data by using the number of times $user_u$ purchased $item_i$ with respect to total transactions of users. Once preference ratings are determined, they are used to formulate user-item rating matrix for collaborative filtering. Major drawbacks of ChoiRec12 are: (a) Sequential purchase pattern is not included from historical or clickstream data during user-item matrix creation. (b) No provision for recommending item to infrequent users.

3. Common Interest Based Approach-ChenRec15 ([11]): It is based on finding the common interest similarity (frequency, duration, and path) between purchase patterns of users to discover the closest neighbors. For the frequency similarity computation, it computes total hits that occurred in an item or category with respect to the total length of users' browsing path. For duration similarity computation, it computes the total time spent on each category with respect to total time spent by users. Finally, for path similarity computation, it uses the longest common subsequence comparison. By selecting Top-N similar users from three indicators, the CF method is used to select close neighbors. The major drawbacks are: (a) It requires domain knowledge for categories, and only supports category level recommendations. (b) Fails to integrate sequential purchase patterns during formation of user-item rating matrix.

4. Historical and Clickstream Based Recommendation-HPCRec18 ([12]): Xiao and Ezeife in [12] proposed HPCRec18 system, which normalizes the purchase frequency matrix to improve rating quality, and mines the session-based consequential bond between clicks and purchases to generate potential ratings to improve the rating quantity. Furthermore, they used frequency of historical purchased items to enrich the user-item matrix from both quantitative (finding possible value for 0 rating), and qualitative (finding more precise value for 1 rating) by using normalization of user-item purchase frequency matrix and consequential bond between clicks and purchases. They also did not integrate mining of sequential patterns of purchases to capture even better customer historical behavior.

3   Proposed Historical Sequential Recommendation (HSPRec) System

The major goal of the proposed HSPRec is to mine frequent sequential patterns from E-commerce historical data to enhance user-item rating matrix from discovered pattern. Thus, HSPRec takes minimum support, historical user-item purchase frequency matrix and consequential bond as input to generate rich user-item matrix as output, (see Algorithm 1).

Algorithm 1. Historical sequential recommendation (HSPRec) system

input	: minimum support(s), historical user-item purchase frequency matrix (M), consequential bond (CB), historical purchase database (DB), historical click database (CDB)
output	: user-item purchase frequency matrix (M_2)
intermediates:	historical sequential purchase database (SDB), weighted purchase pattern (WP), historical sequential click database (SCDB), rule recommended purchase items (RPI), each user u's rating of item i in the matrices is referred to as r_{ui}

1   purchase sequential database (SDB)← SHOD (DB) using Algorithm 2 present in section 3.1;
2   user-item purchase frequency matrix (M_1) ← M modified with Sequential Pattern Rule (SDB) using section 3.2 ;
3   **for** *each user u* **do**
4 | weighted purchase pattern for user u, (WP_u) ← null ;
5   **end**
6   **for** *each user u* **do**
7 | **if** *u has both click and purchase sequences* **then**
8 | compute Click Purchase similarity CPS(click sequence, purchase sequence) from SCDB and SDB using section 3.3;
9 | weighted purchase pattern for user u, (WP_u) ← CPS(click sequence, purchase sequence);
10 | **else**
11 | rule recommended purchase items (RPI) ← Sequential Pattern Rule (SCDB) ;
12 | weighted purchase pattern for user u, (WP_u) ← CPS(click sequence, purchase sequence) using section 3.3;
13 | **end**
14 | rating of item i by user u (r_{ui}) ← Weighted Frequent Purchase Pattern Miner (WP_u) ;
15 | M_2 ← M_1 modified with rating r_{ui};
16   **end**

Steps in the Proposed HSPRec System

1. Convert historical purchase information (Table 3) to user-item purchase frequency matrix (Table 4) by counting the number of each item purchased by each user. For example, User 2 purchased items 1 and 2 twice but other items once.

2. Create daily purchase sequential database (Table 2) of customer purchase (Table 3) by applying sequential historical periodic database generation (SHOD) algorithm presented in Sect. 3.1 to the customer purchase database as input. For example, User 2 daily purchase sequence is $<(1, 4), (3), (2), (1, 2, 5, 6)>$, which shows User 2 purchased item 1 and item 4 together on the same day and purchased item 3 on the next day then purchased item 2 on another day and finally purchased items 1, 2, 5 and 6 together on the next day.

3. Input daily purchase sequential database (Table 2) to Sequential Pattern Rule (SPR) module presented in Sect. 3.2 to generate sequential rule from frequent purchase sequences. For example, 1-frequent purchase sequences $= \{<(1)>, <(2)>, <(3)>, <(5)>, <(6)>, <(7)>\}$. Similarly, some of the 2-frequent purchase sequences $= \{<(6), (5)>, <(3), (6)>, <(3), (5)>, <(2), (7)>, <(2), (6)>, <(2), (5)>\}$ and some of the 3-frequent purchase sequences $= \{<(2), (6), (5)>, <(1), (6), (5)>, <(1), (3), (6)>, <(1), (3), (5)>, <(1), (2), (6)>\}$. Thus, some of the possible sequential purchase pattern rules based on frequent purchase sequences are: (a) 1, 5 → 3 (b) 2, 6 → 1 (c) 2, 6 → 5, where rule (a) states that if user purchases items 1 and 5 together then user will purchase item 3 in next purchase, which will be applied in case of for user 3 in user-item purchase frequency matrix (Table 4).

4. Apply purchase sequential rule in user-item purchase frequency matrix to improve quantity of ratings.

5. For each user, where click happened without a purchase such as user 5 in Table 6, we use consequential bond between clicks and purchases derived as sequential pattern rules mined from the click sequence database. The consequential bond for user 5 in Table 4 is computed by finding frequent click sequential patterns using the SPR (click sequence database) call. From the generated sequential patterns, we filter out only strong rules containing the sequences in user 5 clicks so that they can be used to derive user 5 is possible future purchases.

6. Next, compute Click and Purchase Similarity (click sequence, purchase sequence) using longest common subsequence rate (LCSR) and frequency similarity (FS) Eq. 4 presented of Sect. 3.3.

7. Assign Click Purchase Similarity (click sequence, purchase sequence) value to purchase patterns present in consequential bond (Table 6) to create weighted purchase patterns (Table 7).

8. Input weighted purchase patterns to Weighted Frequent Purchase Pattern Miner (WFPPM) presented in Sect. 3.4 to calculate the weight for each frequent individual item based on its occurrence in weighted purchase patterns.

9. Repeat the steps 5, 6, 7, and 8 if there are more users with clicks but without purchases, otherwise assign computed items weights to modify enhanced user-item frequency matrix (Table 5) and apply collaborative filtering.

Table 4. User-item purchase frequency matrix (M)

User/item	1	2	3	4	5	6	7
User1	1	1	1	?	1	1	1
User2	2	2	1	1	1	1	?
User3	1	1	?	?	1	1	?
User4	?	1	?	?	?	1	1
User5	?	?	?	?	?	?	?

Table 5. Enhanced user-item frequency matrix with Sequential Pattern Rule (M_1)

User/item	1	2	3	4	5	6	7
User1	1	1	1	?	1	1	1
User2	2	2	1	1	1	1	?
User3	1	1	1	?	1	1	?
User4	1	1	?	?	1	1	1
User5	?	?	?	?	?	?	?

Table 6. Consequential bond table showing clicks and purchases from historical data

UID	Click sequence	Purchase sequences
1	<(1,2,3), (7,5,3), (1,6), (6), (1,5)>	<(1,2), (3), (6), (7), (5)>
2	<(1,4), (6,3), (1,2), (1,2,5,6)>	<(1,4), (3), (2), (1,2,5,6)>
3	<(1,5), (6,5,2), (6), (5)>	<(1), (2), (6), (5)>
4	<(2,7), (6,6,7)>	<(2), (6,7)>
5	<(1,5)>	?

3.1 Historical Periodic Sequential Database Generation (SHOD) Module

The proposed sequential historical pattern generation (SHOD) module takes historical (click or purchase database) data as input and produces periodic (daily, weekly, monthly) sequential (click or purchase) database as output as present in Algorithm 2. This algorithm is based on checking timestamp between items to form periodic sequences if the difference between purchase times by the same user are within 24 h, 24 * 7 h respectively.

Example of creating periodic sequential database Check time difference between purchased products.

(a) If the time difference between two product purchases is less than 24 h, it adds itemID to itemset in daily.txt file. In our case, purchased time difference between two products {1, 2} purchased by user {Tuserid = 1} is less than 24 h. So, it adds two items to itemset in daily.txt as: **1, 2**.

(b) However, if the time difference between purchased items is more than 24 h, it adds -I to indicate the end of itemset and adds itemID after -I.
For example, **1, 2 -I 3**.

If user identity is not similar, then add -I and -S after item to indicate the end of itemset and sequence respectively and goto step 2 by updating temporary variable. Repeat steps with each record read until there is no record in the historical database. In our case, the created daily purchase sequential database

Algorithm 2. Historical periodic sequential database (SHOD)

input   : historical click or purchase data
output   : periodic (daily, weekly, monthly) sequential database
intermediate: Tuserid= temporary userid, Ttimestamp= temporary timestamp

1   historical.txt ← extract userid, itemid, timestamp from historical data;
2   read first line from historical.txt and store userid, timestamp into Tuserid,
     Ttimestamp ;
3   **for** *each user u, u ∈ historical.txt* **do**
4   **if** *userid==Tuserid* **then**
5   Tdur ← (Ttimestamp - timestamp) ;
6   **if** *Tdur <= 24 hrs* **then**
7   add itemid to **daily-sequence-database.txt** and goto step 3 ;
8   **else**
9   add -I to indicate end of itemset and goto step 3 ;
10   **end**
11   **if** *Tdur > 24 hrs and Tdur <=168 hrs* **then**
12   add itemid to **weekly-sequence-database.txt** and goto step 3;
13   **else**
14   add -I to indicate end of itemset and goto step 3 ;
15   **end**
16   add -S to indicate end of sequence and goto step 3 ;
17   **else**
18   add -I after itemid to indicate end of itemset, -S to indicate end of
             sequence and update Tuserid and goto step 3 ;
19   **end**
20   **end**

is presented in Table 2 and the same steps can be repeated to generate daily
click sequential database by using click item database as input.

3.2 Sequential Pattern Rule (SPR) Module

Sequential Pattern Rule (SPR) is based on the use of frequent sequential patterns
created from periodic sequential database. Thus, input of SPR is periodic historical sequential database and output is recommended possible purchase items
derived from generated sequential patterns rules. The major steps in SPR are:

1. **Frequent sequences generation:** Generates frequent sequences using Generalized Sequential Pattern Mining (GSP) algorithm [10]. Let us consider
   input: Daily click sequential database in Table 1, **minimum support** = 2
   and **candidate set** (C_1) = $\{1, 2, 3, 4, 5, 6, 7\}$ and **algorithm** = GSP as
   defined in Sect. 1.1. **output**: frequent sequential patterns. Some example frequent sequential patterns are: 1-frequent sequences: $\{<(1)>, <(2)>, <(3)>,$
   $<(5)>, <(6)>\}$. Some of the 2-frequent sequences: $\{<(1), (2)>, <(1), (3)>,$
   $<(3), (6)>, <(5), (6)>, <(1, 5)>\}$. Some of 3-frequent sequences: $\{<(1), (2),$
   $(6)>, <(3), (1), (5)>, <(1), (5), (6)>, <(6), (1, 5)>\}$

2. **Rule generation:** Represent frequent sequences in the form of $U_{click} \rightarrow U_{purchase}$, where the left-hand side of the rule refers to clicked item set while the right-hand side is the recommended item to be purchased. Furthermore, to verify the validity of SPR, confidence of SPR is defined as:

$$Confidence(U_{click} \rightarrow U_{purchase}) = \frac{Support(U_{click} \cup U_{purchase})}{Support(U_{click})} \quad (1)$$

Here, some of the rules from frequent clicks sequences are: (a) $(1,5) \rightarrow (1),(3)$, (b) $(1,5) \rightarrow (3),(1)$, (c) $(1,2) \rightarrow (1,6)$, (d) $(1)(5) \rightarrow (6),(5)$

3. **Rule section:** Assume we are ony interested in rules that satisfy the following criteria: (1) At least two antecedents. (2) Confidence > 50%. (3) Select one rule with highest confidence. Let us consider rule (a) recommends User 5 to purchase item 1 and item 3 when item 1 and item 5 are clicked together.

3.3 Click Purchase Similarity (CPS)

To compute the CPS similarity between click sequence and purchase sequence of each user, we have used sequence similarity and frequency similarity of the two sequences. **Sequence Similarity:** It is based on longest common subsequence rate (LCSR) [2] and presented in Eq. (2).

$$LCSR(X,Y) = \frac{LCS(X,Y)}{max(X,Y)} \quad (2)$$

In our case, X represents click sequence and Y represents purchase sequence and LCS is defined as:

$$LCS(Xi,Yj) = \begin{cases} \phi & if \quad i=0 \quad or \quad j=0 \\ LCS(X_{i-1},Y_{j-1}) \cap X_i & if \quad x_i = y_i \\ longest(LCS(X_i,Y_{j-1}),LCS(X_{i-1},Y_j)) & if \quad x_i! = y_i \end{cases}$$

Frequency Similarity: First, form distinct sets of items from both click and purchase sequential patterns and count the number of items occurring in each sequence to form the vectors specifying the number of times a user clicked or purchased a particular item. Then, apply Cosine frequency similarity (Eq. (3)) to the click and purchase vectors.

$$Cosine(X,Y) = \frac{X_1 * Y_1 + X_2 * Y_2 + ...X_n * Y_n}{\sqrt{X_1^2 + X_2^2 + ... + X_n^2}\sqrt{Y_1^2 + Y_2^2 + ... + Y_n^2}} \quad (3)$$

Thus,

$$CPS = \alpha * LCSR(X,Y) + \beta * FS(X,Y) \quad (4)$$

where $\alpha + \beta = 1, 0 < \alpha, < \beta < 1, \alpha$ and β are weights assigned to reflect the importance of the two sequences of similarity and frequency.

Example of CPS (Click Sequence, Purchase Sequence)

To compute CPS similarity between click sequence (X) = <(2,7), (6,6,7)> and purchase sequence (Y) = <(2), (6,7)>, we take the following steps:

1. Compute the longest common subsequences, LCS(X, Y) between click and purchase sequence. For example, LCS (<(2, 7), (6, 6, 7)>, <(2), (6, 7)>) is 3, because of common subsequence (2), (6, 7).
2. Find the maximum number of items occurring in click or purchase sequence as Max(X, Y). In our case, MAX(X, Y) is 5.
3. Compute sequence similarity of click (X) and purchase (Y) sequences as LCS(X, Y)/Max(X, Y)= 3/5 = 0.6.
4. Compute the frequencies of items in click and purchase sequences. In our case, we have used the format [(item): number of occurrence]. So, frequency counts of clicks is: [(2):1, (6):2, (7):2]. Similarly, frequency counts of purchases is: [(2):1, (6):1, (7):1].
5. Use Cosine similarity function in Eq. 3 to get the frequency similarity between click sequence (X) and purchase sequence (Y) as Cosine (X, Y). In our case, Cosine (X, Y) = 0.96.
6. The Click Purchase Similarity of user click and purchase sequences CPS(X, Y) is presented in Eq. 4. In our case, CPS(X, Y) = 0.8 * 0.6 + 0.2 * 0.96 = 0.67, where $\alpha = 0.8$ and $\beta = 0.2$.

This CPS(X, Y) can be used as weight or probability that user u will purchase the entire sequence as shown in Table 7.

Table 7. Weighted purchased patterns

Purchase	<(1,2), (3), (6), (7), (5)>	<(1,4), (3), (2), (1,2,5,6)>	<(1), (2), (6), (5)>	<(2), (6,7)>	<(1), (3)>
CPS	0.624	0.834	0.636	0.67	0.5

3.4 Weighted Frequent Purchase Pattern Miner (WFPPM) Module

Weighted Frequent Purchase Pattern Miner(WFPPM) takes weighted purchase pattern as input (present in Table 7) and weighted purchase patterns is created by assigning CPS (click sequence, purchase sequence) value to purchase patterns in consequential bond Table 6 to generate frequent purchase patterns with weight under user specified minimum threshold as output. Major steps of WFPPM are:

1. **Count support of items:** Count occurrence of items presented in weighted purchase pattern (Table 7). For example, {support(1):5, support(2):5, support(3):3, support(4):1, support(5):3, support(6):4, support(7):2}.
2. **Calculate weight for individual item:** Compute weight of individual item from weighted purchase pattern (Table 7) using CPS module (Eq. (4)).

$$R, item_i = \frac{\sum_{i=1}^{n} CPS \in item_i}{Support(item_i)} \tag{5}$$

For example, $R_1 = \frac{0.624+0.834+0.834+0.636+0.5}{5} = 0.68$ Similarly, $R_2 = 0.71$, $R_3 = 0.65, R_4 = 0.834, R_5 = 0.698, R_6 = 0.691, R_7 = 0.647$,

3. **Test item weight with minimum threshold:** Define minimum threshold rating for user, here in our case, minimum threshold = 0.2. So, all rating of item are frequent.

3.5   User-Item Matrix Normalization

Normalization in recommendation system helps to predict the level of interest of user on a particular purchased item. Thus, normalization function for a user u's rating of an item i (r_{ui}) takes user-item frequency matrix (Table 8) as input and provides the level of user interest between 0 and 1 using unit vector formula (Eq. (6)).

$$Normalization(r_{ui}) = \frac{r_{ui}}{\sqrt{r_{ui1}^2 + r_{ui2}^2 + ... + r_{uin}^2}} \qquad (6)$$

The normalization of enhanced user-item matrix (Table 8) using Eq. 5 is presented in Table 9. For example, normalization of $User_1$ rating on $Item_1 = \frac{1}{\sqrt{1^2+1^2+1^2+1^2+1^2+1^2}} = 0.40$.

Table 8. Enhanced user-item purchase frequency matrix with rating for user 5 (M_1)

User/item	1	2	3	4	5	6	7
User1	1	1	1	?	1	1	1
User2	2	2	1	1	1	1	?
User3	1	1	1	?	1	1	?
User4	1	1	?	?	1	1	1
User5	0.68	0.71	0.65	0.834	0.698	0.691	0.647

Table 9. Normalized user-item purchase frequency matrix (M_2)

User/item	1	2	3	4	5	6	7
User1	0.40	0.40	0.40	?	0.40	0.40	0.40
User2	0.57	0.57	0.28	0.28	0.28	0.28	?
User3	0.44	0.44	0.44	?	0.44	0.44	?
User4	0.44	0.44	?	?	0.44	0.44	0.44
User5	0.68	0.71	0.65	0.834	0.698	0.691	0.647

4   Experimental Design

We implemented our proposed historical sequential pattern mining based recommendation system, HSPRec with two other existing recommendation systems of ChoiRec12 [3] and HPCRec18 [12] in user-based collaborative filtering to evaluate the performance our proposed system in comparison to others. First, the same data obtained from Amazon product data [8] is converted into user-item matrix for each of the approaches of the three algorithms (ChoiRec12, HPCRec18 and HSPRec) before applying collaborative filtering on the user-item matrix to obtain the recommendations for each of the systems. During this conversion, data is modified into intermediate form, which means, when the value is larger than the minimum threshold; this value would be set to one (highest rating). When the value is less than the minimum threshold, this value would be set to zero (lowest rating). To test user-based collaborative filtering, we have used

Pearson Correlation Coefficient (PCC). Furthermore, 80% of data is used in training and 20% is used in testing performance. To evaluate the performance of the recommendation system, we have used a different number of users and nearest neighbors using three different evaluation parameters (a) mean absolute error (MAE) (b) precision and (c) recall with https://www.librec.net/ LibRec [4] library available in Java.

4.1 Dataset Selection

To perform experiment, we used data available from Amazon product data [8]. The Amazon data sets consists of 23 different categories such as Books, Electronics, Home and Kitchen, Sports and Outdoors, Cell Phones and Accessories, Grocery and Gourmet Food and many more. Data contains 142.8 million transactional records spanning May 1996–July 2014 but in our experiments we have used data of 2013 and 2014.

4.2 Experimental Results

Figure 1 presents results of our experiments. First, we implemented original CF algorithm with explicit rating available from Amazon, and found very low performance. Then, we implemented choiRec12 [3] with rating derived from each

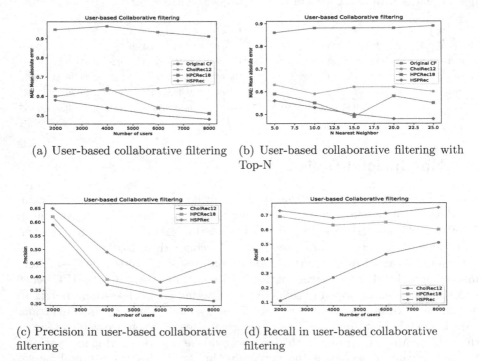

(a) User-based collaborative filtering

(b) User-based collaborative filtering with Top-N

(c) Precision in user-based collaborative filtering

(d) Recall in user-based collaborative filtering

Fig. 1. Experimental result showing evaluation recommendation systems using MAE, precision and recall

user purchases with respect to total purchases by each user and got better result compared to original CF. Then, we implemented HPCRec18 [12] by using user-item frequency first and used consequential bond in user-item frequency matrix and found better result than choiRec12. Finally, for our proposed HSPRec, we constructed user-matrix purchase frequency matrix at first. Then, we discovered frequent sequences of purchased data to create sequential rules and used sequential rule to enhance user-item matrix quantitatively and we applied enhanced user-item frequency matrix to collaborative filtering and found better result compared to choiRec12, and HPCRec18.

5   Conclusions and Future Work

Many recommendation systems neglect sequential patterns during recommendation. Thus, to verify the necessity of sequential patterns in recommendation, we generated sequential patterns from historical E-commerce data and fed them into collaborative filtering to make user-item matrix rich from quantitative and qualitative perspectives. Furthermore, after evaluation with different systems, we got better results with a sequential pattern based recommendation. Thus, some of the possible future works are: (a) Finding more possible ways of integrating sequential patterns to collaborative filtering. (b) Incorporating multiple data sources based sequential pattern with different data schema, and making recommendations based on the overall data set. (c) Finding the more possible ways of integrating sequential pattern in user-item matrix from online data.

References

1. Aggarwal, C.C.: Recommender Systems. Springer, Cham (2016). https://doi.org/10.1007/978-3-319-29659-3
2. Bergroth, L., Hakonen, H., Raita, T.: A survey of longest common subsequence algorithms. In: Seventh International Symposium on String Processing and Information Retrieval, pp. 39–48 (2000)
3. Choi, K., Yoo, D., Kim, G., Suh, Y.: A hybrid online-product recommendation system: combining implicit rating-based collaborative filtering and sequential pattern analysis. Electron. Commer. Res. Appl. 11(4), 309–317 (2012)
4. Guo, G., Zhang, J., Sun, Z., Yorke-Smith, N.: LibRec: a java library for recommender systems. In: Posters, Demos, Late-breaking Results and Workshop Proceedings of the 23rd International Conference on User Modeling, Adaptation and Personalization (2015)
5. Han, J., Pei, J., Kamber, M.: Data Mining: Concepts and Techniques. Elsevier, New York (2011)
6. Kim, Y.S., Yum, B.J., Song, J., Kim, S.M.: Development of a recommender system based on navigational and behavioral patterns of customers in e-commerce sites. Expert Syst. Appl. 28(2), 381–393 (2005)
7. Liu, D.R., Lai, C.H., Lee, W.J.: A hybrid of sequential rules and collaborative filtering for product recommendation. Inf. Sci. 217(20), 3505–3519 (2009)
8. McAuley, J.: Amazon product data (2019). http://jmcauley.ucsd.edu/data/amazon/

9. Schafer, J.B., Konstan, J.A., Riedl, J.: E-commerce recommendation applications. Data Min. Knowl. Discov. **5**(1–2), 115–153 (2001)

10. Srikant, R., Agrawal, R.: Mining sequential patterns: generalizations and performance improvements. In: Apers, P., Bouzeghoub, M., Gardarin, G. (eds.) EDBT 1996. LNCS, vol. 1057, pp. 1–17. Springer, Heidelberg (1996). https://doi.org/10.1007/BFb0014140

11. Su, Q., Chen, L.: A method for discovering clusters of e-commerce interest patterns using click-stream data. Electron. Commer. Res. Appl. **14**(1), 1–13 (2015)

12. Xiao, Y., Ezeife, C.I.: E-Commerce product recommendation using historical purchases and clickstream data. In: Ordonez, C., Bellatreche, L. (eds.) DaWaK 2018. LNCS, vol. 11031, pp. 70–82. Springer, Cham (2018). https://doi.org/10.1007/978-3-319-98539-8_6

Discovering and Visualizing Efficient Patterns in Cost/Utility Sequences

Philippe Fournier-Viger[1]([⊠]), Jiaxuan Li[2], Jerry Chun-Wei Lin[3],
and Tin Truong-Chi[4]

[1] School of Humanities and Social Sciences,
Harbin Institute of Technology (Shenzhen), Shenzhen, China
philfv@hit.edu.cn
[2] School of Computer Sciences, Harbin Institute of Technology (Shenzhen),
Shenzhen, China
jiaxuanliniki@gmail.com
[3] Department of Computing, Mathematics and Physics,
Western Norway University of Applied Sciences (HVL), Bergen, Norway
jerrylin@ieee.org
[4] Department of Mathematics and Informatics, University of Dalat, Dalat, Vietnam
tintc@dlu.edu.vn

Abstract. Many algorithms have been proposed to discover interesting patterns in sequences of events or symbols, to support decision-making or understand the data. In sequential pattern mining, patterns are selected based on criteria such as the occurrence frequency, periodicity, or utility (eg. profit). Although this has many applications, it does not consider the effort or resources consumed to apply these patterns. To address this issue, this paper proposes to discover patterns in cost/utility sequences, in which each event/symbol is annotated with a cost, and where a utility value indicates the benefit obtained by performing each sequence. Such sequences are found in many fields such as in e-learning, where learners do various sequences of learning activities having different cost (time), and obtain different utility (grades). To find patterns that provide a good trade-off between cost and benefits, two algorithms are presented named CEPDO and CEPHU. They integrate many optimizations to find patterns efficiently. Moreover, a visualization module is implemented to let users browse patterns by their skyline and visualize their properties. A case study with e-learning data has shown that insightful patterns are found and that the designed algorithms have excellent performance.

Keywords: Pattern mining · Cost/Utility sequences · Visualization

1 Introduction

Sequences of symbols or events are an important type of data found in many domains [8]. A popular method for analyzing sequences is sequential pattern mining (SPM). It consists of discovering all subsequences of symbols or events that have an occurrence frequency (support) exceeding some minimum frequency

© Springer Nature Switzerland AG 2019
C. Ordonez et al. (Eds.): DaWaK 2019, LNCS 11708, pp. 73–88, 2019.
https://doi.org/10.1007/978-3-030-27520-4_6

threshold [1,4,8]. Albeit discovering sequential patterns is useful to understand data and to support decision making, frequent patterns are not always relevant [15]. To find more interesting patterns, SPM has been recently extended as the problem of High Utility Sequential Pattern Mining (HUSPM) [2,15]. The aim is to find sequences having a high utility, where the utility is a numerical measure of the profit, importance or benefits provided by a pattern. Though, HUSPM has many applications, it focuses on the utility or benefits provided by patterns but ignores the cost or effort for obtaining these benefits. For instance, a pattern $\langle material1, material20, A+ \rangle$ may be found in e-learning data, indicating that reading learning material 1 and then learning material 20 leads to obtaining a high score (A+). But HUSPM ignores the cost of patterns (e.g. the monetary cost of the materials and the effort required to learn using these materials).

Generally, the cost of a pattern could be calculated in terms of different aspects such as the time, money, resources consumed and effort. Because HUSPM does not consider the cost of patterns, it can find numerous patterns that have a high utility but have a high cost. For example, in e-learning, many patterns may lead to a desirable outcome such as a high grade (a high utility) but have a cost that is prohibitive. Moreover, HUSPM may miss numerous patterns that have a lower utility but provides an excellent trade-off between cost and utility. Thus, integrating the concept of cost in HUSPM is desirable but hard since cost and utility may measure different aspects (e.g. profit vs time). Hence, cost cannot be simply subtracted from the utility to apply existing HUSPM algorithms. Moreover, doing so would fail to assess how strong the correlation between cost and utility is for each pattern, while it is desirable to discover patterns that not only have a low cost and high utility but that provide a good trade-off and strong correlation between cost and utility.

This paper addresses this limitation of HUSPM by proposing a novel problem of mining *cost-effective patterns* (CEP) by considering both utility and cost. A CEP is a pattern having a good cost/utility trade-off. Two variations of the problem are defined to handle binary utility values indicating desirable/undesirable outcomes (e.g. *passed* vs *failed*) and numeric values (e.g. exam grades), respectively. Moreover, statistical measures are designed to assess the cost/utility correlation. Two algorithms are proposed to find all CEPs (one for each problem variation). The algorithms rely on optimizations, a lower-bound on the average cost of patterns and an upper-bound on their utility, to find all CEPs efficiently. Moreover, a visualization module is proposed to facilitate pattern analysis. It lets users visualize properties of CEPs, and browse CEPs using their skyline. An experimental study has shown that the algorithms are efficient and that the proposed lower-bound can considerably reduce the search space. In addition, a case study on e-learning data was carried and shows that interesting cost-effective patterns are found using the designed algorithms.

The rest of this paper is organized as follows. Section 2 reviews related work. Section 3 defines the problem of cost-effective pattern mining. Section 4 presents the proposed algorithms. Section 5 presents the performance evaluation. Section 6 describes a case study with e-learning data, and presents the visualization module. Finally, Sect. 7 draws a conclusion and discusses future work.

2   Related Work

High utility sequential pattern mining (HUSPM) has become an important data mining problem in recent years, and many algorithms have been proposed for this problem [2,15]. Although HUSPM has many applications, it was first proposed to mine sequences of purchases that yield a high profit from sequences of customer transactions [15]. The input of HUSPM is a minimum utility threshold *minutil* and a database of quantitative sequences, where each sequence is an ordered list of transactions. A transaction is a set of symbols or events, each annotated with some internal utility value (e.g. a purchase quantity). Moreover, each symbol has a weight called external utility, indicating its relative importance (e.g. unit profit). Table 1 shows an example quantitative sequence database containing four sequences. The first sequence indicates that a customer purchased three units of item a, followed by three units of item b, then two units of item b, and finally one unit of item c. The output of HUSPM is the set of all sequences having a utility that is no less than *minutil* (e.g. all sequences that yield a high profit).

Table 1. A quantitative sequence database containing four sequences.

Quantitative sequences with purchase quantities (internal utility)
sequence 1: $\langle (a,3), (b,3), (c,1), (b,4) \rangle$
sequence 2: $\langle (a,1), (e,3) \rangle$
sequence 3: $\langle (a,6), (c,7), (b,8), (d,9) \rangle$
sequence 4: $\langle (b,3), (c,1) \rangle$
Unit profits (external utility)
$a = 5\$, b = 1\$, c = 2\$, d = 1\$$

Although HUSPM is useful, it does not consider the cost of patterns. Until now, few pattern mining studies have considered a concept of cost. The closest work to the one presented in this paper is in the field of process mining, where cost was considered to perform behavior analysis. The main goal of process mining is to extract models from event logs that characterize a business process. An event log is a list of events with timestamps, ordered by time. Various methods have been used to analyze processes [11,13]. Dalmas et al. integrated the concept of cost in process mining by proposing the TWINCLE algorithm [5]. It was applied to find patterns in event logs of hospitals, where each patient activity has a cost. The algorithm finds sequential rules, where the antecedent and consequent of a rule are events. Rules are selected based on their cost and displayed to the user with the aim of reducing the monetary cost of medical treatments. Although interesting low cost patterns were found using TWINCLE, it ignores the utility of patterns. As a result, TWINCLE can find many patterns representing cheap medical treatments but having a low effectiveness (utility).

To address the limitations of the above studies, the next section proposes to mine patterns in event sequences by considering both a cost and a utility model, and where utility is the sequence outcome and cost is event annotations.

3 Problem Definition

The type of data considered in this paper is a **sequential activity database** (SADB), which contains several sequences, where a sequence is a time-ordered list of activities. Sequences of activities can model various types of data, and are similar to event logs in process mining [3]. The main difference is that this paper considers that each sequence of events includes cost and utility information. Formally, a SADB is a set of sequences $SADB = \{S_1, S_2, \ldots S_n\}$, where each sequence S_s has a unique identifier s ($1 \le s \le n$). A sequence S_s is a list of activities $\langle \{v_1[c_1], v_2[c_2], \ldots v_m[c_m]\} \, |Utility\rangle$, where the notation $v_i[c_i]$ indicates that the activity v_i had a cost c_i (a positive number representing some effort or resources consumed), and u_s represents the utility of the sequence S_s (the benefits obtained after performing the sequence). To be able to address the needs of different applications, the utility is viewed as either binary or numeric values. A SADB containing numeric or binary utility values is called *binary SADB* or *numeric SADB*, respectively.

For instance, Table 2 shows a binary SADB from the e-learning domain containing five sequences, each indicating the learning materials (denoted as a, b, c, d, e) studied by a learner. Each learning material is annotated with a cost value indicating the amount of time that it was studied (in hours), and each sequence has a binary utility indicating whether the student passed (+) or failed (−) the final exam. The first sequence indicates that a student spent 2 h to study a, followed by 4 h to study b, 9 h for c, 2 h for d, and then passed the exam. Table 3 shows a numeric SADB where sequences are represented using the same format but utility is numeric values indicating scores obtained at the final exam. That SADB contains five sequences. The first sequence has a positive utility value of 80.

Table 2. SADB with binary utility

Sid	Sequence (activity[cost])	Utility
1	<(a[2]),(b[4]),(c[9]),(d[2])>	+
2	<(b[1]),(d[12]),(c[10]),(e[1])>	−
3	<(a[5]),(e[4]),(b[8])>	+
4	<(a[3]),(b[5]),(d[1])>	−
5	<(b[3]),(e[4]),(c[2])>	+

Table 3. SADB with numeric utility

Sid	Sequence (activity[cost])	Utility
1	<(a[20]),(b[40]),(c[50])>	80
2	<(b[25]),(d[12]),(c[30])>	60
3	<(a[25]),(e[14]),(b[30])>	50
4	<(a[40]),(b[16]),(d[40])>	40
5	<(b[20]),(e[24]),(c[20])>	70

To find interesting patterns in sequences of activities, this paper consider that a pattern p is an ordered list of activities $\{v_1, v_2, \ldots, v_o\}$. To select interesting patterns, three measures are first considered: support, cost and occupancy. The *support* measure is used to reduce the search space and eliminate rare patterns that may represent noise in the data and may be insignificant. It is used and defined as in SPM [8].

Definition 1. *The **support of a pattern** p is the number of sequences containing p, i.e. $sup(p) = |\{S_s|p \subseteq S_s \wedge S_s \in SADB\}$* [8].

The second measure is the *cost*. It allows evaluating the amount of resources spent to apply a pattern. But because a pattern's cost may not be the same in all sequences where it appears, we propose to calculate the average cost.

Definition 2. *The **cost of a pattern** p in a **sequence** S_s is: $c(p, S_s) = \sum_{v_i \in first(p,S_s)} c(v_i, S_s)$ if $p \subseteq S_s$ and otherwise 0, where $first(p, S_s)$ denotes the first occurrence of p in S_s. The **cost of a pattern** p in a **SADB** is the sum of its cost in all sequences, i.e. $c(p) = \sum_{p \subseteq S_s \wedge S_s \in SADB} c(p, S_s)$. The **average cost** of a pattern p is : $ac(p) = \frac{c(p)}{|sup(p)|}$, and represents the average effort to apply the pattern p.*

The average cost measure is useful as it provides an overview of the resources consumed to apply each pattern. It is to be noted that since a pattern may have multiple occurrences in a sequence, the pattern's cost could be calculated in different ways. A possibility could be to use the sum or the average of the costs of all occurrences. But this is not trivial to calculate and can lead to overestimating the cost because two occurrences may overlap and share events. Thus, it was decided to simply use the first occurrence. The proposed definition can also work for the last occurrence (by processing sequences in backward order).

The third measure is the *occupancy*, which is borrowed from SPM [17]. This measure aims at finding patterns that cover large parts of sequences where they appear. The assumption is that high occupancy patterns represent well these sequences and thus are more likely to have influenced their outcome (utility) than low occupancy patterns.

Definition 3. *The **occupancy of a pattern** p is calculated as: $occup(p) = \frac{1}{sup(p)} \sum_{p \subseteq S_s \wedge S_s \in SADB} \frac{|p|}{|S_s|}$* [17].

For example, consider the database of Table 2 and pattern $p = \{ab\}$. It is found that $c(p, S_1) = 6$, $c(p, S_3) = 13$, $c(p, S_4) = 8$, $sup(p) = 3$, $ac(p) = (6 + 13 + 8)/3 = 9$, and $occup(p) = (2/4 + 2/3 + 2/3)/3 \approx 0.61$.

The above measures are used to ensure that patterns have a low cost, are not infrequent, and are representative of sequences. But those measures cannot evaluate if patterns are cost-efficient (provide a good trade-off between cost and utility) and are correlated to a positive outcome (utility). For this purpose, the following paragraphs propose two measures of cost-efficiency called *cor* and *trade-off*, which are defined to handle binary and numeric SADBs, respectively.

Correlation of a Pattern in a Binary SADB. In a binary SADB, the utility of a sequence is a binary value indicating a positive or negative outcome (e.g. passed/failed an exam). To find patterns that contribute to a positive outcome, the *cor* measure is defined as follows.

Definition 4. *The **correlation** of a pattern p in a binary SADB is:*

$cor(p) = \frac{ac(D_p^+) - ac(D_p^-)}{Std} \sqrt{\frac{|D_p^+||D_p^-|}{|D_p^+ \cup D_p^-|}}$ *where D_p^+ and D_p^- respectively denote the set of positive and negative sequences containing the pattern p, $ac(D_p^+)$ and $ac(D_p^-)$ are the pattern p's average cost in D_p^+ and D_p^-, Std is the standard deviation of p's cost, and $|D_p^+|$ and $|D_p^-|$ are the support of p in D_p^+ and D_p^-, respectively.*

The *cor* measure is a variation of the biserial correlation measure [12] used in statistics to assess the correlation between a binary and a numeric attribute. The *cor* measure values are in the $[-1, 1]$ interval. The greater positive (smaller negative) the *cor* measure is, the more a pattern is correlated with a positive (negative) utility. In the *cor* measure, the term $ac(D_p^+) - ac(D_p^-)$ is used to find patterns that have a large difference in terms of average cost for positive and negative sequences. That cost difference is divided by the standard deviation of the cost to avoid using absolute values in the equation. The term $\sqrt{\frac{|D_p^+||D_p^-|}{|D_p^+ \cup D_p^-|}}$ is used to find patterns that appears more frequently in sequences having a positive outcome than having a negative outcome.

For example, consider the binary SADB of Table 2 and $p = \{ab\}$. It is found that $c(p, S_1) = 6$, $c(p, S_3) = 13$, $c(p, S_4) = 8$, $sup(p, +) = 2$, $sup(p, -) = 1$, $ac(D_p^+) = (6 + 13)/2 = 9.5$ and $ac(D_p^-) = (8)/1 = 8$. Hence, $cor(ab) = \frac{9.5 - 8}{1.06}\sqrt{\frac{2 \times 1}{|3|}} \approx 0.314 > 0$ is correlated with a positive outcome.

Correlation of a Pattern in a Numeric SADB. In a numeric SADB, the utility of a sequence is a positive number (e.g. a grade obtained at an exam) where a high (low) value indicate a good (bad) outcome. To measure the correlation of a pattern with a good outcome represented as numeric utility, the *trade-off* measure is proposed. It evaluates the relationship between cost and utility. The utility of a pattern and its *trade-off* are calculated as follows.

Definition 5. *The **utility** of a pattern p in a numeric SADB is the average of the utility of sequences in which it appears, that is $u(p) = \frac{\sum_{p \subseteq S_s \wedge S_s \in SADB} su(S_s)}{|sup(p)|}$, where $su(S_s)$ denotes the utility of a sequence S_s.*

Definition 6. *The **trade-off** of a pattern p is the ratio of its average cost to its average utility, that is $tf(p) = ac(p)/u(p)$.*

Trade-off values are in the $(0, \infty]$ interval. The trade-off value of a pattern indicates how efficient the pattern is. A pattern having a small trade-off is viewed as being cost-effective as it provides utility at a low cost (it requires a small amount of resources to obtain a desirable outcome). For example, in e-learning, a pattern with a small trade-off may indicate that studying some learning materials (events) typically requires a small amount of time (cost) and is correlated with obtaining high scores (utility). On the other hand, patterns having a larger trade-off may be viewed as being less efficient.

For instance, consider Table 3 and pattern $p = \{ab\}$. It is found that $c(p, S_1) = 60$, $c(p, S_3) = 55$, $c(p, S_4) = 56$, $sup(p) = 3$, $ac(p) = (60+55+56)/3 = 57$, $u(p) = \frac{su(p,S_1)+su(S_3)+su(p,S_4)}{sup(p)} = \frac{80+50+40}{3} \approx 56.67$. The trade-off of pattern p is $tf(p) = ac(p)/u(p) = 57/56.67 \approx 1.01$.

Based on the measures presented in this section, two problems are defined to find cost-effective patterns in binary and numeric SADBs, respectively.

Problem 1. Mining Cost-Effective Patterns in a binary SADB. Given user-specified *minsup*, *minoccup* and *maxcost* thresholds, a pattern p is a cost-effective pattern in a binary SADB if $(\sup(p) \geq minsup) \land (occup(p) \geq minoccup) \land (ac(p) \leq maxcost)$, and $cor(p)$ has a high positive value.

Problem 2. Mining Cost-Effective Patterns in a numeric SADB. Given user-specified *minsup*, *minoccup*, *minutil* and *maxcost* thresholds a pattern p is a cost-effective pattern in a numeric SADB if $(\sup(p) \geq minsup) \land (occup(p) \geq minoccup) \land (ac(p) \leq maxcost) \land (u(p) \geq minutil)$ and $tf(p)$ has a small value.

4   The CEPDO and CEPHU Algorithms

This section presents two algorithms, designed for mining cost-effective patterns leading to a desirable outcome (CEPDO), and cost-effective patterns leading to a high utility (CEPHU), respectively (Problems 1 and 2).

Both algorithms search for patterns using a pattern-growth approach inspired by the PrefixSpan algorithm for SPM [8]. This approach consists of performing a depth-first search starting from patterns of length 1 (containing single activities), and recursively extending each pattern with additional activities to find larger patterns. For each visited pattern p of length k, a projected database is created, and then scanned to find extensions of p of length $(k + 1)$. Formally, a pattern $e = \{r_1, r_2, ..., r_q\}$ is said to be an *extension* of a pattern $p = \{v_1, v_2, ..., v_o\}$ if there exists integers $1 \leq y_1 \leq y_2 \leq ... \leq y_k \leq o < q$ such that $r_{y1} = v_1$, $r_{y2} = v_2, ...,$ $r_{yo} = v_o$. To avoid exploring all possible patterns, it is necessary to use techniques to reduce the search space. Hence, five search space pruning strategies are introduced. Strategys 1, 2, 3, 4 are used in CEPDO, while Strategy 1, 2, 3, 5 are used in CEPHU. Strategies are first described, and then the algorithms. The first strategy calculates an upper-bound on the occupancy measure introduced in [17] to reduce the search space.

Definition 7. *(Upper bound on occupancy).* For a pattern p and sequence S_i, let $psl[S_i]$, $ssl[S_i]$ and $sl[S_i]$ be p's length in S_i, the length of the subsequence after p in S_i, and S_i's length, respectively. An *upper-bound on the occupancy* of a pattern p is: $ou(p) = \frac{1}{sup(p)} \cdot \max_{S_1,...,S_{sup(p)}} \sum_{i=1}^{sup(p)} \frac{psl[S_i]+ssl[S_i]}{sl[S_i]}$ [17].

Strategy 1. *For a pattern p, if $ou(p) < minoccup$, then pattern p and all its extensions have an occupancy lower than minoccup and can be eliminated* [17].

The second strategy relies on the support measure to reduce the search space and is commonly used in SPM [8].

Strategy 2. *For a pattern p, if $sup(p) < minsup$, then pattern p and all its extensions have a support lower than minsup and can be eliminated.*

The third strategy is novel, and based on the cost measure. A challenge for reducing the search space using the average cost is that this measure is neither monotonic nor anti-monotonic (proof is ommitted due to space limitation). For this reason, Strategy 3 relies on a new lower-bound on the average cost.

Definition 8 *(Lower bound on the average cost).* *Let $C(p)$ be the set of costs of a pattern p in a sequence S_s where $p \subseteq S_s$, and $c_i(p)$ denotes the i-th smallest cost of p in $C(p)$. The* Average Minimum Supported Cost *(AMSC) is a lower-bound on $ac(p)$, defined as: $amsc(p) = \frac{1}{minsup} \sum_{i=1,2,...,minsup} c_i(p)$.*

Theorem 1 *(Underestimation property of the AMSC).* *The AMSC of a pattern p is smaller than or equal to its true cost, that is $amsc(p) \leq c(p)$.*

Proof. Let $N = sup(p)$ and $M = minsup$. Without loss of generality, assume that all the cost values in $C(p) = \{c_1, c_2, \ldots c_N\}$ are sorted in ascending order, i.e. $c_1 \leq c_2 \leq \ldots \leq c_M \leq \ldots \leq c_N$. Then, $\frac{1}{M} \sum_{i=1,2,...,M} c_i(p) \leq \frac{1}{N} \sum_{i=1,2,...,N} c_i(p)$, because cost values are in ascending order, and thus $(N - M) \sum_{i=1,2,...,M} c_i(p) \leq (N - M) \cdot M \cdot c_M(p) \leq M \cdot (N - M) \cdot c_{M+1}(p) \leq M \sum_{i=M+1,M+2,...,N} c_i(p)$.

Theorem 2 *(Monotonicity property of the AMSC).* *The AMSC measure is monotonic. Let $p_x \subseteq p_y$ be two frequent patterns. Then, $amsc(p_x) \leq amsc(p_y)$.*

Proof. Assume that $n = sup(p_x), m = sup(p_y)$ and $M = minsup$, then $n \geq m \geq M$. Because $\forall p_x \subseteq p_y$, $c(p_x, S_i) \leq c(p_y, S_i)$ where $p_y \subseteq S_i$. And all the cost values $c(p_x, S_i)$ and $c(p_y, S_i)$ are sorted in ascending order, $c(p_y, S_{i_k}) \geq c(p_x, S_i)$ where $\{i_1, i_2, ..., i_m\} \subseteq \{1, 2, ..., n\}$, therefore $amsc(p_y) = \frac{1}{M} \sum_{i=1,2,...,M} c(p_y, S_i) \geq amsc(p_x) = \frac{1}{M} \sum_{i=1,2,...,M} c(p_x, S_i)$.

Strategy 3. *For a pattern p, if $AMSC(p) > maxcost$, then p and its extensions have an average cost greater than maxcost, and thus can be eliminated.*

The fourth search space pruning strategy is novel and used by CEPDO. It consists of eliminating all patterns that do not exist in at least one sequence having a positive utility when mining patterns in a binary SADB (Problem 1).

Strategy 4. *For a pattern p, if $D_p^+ = \emptyset$, then pattern p and all its extensions can be eliminated.*

The rationale of this strategy is that if a pattern only appears in negative sequences, then this pattern is not interesting as one cannot evaluate if it contributes to the positive outcome. To prune the search space in Problem 2, where utility is a numeric value, a fifth strategy is presented. Because numeric utility is neither monotonic nor anti-monotonic, an upper-bound on the utility is proposed to reduce the search space. Note that this upper-bound is different from those used in HUSPM since in this paper each sequence is annotated with the utility rather than each activity.

Definition 9 *(Upper bound on numeric utility).* *For a pattern p, let $n = sup(p)$ and $M = minsup$. An upper-bound on the numeric utility of p is $upperu = \frac{1}{M} \sum_{i=1,2,\ldots,n} u(p, S_i)$.*

Theorem 3 *(Overestimation and antimonotonicity of the numeric utility).* *Let p_x and p_y be two frequent patterns. Then, $u(p_x) \leq upperu(p_x)$. Moreover, if $p_x \subseteq p_y$, then $upperu(p_x) \geq upperu(p_y)$.*

Proof. *Assume that $n = sup(p_x)$, $m = sup(p_y)$, and $M = minsup$. Because $n \geq m \geq M$, $upperu = \frac{1}{M} \sum_{i=1,2,\ldots,n} u(p, S_i) \geq \frac{1}{n} \sum_{i=1,2,\ldots,n} u(p, S_i)$. Besides, $\frac{1}{M} \sum_{i=1,2,..m,\ldots,n} u(p, S_i) \geq \frac{1}{M} \sum_{i=1,2,\ldots,m} u(p, S_i)$.*

Strategy 5. *For a pattern p, if $upper(p) < minutil$, p and its extensions have a utility lower than $minutil$ and can be eliminated.*

The five strategies are designed to reduce the search space, and thus the time and memory required for finding cost-effective patterns in binary and numeric SADBs. The next paragraphs presents the two proposed algorithms.

The CEPDO algorithm (Algorithm 1) first scans the input SADB to calculate measures of length-1 patterns (Line 1). For a pattern p, if the support is not zero in positive sequences (Strategy 4) (Line 4), then the support, occupancy and average cost of p are compared with $minsup$ (Strategy 2), $minoccup$ and $maxcost$ (Line 5). If the threshold requirements are passed, the correlation of p is calculated, and p is saved as a cost-effective pattern with its correlation (Line 6–7). Then, if p's $AMSC$ is no larger than $maxcost$ (Strategy 3) and the occupancy upper-bound is no less than $minoccup$ (Strategy 1), the algorithm creates p's projected database, and scans it to find extensions of the form $p \cup \{a\}$ where a is an activity (Line 8–9). The CEPDO function is then recursively called to enumerate all cost-effective patterns having p as prefix (line 9) following a depth-first search. Since the algorithm only eliminates non cost-efficient patterns using the designed strategies, all cost-efficient patterns are output.

```
input  : a binary SADB D, minsup, maxcost, minoccup
output: the cost-efficient patterns
1  Scan D once to calculate the support, occupancy, average cost and AMSC of each
   length-1 pattern;
2  def CEPDO():
3  foreach activity p do
4  │   if |D_p^+| ≠ 0 then
5  │   │   if (sup(p) ≥ minsup and o(p) ≥ minoccup and (ac(p) ≤ maxcost) then
6  │   │   │   if |D_p^+| = sup(p) then cor(p):=1;
7  │   │   │   else Calculate(cor(p)) and Save(p);
8  │   │   │   if (amsc(p) ≤ maxcost and ou(p) ≥ minoccup then
9  │   │   │   │   foreach extension q = (p ∪ {a}) do call CEPDO(...);
10 │   │   │   end
11 │   │   end
12 │   end
13 end
```

Algorithm 1. The CEPDO algorithm

The CEPHU algorithm (Algorithm 2) scans the database to calculate the support of each length-1 pattern p (Line 1). For a pattern p, if the support is no less than *minsup* (Strategy 2) (Line 4), then the average cost, utility and occupancy of p are compared with *maxcost*, *minutil* and *minoccup* (Line 5). If the threshold requirements are met, the trade-off of p is calculated, and p is saved as a cost-effective pattern with its trade-off. Then, if p's $AMSC$ is no larger than *maxcost* (Strategy 3), the occupancy upper-bound is no less than *minoccup* (Strategy 1), and the utility upper-bound is no less than *minutil* (Strategy 5), the algorithm creates p's projected database, and scans it to find extensions of the form $p \cup \{a\}$ where a is an activity (Line 6-7). The CEPHU function is then recursively called to find all cost-effective patterns having p as prefix following a depth-first search (line 7). Since CEPHU only eliminates non cost-efficient patterns using its strategies, it finds all cost-efficient patterns.


```
input  : a numeric SADB D, minsup, maxcost, minoccup, minutil
output: the cost-efficient patterns
1  Scan D to calculate the support, occupancy, average cost, utility, upperu and ASMC of
   each length-1 pattern;
2  def CEPHU():
3  foreach activity p do
4  │   if (sup(p) ≥ minsup) then
5  │   │   if (ac(p) ≤ maxcost and u(p) ≥ maxutility and o(p) ≥ minoccup) then
   │   │       Calculate tf(p) and Save(p);
6  │   │   if (amsc(p) ≤ maxcost and upperu(p) ≥ minutil) and ou(p) ≥ minoccup then
7  │   │   │   foreach extension q = (p ∪ {a}) do call CEPHU(...);
8  │   │   end
9  │   end
10 end
```

Algorithm 2. The CEPHU algorithm

The proposed CEPDO and CEPHU algorithms can find cost-effective patterns in binary and numeric SADBs, respectively. To let users conveniently analyze patterns found, a user interface providing various visualizations were designed. This module will be described in the case study (Sect. 6).

5 Performance Evaluation

To evaluate the performance of the proposed CEPDO and CEPHU algorithms, experiments were carried out on a computer having a 64 bit Xeon E3-1270 3.6 Ghz CPU and 64 GB of RAM, running the Windows 10 operating system. Algorithms were implemented in Java, and source code is offered in the SPMF software [6]. Algorithms were evaluated in terms of performance on three standard benchmark datasets used in SPM, namely Bible, BMS and SIGN, obtained from the SPMF website. Those datasets have different characteristics such as dense, sparse, long and short sequences. In these datasets, the cost and utility values were randomly generated using a simulation model as in previous work in HUSPM [15]. The Bible dataset contains 36,369 sequences with 13,905 distinct items and an average sequence length of 44.3. The BMS dataset contains 59,601 sequences with 497 distinct items and an average sequence length of 6.02. The SIGN dataset contains 730 sequences with 267 distinct items and an average sequence length of 104.1. Because CEPDO and CEPHU are the first algorithms for mining cost-effective patterns, they cannot be compared with previous SPM and HUSPM techniques. For this reason, two versions of each algorithm were compared using i. Strategy 3 based on the proposed AMSC lower-bound on the average cost and ii. without using Strategy 3. The execution times of the two versions of CEPDO and CEPHU are shown in Fig. 1A and B for each dataset, while the $maxcost$ threshold is varied. As shown in Fig. 1, using Strategy 3 with the AMSC decreases runtime by up to 4 times. Besides, as the $maxcost$ threshold is increased, the version using AMSC become more efficient. Because of space

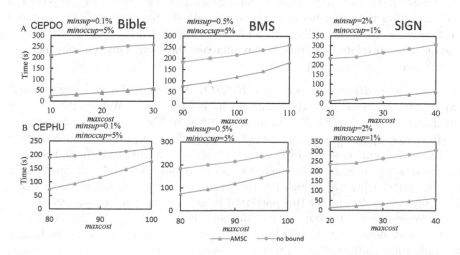

Fig. 1. Runtimes of CEPHU/CEPDO with/without Strategy 3 when $maxcost$ is varied

limitation and that occupancy and support were used in previous papers, results obtained when varying the *minsup* and *minoccup* threshold are not presented. It is found that the AMSC pruning strategy greatly improves efficiency.

6   Case Study in E-Learning

To evaluate how useful the proposed patterns are, the developed algorithms were applied on e-learning data of Vahdat et. al [16] to find cost-effective patterns that can lead to obtaining high scores. This data consists of logs of the Deeds e-learning environment, for learning digital electronics. The environment provides learning materials. Furthermore, learners are asked to solve problems of different levels of difficulty to assess their knowledge. The dataset contains sequence of activities performed by each student in each learning sessions. The time spent for each activity by a student is indicated. Moreover a score is given at the end of each session, and a score is provided for the fnal exam.

Finding Patterns Using CEPDO. CEPDO was first applied to find cost-effective sequences of sessions to pass the final exam. Each sequence consists of several sessions, each annotated with a time cost. The utility of each sequence is the exam score, transformed into binary classes (*passed*/*failed*). CEPDO was applied with $minsup = 0.5$, $minoccup = 0.5$, and $maxcost = 600$. Some patterns providing interesting insights were discovered. For example, the patterns $\{1, 2, 5, 6\}$, $\{1, 2, 6\}$ and $\{1, 5, 6\}$ have a large positive correlation (0.21, 0.20, 0.19), respectively, where numbers represents session IDs. It was also found that some patterns such as $\{4, 5\}$ and $\{5\}$ have a negative correlation (-0.11 and -0.15, respectively). Moreover, it was found that some patterns such as $\{2, 3\}$, $\{3, 4, 5, 6\}$ are barely correlated with the final exam result as their correlation are both 0.001. Overall, based on these patterns, it is found that students who learnt Session 1, Session 2, Session 5 and Session 6 are more likely to *pass* the final exam. On the other hand, if a student only studies Session 5, or Session 4 and Session 5, he is more likely to *fail* the exam. Besides, if a student spends time on other unrelated sessions, it may increase the study time but not increase much his chances of passing the exam.

Visualizing Patterns Found by CEPDO. We designed a visualization system to further investigate the relationship between cost and utility for each pattern found by CEPDO. Charts in the top row of Fig. 2 shows the cost-utility distribution graph of three patterns ($\{1, 2, 5\}$, $\{4, 5\}$ and $\{2, 3\}$). Since a pattern may have the same cost but different outcomes in multiple sequences, the average outcome was calculated for each cost value (a value in $[-1, 1]$). It can be seen that the pattern $\{1, 2, 5\}$, which has the largest positive correlation, has a dense distribution (Fig. 2) and for this pattern a larger cost is spent in sequences leading to the desirable outcome (1) than the negative outcome (-1). Pattern $\{4, 5\}$, has the most dense distribution and most cost is spent in sequences leading to the undesirable outcome (-1), which is reasonable since $\{4, 5\}$ has the largest negative correlation. Pattern $\{2, 3\}$'s distribution does not show a considerable

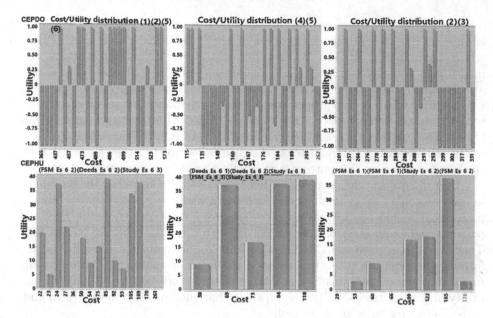

Fig. 2. Distribution graph in terms of cost-utility

difference for all sequences in terms of cost, which is reasonable since its correlation is 0.001. Another observation is that because few students passed the exam for the 60 point threshold, the correlation is on overall small. If that threshold is reduced, correlation values increase.

Finding Patterns Using CEPHU. In another experiment, the designed CEPHU algorithm was applied to the same e-learning database to analyse activities within sessions rather than analyzing the whole learning process. The database contains multiple sequences for the same session (one for each student) in which each activity (learning material) has a time cost. The score of each session exam is the utility of the sequence. The goal is to obtain insights about how to efficiently use learning materials to obtain high scores in a session test. The algorithm was applied to data from six sessions to find patterns that have a small trade-off and lead to a high score. It was found that it's unreasonable to recommend the patterns that only have a small trade-off, because even though some patterns have a small trade-off, they sometimes also have a low utility. Thus, we defined several utility ranges (e.g. $[14, 15)$, $[15, 16)$), and sorted patterns by trade-off in each utility range. By considering ranges, interesting patterns having a relatively small trade-off but a high utility are observed. For example, consider session 6, $minsup = 0.1$, $occup = 0.28$, $minutility = 10$ and $maxcost = 100$. To obtain the average score of 15, the most cost-effective pattern is (Study_Es_6_2)(Deeds_Es_6_2)(Study_Es_6_3), having a trade-off of 1.74. But, the most cost-effective pattern to obtain a high score of 28/40 is (Deeds_Es_6_1)(Deeds_Es_6_2)(Study_Es_6_3)(FSM_Es_6_3) (Study_Es_6_3) with a trade-off of 2.71. Pictures of the bottom row of Fig. 2 show the visualization of the cost/utility distribution of three patterns found by CEPHU, which haves utility of 15, 28, and 10, respectively. Two observations are made from these charts. First, although most effective patterns have a relatively small cost, they can still lead to a relatively high score. Second, we can find what is the proper cost range that should be spent on a pattern to obtain a target score. For example, Fig. 2 (bottom row) shows that (Deeds_Es_6_1)(Deeds_Es_6_2)(Study_Es_6_3)(FSM_Es_6_3)(Study_Es_6_3) generally provides a high score at a low cost compared to the other patterns. And the cost range where this pattern is most effective is approximately [69, 118].

After discovering the patterns, we have also carefully looked at the questions in each session's final exam and compared them with the materials of the mined patterns. This has confirmed that there is indeed a strong correlation between patterns and exam questions, which indicates that the patterns are reasonable. Then, we also applied the CEPDO algorithm for individual sessions by transforming the sequence utility to a binary class based on the average score. It was found that several patterns having a large positive correlation also have a high utility and a low trade-off. This again show that the proposed measures to find cost-effective patterns are reasonable.

Visualizing the Skyline of Patterns Found by CEPDO and CEPHU.
To give users a more clear view of patterns found by the algorithms, a second visualization was implemented based on the concept of skyline [14]. Figure 3 displays the skyline of patterns that have been extracted by CEPDO. Each data point represents the pattern p having the highest correlation for its average cost value $ac(p)$. The skyline is useful as it lets users quickly find the patterns having the highest correlation for each cost value, and thus to more easily browse patterns. Similarly, Fig. 4 displays the skyline of patterns that have been extracted by CEPHU. Each data point represents the pattern p having the lowest trade-

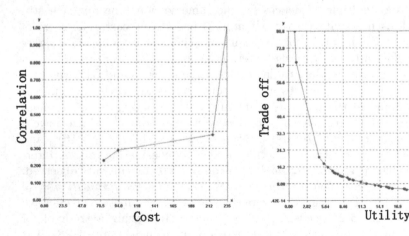

Fig. 3. Skyline patterns of CEPDO   **Fig. 4.** Skyline patterns of CEPHU

off for its utility value. It is interesting to see that patterns scattered in the utility range [16.9, 28.2] are very efficient, while those scattered in [0 to 2.82] have a high cost but lead to a low utility. Note that the designed skyline visualization system for patterns can be used using other pairs of measures such as {*utility*, *occupancy*} or {*support*, *utility*} but this feature is not illustrated due to space limitation.

On overall, Sect. 6 has shown that insightful cost-efficient patterns have been found in e-learning data using the proposed algorithms.

7 Conclusion

This paper proposed two algorithms named CEPDO and CEPHU for mining cost-effective patterns in sequences of activites with cost/utility sequences, where utility is either binary or numeric values. A tight lower bound on the cost and an upper bound on the utility was proposed to efficiently find cost-effective patterns. Furthermore, occupancy and support measures were integrated into the algorithms to eliminate non representative and infrequent patterns. Moreover, a visualization system was developed to show the cost-utility distribution of patterns and browse patterns using their skyline. Experimental results have shown that the algorithms have good performance and a case study has shown that interesting patterns are discovered in e-learning data.

In future work, we will consider optimizing the proposed algorithms, and extending the model to other types of patterns such as periodic patterns [7], peak patterns [10], and itemsets [9]. Moreover, applications of the algorithms in other domains such as for analyzing business procsses will be investigated.

Acknowledgement. This project was funded by the National Science Fundation of China and the Harbin Institute of Technology (Shenzhen).

References

1. Agrawal, R., Srikant, R.: Mining sequential patterns. In: Proceedings of the 11th International Conference on Data Engineering, pp. 3–14. IEEE (1995)
2. Alkan, O.K., Karagoz, P.: Crom and huspext: improving efficiency of high utility sequential pattern extraction. IEEE Trans. Knowl. Data Eng. **27**(10), 2645–2657 (2015)
3. Bogarín, A., Cerezo, R., Romero, C.: A survey on educational process mining. Wiley Interdisc. Rev.: Data Min. Knowl. Disc. **8**(1), e1230 (2018)
4. Choi, D.W., Pei, J., Heinis, T.: Efficient mining of regional movement patterns in semantic trajectories. Proc. VLDB Endowment **10**(13), 2073–2084 (2017)
5. Dalmas, B., Fournier-Viger, P., Norre, S.: Twincle: a constrained sequential rule mining algorithm for event logs. Procedia Comput. Sci. **112**, 205–214 (2017)
6. Fournier-Viger, P., Gomariz, A., Gueniche, T., Soltani, A., Wu, C.W., Tseng, V.S.: SPMF: a Java open-source pattern mining library. J. Mach. Learn. Res. **15**(1), 3389–3393 (2014)

7. Fournier-Viger, P., Li, Z., Lin, J.C.W., Kiran, R.U., Fujita, H.: Efficient algorithms to identify periodic patterns in multiple sequences. Inf. Sci. **489**, 205–226 (2019)
8. Fournier-Viger, P., Lin, J.C.W., Kiran, R.U., Koh, Y.S., Thomas, R.: A survey of sequential pattern mining. Data Sci. Pattern Recogn. **1**(1), 54–77 (2017)
9. Fournier-Viger, P., Chun-Wei Lin, J., Truong-Chi, T., Nkambou, R.: A survey of high utility itemset mining. In: Fournier-Viger, P., Lin, J.C.-W., Nkambou, R., Vo, B., Tseng, V.S. (eds.) High-Utility Pattern Mining. SBD, vol. 51, pp. 1–45. Springer, Cham (2019). https://doi.org/10.1007/978-3-030-04921-8_1
10. Fournier-Viger, P., Zhang, Y., Lin, J.C.W., Fujita, H., Koh, Y.S.: Mining local and peak high utility itemsets. Inf. Sci. **481**, 344–367 (2019)
11. Fani Sani, M., van der Aalst, W., Bolt, A., García-Algarra, J.: Subgroup discovery in process mining. In: Abramowicz, W. (ed.) BIS 2017. LNBIP, vol. 288, pp. 237–252. Springer, Cham (2017). https://doi.org/10.1007/978-3-319-59336-4_17
12. Glass, G.V., Hopkins, K.D.: Statistical Methods in Education and Psychology. Pearson, Harlow (1996)
13. Ranjan, J., Malik, K.: Effective educational process: a data-mining approach. Vine **37**(4), 502–515 (2007)
14. Soulet, A., Raïssi, C., Plantevit, M., Cremilleux, B.: Mining dominant patterns in the sky. In: IEEE 11th International Conference on Data Mining, pp. 655–664. IEEE (2011)
15. Truong-Chi, T., Fournier-Viger, P.: A survey of high utility sequential pattern mining. In: Fournier-Viger, P., Lin, J.C.-W., Nkambou, R., Vo, B., Tseng, V.S. (eds.) High-Utility Pattern Mining. SBD, vol. 51, pp. 97–129. Springer, Cham (2019). https://doi.org/10.1007/978-3-030-04921-8_4
16. Vahdat, M., Oneto, L., Anguita, D., Funk, M., Rauterberg, M.: A learning analytics approach to correlate the academic achievements of students with interaction data from an educational simulator. In: Conole, G., Klobučar, T., Rensing, C., Konert, J., Lavoué, É. (eds.) EC-TEL 2015. LNCS, vol. 9307, pp. 352–366. Springer, Cham (2015). https://doi.org/10.1007/978-3-319-24258-3_26
17. Zhang, L., et al.: Occupancy-based frequent pattern mining. ACM Trans. Knowl. Discov. Data **10**(2), 14 (2015)

Efficient Row Pattern Matching Using Pattern Hierarchies for Sequence OLAP

Yuya Nasu[1], Hiroyuki Kitagawa[2(⊠)], and Kosuke Nakabasami[1,3]

[1] Graduate School of Systems and Information Engineering, University of Tsukuba,
Tsukuba, Japan
yuuya.n@kde.cs.tsukuba.ac.jp
[2] Center for Computational Sciences, University of Tsukuba, Tsukuba, Japan
kitagawa@cs.tsukuba.ac.jp
[3] Railway Technical Research Institute, Tokyo, Japan
nakabasami.kosuke.39@rtri.or.jp

Abstract. *Sequence OLAP* is a variant of OLAP for sequence data analysis such as analysis of RFID log and person trip data. It extracts *pattern occurrences* of the given patterns (e.g., state transition pattern $S_1 \rightarrow S_2$, movement pattern $A \rightarrow B$) on sequence data and executes multi-dimensional aggregate using *OLAP operations* (such as *drill-down* and *roll-up*) and *pattern OLAP operations* (such as *pattern-drill-down* and *pattern-roll-up*). The pattern OLAP operations are specific to Sequence OLAP and involve a hierarchy of multiple patterns. When sequence data is stored in relational databases as sequences of rows, row pattern matching finds all subsequences of rows which match a given pattern. To do Sequence OLAP, especially pattern OLAP operations, on relational databases, it is required to execute row pattern matching for such a hierarchy of multiple patterns and identify parent-child relationships among pattern occurrences. Generally, row pattern matching needs sequential scan of a large table and is an expensive operation. If row pattern matching is executed individually for each pattern, it is very time consuming. Therefore, it is strongly demanded to execute multiple row pattern matching for a given hierarchy of patterns efficiently. This paper formalizes a pattern hierarchy model for Sequence OLAP and proposes a very efficient algorithm to do multiple row pattern matching using SP-NFA (Shared Prefix Nondeterministic Finite Automaton). In experiments, we implement our algorithm in PostgreSQL and evaluate the effectiveness of the proposal.

Keywords: Row pattern matching · Sequence OLAP · Pattern hierarchy

1 Introduction

Various kinds of data have been generated and accumulated in databases due to recent development of IoT and social media. In particular, we have more

C. Ordonez et al. (Eds.): DaWaK 2019, LNCS 11708, pp. 89–104, 2019.
https://doi.org/10.1007/978-3-030-27520-4_7

and more sequence data such as RFID log and person trip data. *Sequence OLAP* [8,9] is a variant of OLAP (Online Analytical Processing) for sequence data analysis. Sequence OLAP extracts *pattern occurrences* of the given patterns (e.g., state transition pattern $S_1 \rightarrow S_2$, movement pattern $A \rightarrow B$) on sequence data and executes multi-dimensional aggregate using *OLAP operations* such as *drill-down* and *roll-up* and *pattern OLAP operations* such as *pattern-drill-down* and *pattern-roll-up*. The pattern OLAP operations are specific to Sequence OLAP and involve a hierarchy of multiple patterns. In addition, Sequence OLAP also involves other pattern related operations such as *pattern-occurrence-drill-down* and *pattern-occurrence-roll-up*, which follow parent-child relationships focusing on some specific pattern occurrences.

Executing Sequence OLAP on relational databases is very desirable since they are commonly used for storing sequence data and integration with other data is easy. Sequence data is typically represented as sequences of rows in relational databases, and row pattern matching finds all subsequences of rows which match a given pattern. Actually, row pattern matching was standardized as SQL/RPR in 2016 [1], and we can extract pattern occurrences of a given pattern using MATCH_RECOGNIZE (MR) clause in SQL/RPR. However, as aforementioned, we need to execute row pattern matching for hierarchies of multiple patterns and identify parent-child relationships among pattern occurrences in Sequence OLAP. Generally, row pattern matching needs sequential scan of a large table and is an expensive operation. If row pattern matching is executed individually for each pattern, it is very time consuming.

To address these problems, we propose a new algorithm to do multiple row pattern matching for a given hierarchy of patterns. Although support for Sequence OLAP has been studied in the past research as discussed in Sect. 7, this work first formalizes a pattern hierarchy for Sequence OLAP and proposes an efficient multiple row pattern matching in relational database contexts. The proposed algorithm employs SP-NFA (Shared Prefix Nondeterministic Finite Automaton), and executes multiple row pattern matching efficiently following a given hierarchy of multiple patterns. In addition, it identifies parent-child relationships among all found pattern occurrences at the same time. In experiments, we implement our proposed algorithm as UDF (User Defined Function) on top of PostgreSQL and verify advantages of the proposal over other baseline methods.

The remaining part of this paper is organized as follows. In Sect. 2, we give an overview of Sequence OLAP and show the need for row pattern matching following a hierarchy of multiple patterns. Section 3 presents our pattern hierarchy model. Section 4 shows how our hierarchical row pattern matching looks like in the context of SQL as an extension of the SQL/RPR standard. Section 5 presents our proposed algorithm to do row pattern matching following a hierarchy of multiple patterns. Section 6 reports experiment using PostgreSQL. Section 7 discusses related work. Section 8 concludes this paper and shows future research issues.

sequence table			
sid	city	venue	time
1	Vancouver	Airport	01/04 10:20
1	New York	Airport	01/04 16:35
1	New York	Hotel	01/04 17:05
1	New York	Restaurant	01/04 18:00
1	New York	Airport	01/05 07:15
1	London	Airport	01/05 14:10
1	Paris	Airport	01/05 22:30
1	Paris	Hotel	01/05 23:05
...	...		...

pattern occurrence table : (X Y)						
sid	X_city	X_venue	X_time	Y_city	Y_venue	Y_time
1	Vancouver	Airport	01/04 10:20	New York	Airport	01/04 16:35
1	Vancouver	Airport	01/04 10:20	New York	Airport	01/05 07:15
1	Vancouver	Airport	01/04 10:20	London	Airport	01/05 14:10
1	Vancouver	Airport	01/04 10:20	Paris	Airport	01/05 22:30
...	...	...	...	...	...	
1	New York	Airport	01/04 16:35	New York	Airport	01/05 07:15
1	New York	Airport	01/04 16:35	London	Airport	01/05 14:10
...	...	...	...	...	...	...

Fig. 1. Sequence table and pattern occurrence table

2 Sequence OLAP

Sequence OLAP is known as an aggregation analysis technique for sequence data. Let us consider travel log data stored in a sequence table on the left-hand side in Fig. 1. To apply Sequence OLAP on this table, we focus on some sequence pattern (row pattern) and extract its pattern occurrences by row pattern matching. A pattern is represented as a regular expression such as (X Y) and (X Y* Z|A) involving a number of variables (called *pattern variables*) corresponding to rows. Pattern variables may need to meet some conditions with respect to their attributes. For instance, pattern (X Y) with conditions *X.venue = Airport* & *Y.venue = Airport* represents a movement between two airports. The row pattern matching results in a *pattern occurrences table*, which stores all pattern occurrences as rows as shown on the right-hand side in Fig. 1. In Sequence OLAP, this pattern occurrence table is used as a *fact table*. By joining this fact table with different dimension tables and apply several aggregate functions, we can analyze sequence data by OLAP operations such as drill-down and roll-up.

Sequence OLAP usually allows us to execute *pattern OLAP operations* in addition to such traditional OLAP operations. They include *pattern-drill-down* and *pattern-roll-up*, which involve a hierarchy of multiple patterns. For example, when the user is analyzing data using the traveling pattern between two airports (X Y), she/he may further want to know the next venue Z after airport Y. In this case, the pattern which she/he is interested in is elaborated from (X Y) to (X Y Z). The pattern-drill-down allows us to such changes in focused patterns. Given a hierarchy of multiple patterns, the pattern-drill-down allows us to move from a parent pattern to a child pattern to see more details using a more elaborated pattern. The pattern-roll-up allows us to go up the pattern hierarchy.

In addition to pattern-drill-down and pattern-roll-up, Sequence OLAP also involves *pattern-occurrence-drill-down* and *pattern-occurrence-roll-up*, which follow parent-child relationships focusing on some specific pattern occurrences. For example, let us assume that occurrences of the pattern (X Y) are extracted in the left-hand side table in Fig. 2. The user would further want to know the next venue after moving from New York airport to London airport. In this case, we need to show pattern occurrences of (X Y Z) extending the original movement from New York airport to London airport as shown in Fig. 2. This operation is called pattern-occurrence-drill-down. Its inverse is pattern-occurrence-roll-up. To respond to such requests, it is necessary to identify parent-child relationships among pattern occurrences of (X Y) and (X Y Z).

aggregation result of (X Y)

X_city	X_venue	Y_city	Y_venue	count
Vancouver	Airport	New York	Airport	5
Vancouver	Airport	London	Airport	2
...	...	...	...	...
New York	Airport	London	Airport	3
...	...	...	...	...

pattern occurrences of (X Y Z)

sid	X_city	X_venue	Y_city	Y_venue	Z_city	Z_venue
1	New York	Airport	London	Airport	Paris	Airport
1	New York	Airport	London	Airport	Paris	Hotel
2	New York	Airport	London	Airport	London	Cafe
3	New York	Airport	London	Airport	London	Hotel

Fig. 2. Pattern-occurrence-drill-down

To summarize, for efficient Sequence OLAP on relational databases, we need to support row pattern matching for a hierarchy of multiple patterns and identification of parent-child relationships among all found pattern occurrences. In SQL/RPR standardized in 2016 [1], row pattern recognition can be specified using MR clause. It directly generates a pattern occurrence table from the input table as shown in Fig. 1, which serves as a fact table in Sequence OLAP. However, since MR clause can specify only one pattern, we need to execute multiple SQL/RPR queries repeatedly to do Sequence OLAP using a hierarchy of multiple patterns. In addition, SQL/RPR does not support identification of parent-child relationships among all found pattern occurrences. To do this in the current framework of SQL, we need to join multiple pattern occurrence tables for multiple patterns. Our motive for this research is to address these problems by proposing a new framework and algorithm for row pattern matching for a hierarchy of multiple patterns and identification of parent-child relationships among all found pattern occurrences.

3 Pattern Hierarchy Model

In this section, we formalize several basic concepts and introduce pattern hierarchy model used for row pattern matching in this paper. Then, we define row pattern matching with a given pattern hierarchy.

Definition 1 (Pattern element). *A **pattern element** p is a pattern variable v (one character) or a regular expression consisting of pattern elements. We consider the following regular expressions:*

- *sequence (denoted by $(p_1 \ldots p_n)$)*
- *Kleene closure (denoted by $p*$)*
- *alternative (denoted by $p_1|p_2$)*

*where p or p_i is a pattern element, and each pattern element does not contain multiple occurrences of the same pattern variable. We call a pattern element of the first case a **sequence pattern element**.*

Definition 2 (Pattern). *A* **pattern** *P is a sequence pattern element, denoted by* $P = (p_1 \ldots p_n)$.

Definition 3 (Parent-child relationship of patterns). *If a pattern* P_1 *can be derived from* P_2 *by one of the following insert operations,* P_1 *is a* **child** *of* P_2 *(*P_2 *is a* **parent** *of* P_1*). (1) Insert a new pattern element* p_{new} *before* P_2 *(*$P_1 = (p_{new} \ P_2)$*), (2) Insert a new pattern element* p_{new} *after* P_2 *(*$P_1 = (P_2 \ p_{new})$*), and (3) Insert a new element* p_{new} *in between pattern elements of* P_2.

Definition 4 (Pattern hierarchy). *If a set of patterns forms a rooted tree with respect to their parent-child relationships, it is called a* **pattern hierarchy**.

Definition 5 (Prefix and suffix match). *Given a pair of pattern* P_{parent} *and child pattern* P_{child}, *if the (top level) pattern elements of* P_{parent} *and* P_{child} *are equal except for the last one, the parent-child relationship corresponds to* **prefix match**. *Otherwise, the parent-child relationship corresponds to* **suffix match**.

Example 1 **(Pattern hierarchy).** Figure 3 shows an example of a pattern hierarchy. Its root is pattern (X Y). Edges represent parent-child relationships and edge labels shows classification into prefix and suffix match cases. For simplicity, we assume that pattern variables newly included in descendant patterns are distinct with each other. For example, it is not allowed to add (X W Y Z) as a child of (X W Y), since pattern variable Z already appears in (X Y Z) in the pattern hierarchy in Fig. 3.

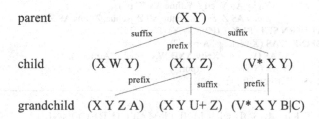

Fig. 3. Pattern hierarchy

Definition 6 (Pattern occurrence). *A pattern occurrence* o_P *for pattern P is a sequence of rows, denoted* $o_P = (r_1, \ldots, r_m)$, *which meet both the chronological pattern P and its associated attribute value conditions.*

In this paper, we employ *Skip-Till-Any-Match* [2,15,18] semantics for row pattern matching, since it is flexible and often assumed in the context of Sequence OLAP. Skip-Till-Any-Match detects pattern occurrences of all combinations that satisfy the pattern conditions as shown in Fig. 1. For example, let us assume that row pattern matching with (X Y) is done for the sequence of rows $(r_1, \ldots, r_4)$. If no attribute condition for X and Y is specified, we get pattern occurrences (r_1, r_2), (r_1, r_3), (r_1, r_4), (r_2, r_3), (r_2, r_4), (r_3, r_4) in Skip-Till-Any-Match semantics.

Definition 7 (Parent-child relationship of pattern occurrences). *Given a pair of child pattern P_{child} and parent pattern P_{parent}, an occurrence $o_{P_{child}}$ of P_{child} is a* **child pattern occurrence** *of a parent pattern occurrence $o_{P_{parent}}$ of P_{parent}, if $\forall r_i(r_i \in o_{P_{parent}} \Rightarrow r_i \in o_{P_{child}})$.*

Definition 8 (Row pattern matching with a pattern hierarchy). *Row pattern matching with a given pattern hierarchy H on a sequence table is to extract pattern occurrences for all patterns included in H and identify all parent-child relationships among the extracted pattern occurrences.*

4   MULTI_MATCH_RECOGNIZE in SQL

As mentioned before, SQL/RPR has MR clause to specify row pattern matching using a single pattern. Here, we show how SQL queries look like if we extend MR clause to allow row pattern matching with pattern hierarchies. Our extension of MR clause is MULTI_MATCH_RECOGNIZE (MMR) clause. A sample SQL query using MMR clause is shown below.

```
SELECT * FROM sequence
MULTI MATCH RECOGNIZE (
PARTITION BY sid ORDER BY time
MEASURES X.city AS X_city, X.time AS X_time,
         W.city AS W_city, W.venue AS W_venue, W.time AS W_time,
         Y.city AS Y_city, Y.time AS Y_time,
         Z.city AS Z_city, Z.venue AS Z_venue, Z.time AS Z_time
PATTERN SET (P1;X Y (P2;X W Y) (P3;X Y Z))
DEFINE X AS (X.venue = "Airport"),
       W AS (W.city = Xcity AND W.venue != "Airport"),
       Y AS (Y.venue = "Airport")
WINDOW [time 24h] ) ;
```

Fig. 4. SQL with MULTI_MATCH_RECOGNIZE

The main difference from MR is PATTERN SET clause. In MR, the user can specify only one pattern in PATTERN clause. In PATTERN SET clause, the user can specify a pattern hierarchy rather than a single pattern. In this example, pattern (X Y) named P1 is the root, and two child patterns (X W Y) named P2 and (X Y Z) named P3 are specified. In PATTERN SET clause, the pattern hierarchy is represented in a nested list. SQL/RPR takes Strict-Contiguity semantics, which is more strict than Skip-Till-Any-Match. Due to this, we also introduce WINDOW clause to specify the maximum pattern occurrence length. Except for some minor changes caused by this difference in semantics, the other components in MMR clause are same as in MR clause. PARTITION BY specifies key attributes and ordering attributes to form sequences. MEASURES clause specifies attributes to be included in the output pattern occurrence table. DEFINE clause gives conditions on pattern variables.

Figure 5 show the pattern occurrence table obtained by this example query applied to the input sequence table in Fig. 1. Each row corresponds to a pattern occurrence of one of the patterns in the given pattern hierarchy. Attributes specified in PARTITION BY and MEASURES clauses appear in the output pattern occurrence table as in the original MR. In the case of MMR, some additional attributes are automatically added to the output table. Attribute **oid** is the primary key of the output pattern occurrence table. Attribute **pname** gives the name of the pattern whose occurrence is represented by the current row. Attribute <**pattern_name**>_oid gives an occurrence identifier. If <pattern_name> is the name of the pattern an occurrence of which the row represents, <pattern_name>_oid designates the occurrence. Otherwise, <pattern_name>_oid designates a parent or ancestor pattern occurrence corresponding to the pattern <pattern_name>.

In Fig. 5, rows with oid = 1, 5, 6 represent pattern occurrences of pattern P1: (X Y). Rows with oid = 7, 8 represent pattern occurrences of pattern P2:(X W Y), and their parent pattern occurrences of P1 are given in P1_oid. Similarly, rows with oid = 2, 3, 4 represent pattern occurrences of pattern P3: (X Y Z), and their parent pattern occurrences of P1 are given in P1_oid.

pattern occurrence table : P1 (X Y), P2 (X W Y), P3 (X Y Z)

oid	sid	X_city	X_time	W_city	W_venue	W_time	Y_city	Y_time	Z_city	Z_venue	Z_time	pname	P1_oid	P2_oid	P3_oid
1	1	Vancouver	01/04 10:20				New York	01/04 16:35				P1	1		
2	1	Vancouver	01/04 10:20				New York	01/04 16:35	New York	Hotel	01/04 17:05	P3	1		1
3	1	Vancouver	01/04 10:20				New York	01/04 16:35	New York	Restaurant	01/04 18:00	P3	1		2
4	1	Vancouver	01/04 10:20				New York	01/04 16:35	New York	Airport	01/05 07:15	P3	1		3
5	1	Vancouver	01/04 10:20				New York	01/05 07:15				P1	2		
6	1	New York	01/04 16:35				New York	01/05 07:15				P1	3		
7	1	New York	01/04 16:35	New York	Hotel	01/04 17:05	New York	01/05 07:15				P2	3	1	
8	1	New York	01/04 16:35	New York	Restaurant	01/04 18:00	New York	01/05 07:15				P2	3	2	
...	...	...	...	...	...	...	...	...	...	...	...	...	...	...	...

Fig. 5. Pattern occurrence table derived by MULTI_MATCH_RECOGNIZE

5 Query Processing Algorithms

This section explains how to execute queries including row pattern matching with pattern hierarchies. In particular, we focus on how to do row pattern matching. The essential point is to efficiently extract pattern occurrences for all patterns included in a given pattern hierarchy from a sequence of rows and identify all parent-child relationships among the extracted pattern occurrences. For this purpose, we use SP-NFA (Shared Prefix Nondeterministic Finite Automaton), which allows pattern matching for multiple patterns sharing prefixes at the same time. Section 5.1 shows how to construct SP-NFA. Section 5.2 give an algorithm to extract pattern occurrences for all patterns included in a given pattern hierarchy using SP-NFA. Section 5.3 shows an algorithm to identify all parent-child relationships among the extracted pattern occurrences.

5.1   Construction of SP-NFA

SP-NFA which allows pattern matching for multiple patterns sharing prefixes at the same time. We traverse the given pattern hierarchy from the root in a depth first manner. If the relationship between the parent and the child patterns is prefix match, then the state transition path in SP-NFA is expanded for the child pattern sharing the prefix with the parent. Otherwise, a new path is added to SP-NFA for the child pattern. In this case, the new path will share prefix of some existing path as much as possible. If no prefix path sharing is possible, a new branch from the initial state q_0 is created for the child pattern. Using SP-NFA, multiple patterns with common prefixes can share intermediate results, and we can easily identify parent-child relationships among occurrences of parent-child patterns of prefix match.

Algorithm 1 shows construction of SP-NFA. SP-NFA is developed based on the descriptions in PATTERN SET, DEFINE, WINDOW clauses in MMR. Algorithm 1 scans the pattern hierarchy specified in PATTERN SET, and expands the SP-NFA from the initial state q_0 as explained above. Actions and state transition conditions are given to edges between states, which are decided based on the descriptions in DEFINE and WINDOW. We use two actions *take* (hold the current row) and *ignore* (skip the current row) as in SASE [15]. For simplicity, Algorithm 1 assumes that all pattern elements are single pattern variables.

Algorithm 1. Construction of SP-NFA

1: **procedure** CONSTSPNFA(pattern hierarchy H, conditions C)
2:      $A \leftarrow$ new SP-NFA having only the initial state
3:      Expand(A, H, C)
4: **end procedure**
5:
6: **procedure** EXPAND(SP-NFA A, pattern hierarchy H, conditions C)
7:      $r \leftarrow$ root of H
8:      $cond \leftarrow$ conditions relevant to r in C
9:      Add($A, r, cond$)
10:      **for each** $child \in r.children$ **do**
11:          Expand(A, subtree of H rooted by $child$, C)
12:      **end for**
13: **end procedure**

Figure 6 shows an SP-NFA for the MMR example in Fig. 4. Note that, (X, Y), (X, W, Y), (X, Y, Z) share the prefix up to the pattern variable X, and (X, Y), (X, Y, Z) also share the prefix up to the pattern variable Y.

5.2   Row Pattern Matching with Pattern Hierarchies

Algorithm 2 shows the main algorithm to do row pattern matching with pattern hierarchies. Algorithm 2 inputs a sequence table as shown in Fig. 1, and outputs a pattern occurrence table as shown in Fig. 5.

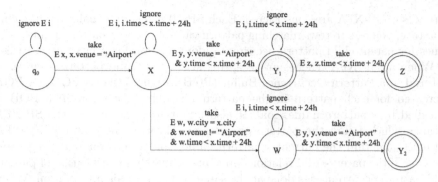

Fig. 6. SP-NFA

Algorithm 2. Row pattern matching with pattern hierarchies

Require: rows R, SP-NFA A, Pattern Occurrence Buffers B
1: $patternOccurrenceTable \leftarrow \emptyset$
2: $instances \leftarrow \{newInstance(A)\}$
3: **for each** $r \in R$ **do**
4:     $newInstances \leftarrow \emptyset$
5:     **for each** $i \in instances$ **do**
6:         $edges \leftarrow A.getTransitableEdges(i, r)$
7:         **for each** $e \in edges$ **do**
8:             $ic \leftarrow newInstance(A)$
9:             $ic \leftarrow i.copy()$
10:             $ic.state \leftarrow e.destination$
11:             **if** e.action is "take" **then**
12:                 insert r into $ic.matchBuffer$
13:                 **if** e.destination is an acceptance state **then**
14:                     $pushPatternOccurrence(ic.matchBuffer, B)$
15:                 **end if**
16:             **end if**
17:             $newInstances \leftarrow newInstances \cup \{ic\}$
18:         **end for**
19:     **end for**
20:     $identifyParentChildRelationships(B)$
21:     $pushPatternOccurrences(B, patternOccurrenceTable)$
22:     $instances \leftarrow newInstances$
23: **end for**
24: **return** $patternOccurrenceTable$

SP-NFA is constructed before executing Algorithm 2. Algorithm 2 reads rows
in the input sequence table one by one. At the beginning, an instance of SP-NFA
is created and its current state is the initial state q_0. Whenever a state transition corresponding to an edge occurs, a new instance of SP-NFA is constructed
and the internal status of the original instance is inherited by the new instance.
When state transitions following multiple edges occur at the same time, multiple

instances of SP-NFA are constructed. Each SP-NFA instance maintains the current state, pointers to rows matching pattern variables, and <pattern_name>_oid values for parent-child pattern occurrence identification for prefix match cases.

When an SP-NFA reaches an accepting state, the found pattern occurrence is stored in a Pattern Occurrence Buffer (POB). Figure 7 shows POBs. A POB is created for each pattern in the pattern hierarchy. Each row in a POB is identified by <pattern_name>_oid. It also maintains a pointer to the SP-NFA instance which found the pattern occurrence. Through the pointers, SP-NFA instances maintain the current <pattern_name>_oid values for relevant parent patterns. If the parent-child relationship is prefix match, identification of parent-child pattern occurrences is done at the same time using this information. When all the processing triggered by a new row finishes, the parent-child relationship identification among pattern occurrences for suffix match cases (See Sect. 5.3) is invoked. Finally, pattern occurrences stored in POBs are output and the contents of POBs are cleared. Figure 7 shows the snapshot of POBs when the first five rows in the sequence table in Fig. 1 are processed.

P1

count	instance	X	Y	P1_oid
3	a^3	$[x_1]$	$[y_5]$	2
	a^6	$[x_2]$	$[y_5]$	3

P2

count	instance	X	W	Y	P1_oid	P2_oid
2	a^7	$[x_2]$	$[w_3]$	$[y_5]$	2	1
	a^8	$[x_2]$	$[w_4]$	$[y_5]$	2	2

P3

count	instance	X	Y	Z	P1_oid	P3_oid
3	a^5	$[x_1]$	$[y_2]$	$[z_5]$	1	3

Fig. 7. Pattern occurrence buffers

5.3 Parent-Child Relationship Identification of Pattern Occurrences

Parent-child relationships can be classified into prefix match and suffix match cases as explained in Sect. 3. In prefix match cases, parent-child relationship identification of pattern occurrences is done when a child pattern occurrence is

---

Algorithm 3. Identify parent-child relationships

Require: Pattern Occurrence Buffers B, identification plan P
1: **for each** $(parent, child) \in P$ **do**
2:     **if** $B[parent] \neq \phi$ **then**
3:         $parents \leftarrow parents \cup \{parent\}$
4:     **end if**
5: **end for**
6: sort $parents$ in the traversing order
7: **for each** $parent \in parents$ **do**
8:     $hashTables \leftarrow newHashTable(B[parent])$
9:     **for each** $child \in parent.children$ **do**
10:         $HashJoin(parent, child)$
11:         output parent occurrences oids for child occurrences
12:     **end for**
13: **end for**

found in SP-NFA as explained in the previous subsection. In suffix match cases, this identification is done when all the processing triggered by a new row finishes.

This identification process is invoked at line 20 of Algorithm 2. In suffix match cases, the last pattern variables in the parent and child patterns are same. Then, the parent pattern occurrences and the child pattern occurrences are inserted to POBs when the same row is processed. Therefore, the occurrences and occurrence identifiers of the parent pattern can be always found in POBs. In general, when a row is processed, n parent pattern occurrences and m child pattern occurrences reside in POBs.

Algorithm 3 shows how parent-child relationship identification of pattern occurrences is done in suffix match cases. It employs hash join to find real parent-child pairs among the $n \times m$ combinations. When a pattern occurrence of a pattern P is found, its ancestor occurrences as well as its parent may be found at the same time in suffix match cases. To identify parent-child relationships efficiently, the join order is controlled by the identification plan, which is derived when analyzing queries. Due to this, we can get occurrence identifiers (<pattern_name>_oid) of ancestor pattern occurrences which need to be inserted in the pattern occurrence table as well.

6 Experiment

6.1 Experimental Setting

To evaluate the performance of the proposed method, we implemented our the proposed method on top of PostgreSQL and compared its processing time with some baseline methods. All experiments were done on Ubuntu 16.04.3 LTS, Intel (R) Core (TM) i7-4790 CPU @ 3.60 GHz, 32 GB RAM running PostgreSQL 9.5[1]. The synthetic sequence table is a relation with sid (sequence id), flag (random 1 or 0 value) and time (sequence order). The table has 1000 sequences, each of which contains 10,000 rows, resulting in 10,000,000 rows in total. We use different pattern hierarchies which are categorized as shown in the Fig. 8. In each category, we test pattern hierarchies consisting of patterns P1 through Pn. The maximum length of pattern occurrence of each query fixed to 6 (WINDOW [time 6]). In addition, we change the condition in DEFINE clause, and control the extracted pattern occurrence ratio (ratio of the extracted pattern occurrences to all the

Prefix match cases	Suffix match cases	Complex pattern cases
(P1; A B	(P1; A B	(P1; A B
(P2; A B C	(P2; A C B	(P2; A B C (P5; A B F C (P8; I A B F C)))
(P3; A B C D	(P3; D A C B	(P3; A D* B (P6; G A D* B (P9; G A D* B J)))
(P4; A B C D E	(P4; D A E C B	(P4; E A B (P7; E A B H+ (P10; E A B K H+))))
(P5; A B C D E F)))))	(P5; F D A E C B)))))	

Fig. 8. Categories of pattern hierarchies

---

[1] https://www.postgresql.org/.

pattern occurrences when the condition in DEFINE clause is TRUE) as 5%, 10%, 20%.

6.2 UDF Implementation of the Proposed and Comparative Methods

We implemented the proposed method and some baseline methods using UDF (User Defined Function) on PostgreSQL. PostgreSQL allows users to define table functions which output a table. When the table function is called in FROM clause, and the table output by the function will be processed according to the semantics of SQL. The codes of table functions were written in C.

Proposed Method. We provided a table function implementing the table generation in MMR. Our table function receives description corresponding to MMR clause as set of string parameters and executes row pattern matching with a pattern hierarchy. In Fig. 9, the right-hand side is SQL statement with MMR and the left-hand side is UDF implementation on PostgreSQL.

Collective Identification. One of the baseline methods is Collective Identification. It performs multiple row pattern matching using SP-NFA as in the proposed method, but does not perform parent-child relationship identification of pattern occurrences. Instead, it executes the parent-child relationship identification after all pattern occurrences are found for the input sequence table. The UDF for Collective Identification receives the same arguments as the UDF for the proposed method.

MATCH_RECOGNIZE. Another baseline method uses the original MR. It invokes multiple queries to derive a pattern occurrence table for each pattern, join the multiple pattern occurrence tables, and takes union to get our target table. We use the UDF for the proposed method under the constraint that only a single pattern is given to simulate MR.

UDF for MULTI_MATCH_RECOGNIZE in PostgreSQL	MULTI_MATCH_RECOGNIZE

SELECT * FROM
multi_match_recognize(
'SELECT * FROM ex_data ORDER BY sid, time;',
'A.time AS A_time,B.time AS B_time,C.time AS C_time'
'(P1; A B (P2; A B C))'
'A AS A.time % 5 = 0,B AS B.time < A.0 + 6,C AS C.time <
   A.0 + 6');
	SELECT * FROM ex_data
	MULTI MATCH RECOGNIZE (
	PARTITION BY sid ORDER BY time
	MEASURES A.time AS A_time,B.time AS B_time,
	   C.time AS C_time
	PATTERN SET (P1; A B (P2; A B C))
	DEFINE A AS (A.time % 5 = 0)
	WINDOW [time 6]);

Fig. 9. Implementation of MULTI_MATCH_RECOGNIZE using UDF

6.3 Experimental Results

Figures 10 and 11 show performance of our proposal and two baselines for cases of prefix match and suffix match for different parameter settings, respectively.

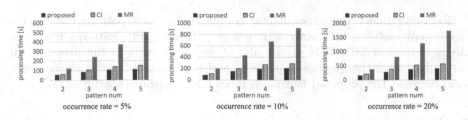

Fig. 10. Experimental results for prefix match cases

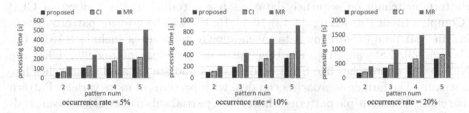

Fig. 11. Experimental results for suffix match cases

Overall, the proposal and Collective Identification clearly exceed MR. When we increase the number of patterns in the pattern hierarchies, the advantage becomes more clear. The reason is that MATCH_REC-OGNIZE needs to repeat scanning the input sequence table for the number of patterns. When we compare processing time of the proposal for prefix match and suffix match cases, the former needs less time. The reason will be that more pattern variables are shared in SP-NFAs in the former cases.

We can see that the proposal exceeds Collective Identification. In particular, the proposal is more advantageous in prefix match cases. The basic pattern matching scheme is the same in both methods. In prefix match cases, the proposal can identify parent-child pattern occurrences within the processing flow of SP-NFAs. In contrast, Collective Identification needs to do join of parent and child occurrences. In suffix cases, the proposal needs to do the Parent-child Relationship Identification as explained in Sect. 5.3. This process involves only pattern occurrences newly found for an incoming row. In contrast, join in Collective Identification needs to deal with all pattern occurrences.

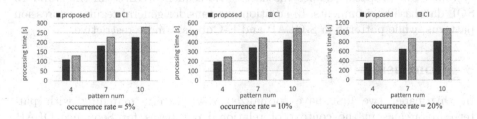

Fig. 12. Experimental results for complex pattern cases

Figure 12 shows the experimental results in complex pattern cases. This experiment excludes MR, since it is not easy to deal with patterns including kleene closure in MR. This results also shows advantages of the proposal. In particular, when the number of patterns and the extracted pattern occurrence ratio increase, the proposed method is more advantageous (Maximum 25%).

7   Related Work

Pattern matching on sequence data has been studied in the context of CEP (Complex Event Processing) [3,4,6,10,11,15,17], which performs pattern matching in real time on stream data. Well-known approaches include SASE [15], Cayuga [3], ZStream [10]. Existing pattern matching methods in CEP are NFA-based approaches like SASE and Cayuga and stack-based approaches like ZStream. The former approaches can deal with patterns of more variety. Pattern matching with multiple patterns which share partial sub-patterns has been studied [5,12,16,19]. The context of those studies is stream processing rather than sequence OLAP. In addition, identification of parent-child pattern occurrences has not been considered in these studies.

The work on sequence databases [13,14] proposed pattern matching as an extension of SQL. However, SEQUIN [14] cannot specify Kleene closure. SQL-TS [13] supports pattern matching with Kleene closure, but it follows Strict-Contiguity semantics. SQL/RPR (Row Pattern Recognition), which executes row pattern matching on relations, was standardized in SQL 2016 [1]. In SQL/RP-R, a pattern can be defined in MR and its pattern occurrences can be extracted. It also follows Strict-Contiguity semantics. None of those methods support multiple row pattern matching.

S-OLAP [9] executes OLAP analysis on sequence data. S-OLAP does pattern matching on sequence data and performs sequence OLAP analysis. S-OLAP keeps the information of the pattern occurrences which are queried in the inverted index. Furthermore, changes of the target patterns in a query can be handled efficiently by utilizing another inverted index for pattern matching. E-Cube [7,8] is known as a method for sequence OLAP analysis on real time data. By partially sharing the aggregation result of the defined parent-child patterns, E-Cube can efficiently execute pattern matching for multiple patterns. However, both S-OLAP and E-Cube are based on their own query languages and systems. Our proposal enables the user to do sequence OLAP in the standard SQL database environments. In addition, we consider general regular expression patterns, while patterns in S-OLAP and E-Cube are more restrictive.

8   Conclusion

In this paper, we first have formalized row pattern matching with pattern hierarchies in the context of relational databases for Sequence OLAP. Then, we have proposed algorithms using SP-NFA to efficiently execute row patterns matching with pattern hierarchies. Also, the SQL extension

MULTI_MATCH_RECOGNIZ-E clause for the formalized multiple row patterns matching has been proposed. The proposed method has been implemented on top of PostgreSQL using UDF, and the experiments proved the advantages of the proposal over some baseline methods. Future research issues include application of the proposed framework for stream data and development of sequence OLAP systems utilizing the proposed query processing methods.

Acknowledgement. This work was partly supported by Grant-in-Aid for Scientific Research (B) (#19H04114) from JSPS.

References

1. ISO, IEC 2016: Information technology — database languages — sql technical reports — part 5:row pattern recognition in sql. Technical report, ISO copyright office (2016)
2. Agrawal, J., Diao, Y., Gyllstrom, D., Immerman, N.: Efficient pattern matching over event streams. In: Proceedings of the 2008 ACM SIGMOD International Conference on Management of Data, pp. 147–160. ACM (2008)
3. Demers, A.J., et al.: Cayuga: a general purpose event monitoring system. In: CIDR, vol. 7, pp. 412–422 (2007)
4. Gyllstrom, D., Agrawal, J., Diao, Y., Immerman, N.: On supporting kleene closure over event streams. In: IEEE 24th International Conference on Data Engineering. ICDE 2008, pp. 1391–1393. IEEE (2008)
5. Hong, M., Riedewald, M., Koch, C., Gehrke, J., Demers, A.: Rule-based multi-query optimization. In: Proceedings of the 12th International Conference on Extending Database Technology: Advances in Database Technology, pp. 120–131. ACM (2009)
6. Kolchinsky, I., Sharfman, I., Schuster, A.: Lazy evaluation methods for detecting complex events. In: Proceedings of the 9th ACM International Conference on Distributed Event-Based Systems, pp. 34–45. ACM (2015)
7. Liu, M., et al.: E-cube: multi-dimensional event sequence processing using concept and pattern hierarchies. In: 2010 IEEE 26th International Conference on Data Engineering (ICDE), pp. 1097–1100. IEEE (2010)
8. Liu, M., et al.: E-cube: multi-dimensional event sequence analysis using hierarchical pattern query sharing. In: Proceedings of the 2011 ACM SIGMOD International Conference on Management of Data, pp. 889–900. ACM (2011)
9. Lo, E., Kao, B., Ho, W.-S., Lee, S.D., Chui, C.K., Cheung, D.W.: Olap on sequence data. In: Proceedings of the 2008 ACM SIGMOD International Conference on Management of Data, pp. 649–660. ACM (2008)
10. Mei, Y., Madden, S.: Zstream: a cost-based query processor for adaptively detecting composite events. In: Proceedings of the 2009 ACM SIGMOD International Conference on Management of Data, pp. 193–206. ACM (2009)
11. Poppe, O., Lei, C., Ahmed, S., Rundensteiner, E.A.: Complete event trend detection in high-rate event streams. In: Proceedings of the 2017 ACM International Conference on Management of Data, pp. 109–124. ACM (2017)
12. Ray, M., Lei, C., Rundensteiner, E.A.: Scalable pattern sharing on event streams. In: Proceedings of the 2016 International Conference on Management of Data, pp. 495–510. ACM (2016)

13. Sadri, R., Zaniolo, C., Zarkesh, A., Adibi, J.: Expressing and optimizing sequence queries in database systems. ACM Trans. Database Syst. (TODS) **29**(2), 282–318 (2004)
14. Seshadri, P., Livny, M., Ramakrishnan, R.: Seq: design and implementation of a sequence database system. Technical report, University of Wisconsin-Madison Department of Computer Sciences (1996)
15. Wu, E., Diao, Y., Rizvi, S.: High-performance complex event processing over streams. In: Proceedings of the 2006 ACM SIGMOD International Conference on Management of Data, pp. 407–418. ACM (2006)
16. Yan, Y., Zhang, J., Shan, M.-C.: Scheduling for fast response multi-pattern matching over streaming events. In: 2010 IEEE 26th International Conference on Data Engineering (ICDE), pp. 89–100. IEEE (2010)
17. Zhang, H., Diao, Y., Immerman, N.: Recognizing patterns in streams with imprecise timestamps. Proc. VLDB Endowment **3**(1–2), 244–255 (2010)
18. Zhang, H., Diao, Y., Immerman, N.: On complexity and optimization of expensive queries in complex event processing. In: Proceedings of the 2014 ACM SIGMOD International Conference on Management of Data, pp. 217–228. ACM (2014)
19. Zhang, S., Vo, H.T., Dahlmeier, D., He, B.: Multi-query optimization for complex event processing in SAP ESP. In: 2017 IEEE 33rd International Conference on Data Engineering (ICDE), pp. 1213–1224. IEEE (2017)

Statistically Significant Discriminative Patterns Searching

Hoang Son Pham[1]([✉]), Gwendal Virlet[2], Dominique Lavenier[2],
and Alexandre Termier[2]

[1] ICTEAM, UCLouvain, Louvain-la-Neuve, Belgium
hoang.s.pham@uclouvain.be
[2] Univ Rennes, Inria, CNRS, IRISA, Rennes, France

Abstract. In this paper, we propose a novel algorithm, named SSDPS,
to discover patterns in two-class datasets. The SSDPS algorithm owes
its efficiency to an original enumeration strategy of the patterns, which
allows to exploit some degrees of anti-monotonicity on the measures of
discriminance and statistical significance. Experimental results demonstrate that the performance of the SSDPS algorithm is better than others. In addition, the number of generated patterns is much less than the
number of the other algorithms. Experiment on real data also shows that
SSDPS efficiently detects multiple SNPs combinations in genetic data.

Keywords: Discriminative patterns · Discriminative measures ·
Statistical significance · Anti-monotonicity

1 Introduction

Recently, the use of discriminative pattern mining (also known under other
terms such as emerging pattern mining [1], contrast set mining [2]) has been
investigated to tackle various applications such as bioinformatics [3], data classification [4]. In this paper, we propose a novel algorithm, named SSDPS, that
discovers discriminative patterns in two-class datasets. This algorithm aims at
searching patterns satisfying both discriminative scores and confidence intervals
thresholds. These patterns are defined as **statistically significant discriminative patterns**. The SSDPS algorithm is based on an enumeration strategy
in which discriminative measures and confidence intervals can be used as anti-
monotonicity properties. These properties allow the search space to be pruned
efficiently. All patterns are directly tested for discriminative scores and confidence intervals thresholds in the mining process. Only patterns satisfying both
of thresholds are considered as the target output. According to our knowledge,
there doesn't exist any algorithms that combine discriminative measures and
statistical significance as anti-monotonicity to evaluate and prune the discriminative patterns.

The SSDPS algorithm has been used to conduct various experiments on both
synthetic and real genomic data. As a result, the SSDPS algorithm effectively

© Springer Nature Switzerland AG 2019
C. Ordonez et al. (Eds.): DaWaK 2019, LNCS 11708, pp. 105–115, 2019.
https://doi.org/10.1007/978-3-030-27520-4_8

deploys the anti-monotonic properties to prune the search space. In comparison with other well-known algorithms such as SFP-GROWTH [5] or CIMCP [6], the SSDPS obtains a better performance. In addition the proportion of generated patterns is much smaller than the amount of patterns output by these algorithms.

The rest of this paper is organized as follows: Sect. 2 precisely defines the concept of statistically significant discriminative pattern, and Sect. 3 presents the enumeration strategy used by the SSDPS algorithm. In Sect. 4, the design of the SSDPS algorithm is described. Section 5 is dedicated to experiments and results. Section 6 concludes the paper.

2   Problem Definition

The purpose of discriminative pattern mining algorithms is to find groups of items satisfying some thresholds. The formal presentation of this problem is given in the following:

Let I be a set of m items $I = \{i_1, ..., i_m\}$ and S_1, S_2 be two labels. A *transaction* over I is a pair $t_i = \{(x_i, y_i)\}$, where $x_i \subseteq I$, $y_i \in \{S_1, S_2\}$. Each transaction t_i is identified by an integer i, denoted *tid* (*transaction identifier*). A set of transactions $T = \{t_1, ..., t_n\}$ over I can be termed as a *transaction dataset* D over I. T can be partitioned along labels S_1 and S_2 into $D_1 = \{t_i \mid t_i = (x_i, S_1) \in T\}$ and $D_2 = \{t_i \mid t_i = (x_i, S_2) \in T\}$. The associated tids are denoted $D_1.tids$ and $D_2.tids$.

For example, Table 1 presents a dataset including 9 transactions (identified by 1..9) which are described by 10 items (denoted by $a..j$). The dataset is partitioned into two classes (class label 1 or 0).

Table 1. Two-class data example

Tids	Items										Class
1	a	b	c			f			i	j	1
2	a	b	c		e		g		i		1
3	a	b	c			f		h		j	1
4		b		d	e		g		i	j	1
5				d		f	g	h	i	j	1
6		b	c		e		g	h		j	0
7	a	b	c			f	g	h			0
8		b	c	d	e			h	i		0
9	a			d	e		g	h		j	0

A set $p \subseteq I$ is called an *itemset* (or pattern) and a set $q \subseteq \{1..n\}$ is called a *tidset*. For convenience we write a tidset $\{1, 2, 3\}$ as 123, and an itemset $\{a, b, c\}$ as abc. The number of transactions in D_i containing p is denoted by $|D_i(p)|$.

The *relational support* of p in class D_i (denoted $sup(p, D_i)$), and *negative support* of p in D_i (denoted $\overline{sup}(p, D_i)$), are defined as follows:

$$sup(p, D_i) = \frac{|D_i(p)|}{|D_i|}; \quad \overline{sup}(p, D_i) = 1 - sup(p, D_i)$$

To evaluate the discriminative score of pattern p in a two-class dataset D, different measures are defined over the relational supports of p. The most popular discriminative measures are *support difference, grown rate support* and *odds ratio support* which are calculated by formulas 1, 2, 3 respectively.

$$SD(p, D) = sup(p, D_1) - sup(p, D_2) \tag{1}$$

$$GR(p, D) = \frac{sup(p, D_1)}{sup(p, D_2)} \tag{2}$$

$$ORS(p, D) = \frac{sup(p, D_1)/\overline{sup}(p, D_1)}{sup(p, D_2)/\overline{sup}(p, D_2)} \tag{3}$$

A pattern p is discriminative if its score is not less than a given threshold α. For example, let $\alpha = 2$ be the threshold of growth rate support. Pattern abc is discriminative since $GR(abc, D) = 2.4$.

Definition 1 *(Discriminative pattern). Let α be a discriminative threshold, $scr(p, D)$ be the discriminative score of pattern p in D. The pattern p is discriminative if $scr(p, D) \geq \alpha$.*

In addition to the discriminative score, to evaluate the statistical significance of a discriminative pattern we need to consider the confidence intervals (CI). Confidence intervals are the result of a statistical measure. They provide information about a range of values (lower confidence interval (LCI) to upper confidence interval (UCI)) in which the true value lies with a certain degree of probability. CI is able to assess the statistical significance of a result [7]. A confidence level of 95% is usually selected. It means that the CI covers the true value in 95 out of 100 studies.

Let $a = |D_1(p)|$, $b = |D_1| - |D_1(p)|$, $c = |D_2(p)|$, $d = |D_2| - |D_2(p)|$, the 95% LCI and UCI of GR are estimated as formulas 4 and 5 respectively.

$$LCI_{GR} = e^{\left(ln(GR) - 1.96\sqrt{\frac{1}{a} - \frac{1}{a+b} + \frac{1}{c} - \frac{1}{c+d}}\right)} \tag{4}$$

$$UCI_{GR} = e^{\left(ln(GR) + 1.96\sqrt{\frac{1}{a} - \frac{1}{a+b} + \frac{1}{c} - \frac{1}{c+d}}\right)} \tag{5}$$

Similarly, the 95% LCI and UCI of OR are estimated as formulas 6 and 7 respectively.

$$LCI_{ORS} = e^{\left(ln(ORS) - 1.96\sqrt{\frac{1}{a} + \frac{1}{b} + \frac{1}{c} + \frac{1}{d}}\right)} \tag{6}$$

$$UCI_{ORS} = e^{\left(ln(ORS) + 1.96\sqrt{\frac{1}{a} + \frac{1}{b} + \frac{1}{c} + \frac{1}{d}}\right)} \tag{7}$$

For example, consider the pattern abc in the previous example, the 95%CI of GR are $LCI_{GR} = 0.37$, $UCI_{GR} = 16.60$. Thus the GR score of abc is statistically significant because this score lies between LCI and UCI values.

Definition 2 *(Statistically significant discriminative pattern).* *Given a discriminance score scr $\in \{GR, ORS\}$, a discriminative threshold α and a lower confidence interval threshold β, the pattern p is statistically significant discriminative in D if $scr(p, D) \geq \alpha$ and $lci_{scr}(p, D) > \beta$.*

Problem statement: Given a two-class dataset D, a discriminance score scr and two thresholds α and β, the problem is to discover the complete set of patterns P that are statically significant discriminative for dataset D, discriminative measure scr, discriminative threshold α and lower confidence interval threshold β.

3   Enumeration Strategy

The main practical contribution of this paper is SSDPS, an efficient algorithm for mining statistically significant discriminative patterns. This algorithm will be presented in the next section (Sect. 4).

SSDPS owes its efficiency to an original enumeration strategy of the patterns, which allows to exploit some degree of anti-monotonicity on the measures of discriminance and statistical significance. In pattern mining enumeration strategies, *anti-monotonicity properties* is an important component. When enumerating frequent itemsets, one can notice that if an itemset p is unfrequent ($sup(p, D) < min\_sup$), then no super-itemsets $p' \supset p$ can be frequent (necessarily $sup(p', D) < sup(p, D) < min\_sup$). This allows to stop any further enumeration when an unfrequent itemset p is found, allowing a massive reduction in the search space [8]. As far as we know, no such anti-monotonicity could be defined on measures of discriminance or statistical significance.

The enumeration strategy proposed in SSDPS also builds an enumeration tree. However, it is based on the tidsets and not the itemsets. Each node of the enumeration tree is a tidset (with the empty tidset at the root). For example, consider the node represented by $\boxed{12 : 8}$ in Fig. 1: this node corresponds to the tidset 128 in which $12 \subset D_1.tids$, and $8 \subset D_2.tids$.

Before presenting details of the enumeration strategy we first explain how to recover the itemsets from the tidsets. This is a well known problem: itemsets and tidsets are in facts dual notions, and they can be linked by two functions that form a *Galois connection* [9]. The main difference in our definition is that the main dataset can be divided into two parts ($D = D_1 \cup D_2$), and we want to be able to apply functions of the Galois connection either in the complete dataset D or in any of its parts D_1 or D_2.

Definition 3 (Galois connection). *For a dataset $D = D_1 \cup D_2$:*

– *For any tidset $q \subseteq \{1..n\}$ and any itemset $p \subseteq I$, we define:*

$$f(q, D) = \{i \in I \mid \forall k \in q \; i \in t_k\}; \quad g(p, D) = \{k \in \{1..n\} \mid p \subseteq t_k\}$$

– *For any tidset $q_1 \subseteq D_1.tids$ and any itemset $p \subseteq I$, we define:*

$$f_1(q_1, D_1) = \{i \in I \mid \forall k \in q_1 \; i \in t_k\}; \quad g_1(p, D_1) = \{k \in D_1 \mid p \subseteq t_k\}$$

– *For any tidset $q_2 \subseteq D_2.tids$ and any itemset $p \subseteq I$, we define:*

$$f_2(q_2, D_2) = \{i \in I \mid \forall k \in q_2 \; i \in t_k\}; \quad g_2(p, D_2) = \{k \in D_2 \mid p \subseteq t_k\}$$

Note that this definition marginally differs from the standard definition presented in [9]: here for convenience we operate on the set of tids $\{1..n\}$, whereas the standard definition operates on the set of transaction $\{t_1, ..., t_n\}$.

In Fig. 1, under each tidset q, its associated itemset $f(q, D)$ is displayed. For example for node $\boxed{12{:}8}$, the itemset $f(128, D) = bci$ is displayed. One can verify in Table 1 that bci is the only itemset common to the transactions t_1, t_2 and t_8. A *closure operator* can be defined over the use of the Galois connection.

Definition 4 (Closure operator). *For a dataset D and any tidset $q \subseteq \{1..n\}$, the closure operator is defined as: $c(q, D) = g \circ f(q, D)$. The output of $c(q, D)$ is the tidset of the closed itemset having the smallest tidset containing q.*

We can similarly define $c_1(q_1, D_1) = g_1 \circ f_1(q_1, D_1)$ for $q_1 \subseteq D_1.tids$ and $c_2(q_2, D_2) = g_2 \circ f_2(q_2, D_2)$ for $q_2 \subseteq D_2.tids$.

The basics of the enumeration have been given: the enumeration proceeds by augmenting tidsets (starting from the empty tidset), and for each tidset function f of the Galois connection gives the associated itemset. The specificity of our enumeration strategy is to be designed around statistically significant discriminative patterns. This appears first in our computation of closure: we divide the computation of closure in the two sub-datasets D_1 and D_2. This intermediary step allows some early pruning. Second, most measures of discriminance require the pattern to have a non-zero support in D_2 (GR and ORS). The same condition apply for measures of statistical significance: in both cases we need to defer measures of interest of patterns until it has some tids in D_2. Our enumeration strategy thus operates in two steps:

1. From the empty set, it enumerates closed tidsets containing only elements of D_1 (case group).
2. For each of those tidset containing only tids of D_1, augmentations using only tids of D_2 are generated and their closure is computed. Any subsequent augmentation of such nodes will only be allowed to be augmented by tids of D_2.

More formally, let $q \subseteq \{1..n\}$ be a tidset, with $q = q^+ \cup q^-$, where $q^+ \subseteq D_1.tids$ and $q^- \subseteq D_2.tids$. Then the possible augmentations of q are:

– (*Rule 1*) if $q^- = \emptyset$: q can either:
  - (*Rule 1a*) be augmented with $k \in D_1.tids$ such that $k < min(q^+)$
  - (*Rule 1b*) be augmented with $k \in D_2.tids$
– (*Rule 2*) if $q^- \neq \emptyset$: q can only be augmented with tid $k \in D_2.tids$ such that $k < min(q^-)$

This enumeration strategy allows to benefit from an anti-monotonicity property on the measures of statistical significance and discriminance.

Theorem 1 (Anti-monotonicity). *Let q_1 and q_2 be two tidsets such as: $q_1^+ = q_2^+$ and $q_1^- \subset q_2^-$ (we have $q_1^+ \neq \emptyset$ and $q_2^- \neq \emptyset$). Let $p_1 = f(q_1, D)$ and $p_2 = f(q_2, D)$. Then:*

1. *$scr(p_1, D) > scr(p_2, D)$ with scr a discriminance measure in $\{SD, GR, ORS\}$.*
2. *$lci(p_1, D) > lci(p_2, D)$ with lci a lower confidence interval in $\{LCI_{ORS}, LCI_{GR}\}$.*

Please refer to the full paper at https://arxiv.org/abs/1906.01581 for the detailed demonstration of this part.

This theorem provides pruning by anti-monotonicity in our enumeration strategy: for a node having a tidset with tids both from $D_1.tids$ and $D_2.tids$, if the discriminance or statistical significance measures are below a given threshold, then necessarily its augmentations will also be under the threshold. Hence this part of the enumeration tree can be pruned. For example, node $\boxed{2{:}8}$ has associated itemset $bcei$, and $ORS(bcei, D) = 3/4$. Suppose the ORS threshold $\alpha = 2$ this node can be pruned and its augmentations need not be computed. This allows to significantly reduce the search space.

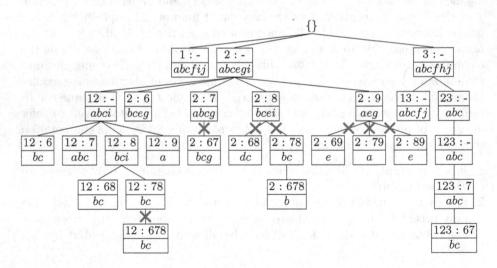

Fig. 1. Tidset-itemset search tree

4   SSDPS: Algorithm Design

This section presents the SSDPS algorithm which exploits the enumeration strategy presented in the Sect. 3.

As mentioned in the previous section, our algorithm is based on an enumeration of the tidsets. It discovers statistically significant discriminative closed patterns. The main procedure for enumerating tidsets is given in Algorithm 1. This procedure calls the recursive procedure `positiveExpand` (Algorithm 2) to find closed frequent itemsets in the positive class. Computing discriminative patterns relies on the recursive procedure `negativeExpand` (Algorithm 3).

Algorithm 1. SSDPS algorithm

---

    input : D, α, β
    output: a set of statistically significant discriminative patterns
1 $t = \emptyset$
2 **for** *each transaction id e in positive_class* **do**
3   ⌊ `positiveExpand`(t, e, D, α, β)

Delving more into details, `positiveExpand` (Algorithm 2) is based on the principles of the LCM algorithm [10], the state of the art for mining closed frequent itemsets. `positiveExpand` takes as input the tidset t of a pattern that is closed in D_1 and a tid $e \in D_1.tids$ that can be used to augment t. This augmentation is performed on line 1, and the pattern p associated to the augmented tidset $t^+ = t \cup \{e\}$ is computed in line 2. If $p = \emptyset$, there are no items common to all transactions of t^+ so the enumeration can stop (test of line 3). Else, we can continue the enumeration by applying *Rule 1* of enumeration presented in Sect. 3. Lines 4 to 9 apply the LCM principles of enumerating closed itemsets without redundancies (the interested reader in referred to [11] Section 3.2 for a recent description of these principles). At this step of the enumeration, the closure is computed in D_1 (line 4). The test of line 5 verifies if the closure actually extends the tidset, requiring a further verification in line 6, and the construction of the new extended tidset (line 7).

Lines 8 to 10 implement *Rule 1a* of enumeration, allowing to grow the positive part of the tidset. Lines 11 to 12 implement *Rule 1b* of enumeration, stopping the growth of the positive part and starting to grow the negative part of the tidset. The same logic is followed in lines 14 to 18, in the case where the tidset is not extended by the closure (test of line 5 is false).

Algorithm 2. positiveExpand Function

```
1  t⁺ ← t ∪ {e}
2  p ← f(t⁺, D)
3  if p is not empty then
4  │   t_ext⁺ ← c₁(t⁺, D₁)
5  │   if t_ext⁺ ≠ t⁺ then
6  │   │   if max(t_ext⁺) < e then
7  │   │   │   q ← t⁺ ∪ t_ext⁺
8  │   │   │   for each e⁺ in D₁.tids \ q do
9  │   │   │   │   if e⁺ < e then
10 │   │   │   │   └   positiveExpand(q, e⁺, D, α, β)
11 │   │   │   for each e⁻ in D₂.tids do
12 │   │   │   └   negativeExpand(q, e⁻, D, α, β)
13 │   else
14 │   │   for each e⁺ in D₁.tids do
15 │   │   │   if e⁺ < min(t⁺) then
16 │   │   │   └   positiveExpand(t⁺, e⁺, D, α, β)
17 │   │   for each e⁻ in D₂.tids do
18 │   │   └   negativeExpand(t⁺, e⁻, D, α, β)
```

The final expansion of the tidset is handled by **negativeExpand** (Algorithm 3), that can only perform augmentations with negative tidsets. It is very similar to **positiveExpand**, with several key differences. The first obvious one is that the closure is this time computed in D_2 (line 4). The second one is that only *Rule 2* of enumeration can be applied (lines 13 and 20). The third and most important difference is that because we have tidsets with positive and negative tids, we can compute discriminance as well as statistical significance measures. Hence, Theorem 1 can be applied to benefit from pruning by anti-monotonicity. This is done in line 3.

Algorithm 3. negativeExpand Function

---

1 $t^- \leftarrow t \cup \{e\}$; $p \leftarrow f(t^-, D)$
2 **if** p *is not empty* **then**
3   **if** $check\_significance(p, D, \alpha, \beta)$ *is true* **then**
4          $t\_ext^- \leftarrow c_2(t^-, D_2)$
5   **if** $t\_ext^- \neq t^-$ **then**
6   **if** $max(t\_ext^-) < e$ **then**
7                  $q \leftarrow t^- \cup t\_ext^-$; $q\_ext \leftarrow c(q, D)$; $p' \leftarrow f(q, D)$
8   **if** q_ext *is empty* **then**
9   **if** $check\_significance(p', D, \alpha, \beta)$ *is true* **then**
10   output: p'
11   **for** *each* $e^- \in D_2.tids \setminus q$ **do**
12   **if** $e^- < e$ **then**
13   negativeExpand$(q, e^-, D, \alpha, \beta)$
14   **else**
15             $t\_ext \leftarrow c(t^-, D)$
16   **if** t_ext *is empty* **then**
17   output: p
18   **for** *each* $e^- \in D_2.tids \setminus t^-$ **do**
19   **if** $e^- < e$ **then**
20   negativeExpand$(t^-, e^-, D, \alpha, \beta)$

---

5   Experimental Results

This section presents various experiments to evaluate the performance of the SSDPS algorithm. In addition, we apply the SSDPS to discover multiple SNPs combinations in a real genomic dataset. The details of the result was presented in the full paper which was published at https://arxiv.org/abs/1906.01581. All experiments have been conducted on a laptop with Core i7-4600U CPU @ 2.10G Hz, 16 GB memory and Linux operating system.

A synthetic two-class data was created to evaluate the pruning strategy as well as compare SSDPS with other algorithms. This dataset includes 100 transactions (50 transactions for each class). Each transaction contains 262 items which are randomly set by value 0 or 1. The density of data is set up to 33%.

5.1   Pruning Efficiency Evaluation

To evaluate the pruning efficiency of the SSDPS algorithm, we executed 2 setups on the synthetic dataset.

- Setup 1: use OR as discriminative measure; the discriminative threshold $\alpha = 2$.
- Setup 2: use OR as discriminative measure and LCI of OR as statistically significant testing; the discriminative threshold $\alpha = 2$, and LCI threshold $\beta = 2$.

As the result, the running time and the number of output patterns significantly reduce when applying LCI_{ORS}. In particular, with the setup 1, the SSDPS algorithm generates 179,334 patterns in 38.69 s while the setup 2 returns 18,273 patterns in 9.10 s. This result shows that a large amount of patterns is removed by using statistically significant testing.

5.2 Comparison with Existing Algorithms

We compare the performance of the SSDPS algorithm with two well-known algorithms: CIMCP [6] and SFP-Growth [5]. Note that these algorithms deploy discriminative measures which are different from the measures of SSDPS. In particular, CIMCP uses one of measures such as chi-square, information-gain and gini-index as a constraint to evaluate discriminative patterns while SFP-GROWTH applies $-log(p\_value)$. For this reason, the number of output patterns and the running times of these algorithms should be different. It is hence not fair to directly compare the performance of SSDPS with these algorithms. However, to have an initial comparison of the performance as well as the quantity of discovered patterns, we select these algorithms.

We ran three algorithms on the same synthetic data. The used parameters and results are given in Table 2.

Table 2. Used parameters and results of 3 algorithms

Algorithms	Measure	Threshold	#Patterns	Time (seconds)
SSDPS	OR, LCI_ORS	$\alpha = 2, \beta = 2$	49,807	73.69
CIMCP	Chi-square	2	5,403,688	143
SFP-GROWTH	$-log(p\_value)$	3	*	>172 (out of memory)

As the result, the SSDPS algorithm finds 49,807 patterns in 73.69 s; CIMCP discovers 5,403,688 patterns in 143 s. The SFP-GROWTH runs out of storage memory after 172 s. Hence the number of patterns isn't reported in this case.

In comparison with these algorithms the SSDPS gives a comparable performance, while the number of output patterns is much smaller. The reason is that the output patterns of SSDPS are tested for statistical significance by CI while other algorithms use only the discriminative measure. However, this amount of patterns is also larger for real biological analysis. Thus, searching for a further reduced number of significant patterns should be taken into account.

6   Conclusion and Perspectives

In this paper we propose a novel algorithm, called SSDPS, that efficiently discover statistically significant discriminative patterns from a two-class dataset. The algorithm directly uses discriminative measures and confidence intervals as anti-monotonic properties to efficiently prune the search space. Experimental results show that the performance of the SSDPS algorithm is better than other discriminative pattern mining algorithms. However, the number of patterns generated by SSDPS is still large for manual analysis. To reduce the amount of patterns our first perspective is to investigate a heuristic approach. Another perspective is to apply statistical techniques such as minimum description length or multiple hypothesis testing in order to further remove uninteresting patterns.

References

1. Dong, G., Li, J.: Efficient mining of emerging patterns: discovering trends and differences. In: Fifth ACM SIGKDD, KDD 1999, pp. 43–52. ACM, New York (1999)
2. Bay, S., Pazzani, M.: Detecting group differences: mining contrast sets. Data Min. Knowl. Discov. 5(3), 213–246 (2001)
3. Cheng, H., Yan, X., Han, J., Yu, P.S.: Direct discriminative pattern mining for effective classification. In: ICDE 2008, pp. 169–178. IEEE Computer Society, Washington, DC (2008)
4. García-Borroto, M., Martínez-Trinidad, J., Carrasco-Ochoa, J.: A survey of emerging patterns for supervised classification. Artif. Intell. Rev. 42(4), 705–721 (2014)
5. Ma, L., Assimes, T.L., Asadi, N.B., Iribarren, C., Quertermous, T., Wong, W.H.: An "almost exhaustive" search-based sequential permutation method for detecting epistasis in disease association studies. Genet. Epidemiol. 34(5), 434–443 (2010)
6. Guns, T., Nijssen, S., De Raedt, L.: Itemset mining: a constraint programming perspective. Artif. Intell. 175(12), 1951–1983 (2011)
7. Morris, J.A., Gardner, M.J.: Statistics in medicine: calculating confidence intervals for relative risks (odds ratios) and standardised ratios and rates. Br. Med. J. 296(6632), 1313–1316 (1988)
8. Agrawal, R., Imieliński, T., Swami, A.: Mining association rules between sets of items in large databases. SIGMOD Rec. 22(2), 207–216 (1993)
9. Pasquier, N., Bastide, Y., Taouil, R., Lakhal, L.: Discovering frequent closed itemsets for association rules. In: Beeri, C., Buneman, P. (eds.) ICDT 1999. LNCS, vol. 1540, pp. 398–416. Springer, Heidelberg (1999). https://doi.org/10.1007/3-540-49257-7_25
10. Uno, T., Kiyomi, M., Arimura, H.: LCM ver. 2: efficient mining algorithms for frequent/closed/maximal itemsets. In: Workshop Frequent Item Set Mining Implementations (2004)
11. Leroy, V., Kirchgessner, M., Termier, A., Amer-Yahia, S.: TopPI: an efficient algorithm for item-centric mining. Inf. Syst. 64, 104–118 (2017)

RDF and Streams

Multidimensional Integration of RDF Datasets

Jam Jahanzeb Khan Behan[1,2]([✉]), Oscar Romero[1], and Esteban Zimányi[2]

[1] Universitat Politècnica de Catalunya,
Calle Jordi Girona, 1-3, 08034 Barcelona, Spain
{behan,oromero}@essi.upc.edu
[2] Université libre de Bruxelles,
Avenue Franklin Roosevelt 50, 1050 Bruxelles, Belgium
{jbehan,ezimanyi}@ulb.ac.be

Abstract. Data providers have been uploading RDF datasets on the web to aid researchers and analysts in finding insights. These datasets, made available by different data providers, contain common characteristics that enable their integration. However, since each provider has their own data dictionary, identifying common concepts is not trivial and we require costly and complex entity resolution and transformation rules to perform such integration. In this paper, we propose a novel method, that given a set of independent RDF datasets, provides a multidimensional interpretation of these datasets and integrates them based on a common multidimensional space (if any) identified. To do so, our method first identifies potential dimensional and factual data on the input datasets and performs entity resolution to merge common dimensional and factual concepts. As a result, we generate a common multidimensional space and identify each input dataset as a cuboid of the resulting lattice. With such output, we are able to exploit open data with OLAP operators in a richer fashion than dealing with them separately.

Keywords: Entity resolution ·
Resource Description Framework (RDF) · Data integration ·
On-line analytical processing (OLAP) · Multidimensional modeling

1 Introduction

Data availability on the Web is ensured as users constantly upload data. Since multiple users can share the same entity, data duplication and unconnected related data grew on the Web. As a consequence, integration of web sources became a necessity and the Web of Linked Data was obtained. Linked Open Data (LOD) enables the sharing of information, structured querying formats, and facilitates access to data by means of Uniform Resource Identifiers (URIs). Yet, due to the heterogeneity of the Web of Linked Data, it is still problematic to develop Linked Data (LD) applications. Nowadays, we cannot assume that all URI aliases have been explicitly stated as links and therefore data integration is

© Springer Nature Switzerland AG 2019
C. Ordonez et al. (Eds.): DaWaK 2019, LNCS 11708, pp. 119–135, 2019.
https://doi.org/10.1007/978-3-030-27520-4_9

still an open issue. Nevertheless, the size of LOD has been increasing exponentially. A study released in April 2014 highlights that the LD cloud has grown to more than 1000 datasets from just 12 datasets cataloged in 2007 [15] having more than 500 million explicit links between them.

The Resource Description Framework (RDF) models information that describes LD in the form of RDF triples. Each triple contains a subject, a predicate, and an object defined by a URI. Data providers publish RDF datasets using a personalized dictionary, hence a single entity has multiple definitions. Yet, most RDF datasets have common characteristics indicating that they share common schemata information, that may enable a tighter integration. The aim of this project is to facilitate the user in querying similar RDF datasets on an integrated, multidimensional fashion (which we refer to as **Integrated Multidimensional Dataset** (IMD)). IMD is annotated as a cube with QB4OLAP vocabulary [5], that enables OLAP functionalities and resolves compatibility issues such as different resource names or granularity details. We claim that performing OLAP analysis on top of disparate RDF datasets allows a richer analysis than regular analysis on each independent or–manually glued together– dataset(s).

To showcase our approach, we use RDF datasets available at the US, the UK and the Eurostat (Linked) data portals as running examples. We focus on carbon emission data, as it is a major factor for air quality degradation. Additionally, we use QBOAirbase[1], a QB4OLAP-compliant dataset, as the basis of our model as it contains several factual data with varying dimensional granularity. We refer to the carbon datasets obtained from the US[2], the UK[3] and the Eurostat[4] data portals as C_{US}, C_{UK}, and C_{EU} respectively. The comparison of these datasets is not straightforward, however, taking a close look, one may identify them as cuboids of the same multidimensional space. For example, C_{US} and C_{UK} contain *location* details on a finer granularity (property/state level), whereas C_{EU} contains higher level data (country level). We perform certain operations to generate a single integrated dimension to enable data compatibility.

In this paper, we propose a framework to integrate and cross external RDF datasets into a unified multidimensional view. Our framework consists of four modules to: (i) identify potential dimensional and factual resources, (ii) independently perform entity resolution (ER) on dimensional and factual resources, (iii) and identify new dimensional and (iv) factual resources to be added in the current IMD (from new links identified in ER module (ii)). Our proposal is an incremental approach that builds the IMD schema by unifying RDF datasets at hand. As a result, based on the granularity level of the measure in the input RDF datasets, each dataset represents a cuboid of the IMD lattice. We thus enable the capability, using OLAP processing, to answer queries at a certain granularity, by means of rolling-up appropriate RDF datasets that are at a lower

---

[1] The European Air Quality RDF Database: http://qweb.cs.aau.dk/qboairbase/.
[2] https://catalog.data.gov/dataset/2015-greenhouse-gas-report-data.
[3] https://opendata.camden.gov.uk/resource/4txj-pb2i.
[4] http://estatwrap.ontologycentral.com/page/t2020_rd300.

granularity level. Thus, the input RDF datasets play the role of materialized views of the IMD lattice and, combining these views, we provide a "bigger picture" of the data. However, since our approach is purely data-driven, the resulting lattice might be highly sparse and there might be no input RDF datasets to compute the granularity for certain cuboids. Figure 1 shows the structure of the IMD lattice obtained after integrating the running example datasets.

The rest of the paper is as follows: Sect. 2 contains the description of our framework and the steps we take to integrate datasets. Section 3 provides the technical details for implementing our framework. Section 4 illustrates the experiments we performed and the results obtained. Section 5 explains other ER frameworks for LD and we finally present our conclusion in Sect. 6.

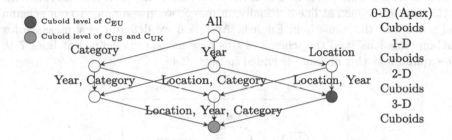

Fig. 1. Lattice of the IMD created for C_{EU}, C_{US}, and C_{UK}

2   Framework

In this section, we present an overview of our framework and how we obtain the IMD by following a purely data-driven approach. Our process is conducted pair-wise, where one input is the new RDF dataset we want to integrate with IMD and the other input is the IMD itself. Figure 2 depicts the data flow and the components of the framework, note that in the first iteration the flow is reduced to integrating two datasets and generating IMD for the first time.

In Sect. 2.1 we propose a supervised learning approach to analyze the RDF dataset D_I and identify resources as dimensions or measures. We refer to the resulting RDF annotated with MD concepts as D_A. We perform ER between same label resources in D_A, thus, obtaining the graph D_R (further explained in Sect. 2.2). In Sect. 2.3, and after performing ER, we enrich IMD and its schema with potentially new dimensions, hierarchies, and dimensional values. Finally, as elaborated in Sect. 2.4, we align factual data by considering potential unit misalignments.

We define the resulting IMD dataset using the QB[5] and QB4OLAP [5] vocabularies. An excerpt defining in QB4OLAP the basic structure of IMD for our running example follows:

---

[5] https://www.w3.org/TR/vocab-data-cube/.

```
1   schema:IMD rdf:type qb:DataStructureDefinition;
2    dct:conformsTo <http://purl.org/qb4olap/cubes_v1.3>;
3   data:IMD rdf:type qb:DataSet;
4    qb:structure schema:IMD;
5    dct:title "Integrated Multidimensional Dataset for carbon emissions"@en.
6   qb:component [qb:measure schema:C ; qb4o:aggregateFunction qb4o:avg];
7   schema:C rdf:type qb:MeasureProperty;
8    rdfs:label "Carbon emission quantity"@en;
9    rdfs:range xsd:float.
```

At line 1, we define the schema of the IMD and state that it follows a Data Cube structure. Line 2 states that this schema has an established standard to which the described resources can conform to. At line 3, we state that the data in the IMD schema is represented as a collection of observations, that can be organized into various slices, and thereby conforming to some common dimensional structure. Line 4 indicates that the structure of the IMD schema conforms to the dataset defined at line 3. Finally, in line 5, we give a name to our schema and state that the name is in English. At line 6 we define a new measure for "carbon emissions" and the schema of this new measure is defined at lines 7–9. We explain how this measure is added in Sect. 2.4.

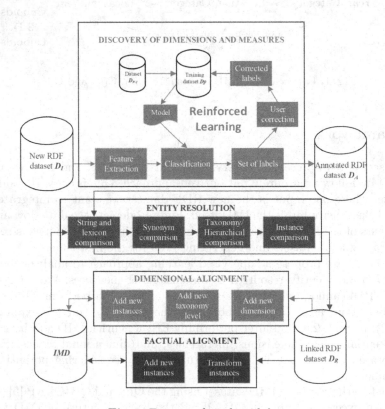

Fig. 2. Framework and modules

2.1   Discovery of Dimensions and Measures

Performing ER at a large scale is rather costly [17]. Thus, we propose a first step to automatically identify potential multidimensional concepts out of the input RDF datasets. Using this technique, the subsequent ER steps will be performed between elements labeled as of the same MD class (i.e., between dimensions or between measures). RDF metadata information, which represents a fair amount of triples, is removed from the process as they do not have a MD meaning.

We automatically label resources in the input RDF dataset as dimensions, measures or metadata using a Decision Tree (DT). This DT is trained on a user labeled dataset, where each resource contains data features and a data label. We use a DT over other ML techniques since a DT provides "rules" used for *explanatory analysis* of the resource labeling. Since the size of the search space is well-limited, as discussed in Sect. 4, using a DT shows good classification results. Then, we give the output to the user to validate the annotations and correct them if needed. Using a reinforcement method, the DT is continuously retrained by means of the user provided corrections. This approach follows a pay-as-you-go model that learns as new datasets come. Our tests show that few iterations are needed to obtain good recall and accuracy. We perform the following steps each time a new dataset is to be integrated in IMD:

Feature Extraction. For each resource in D_I, the following features are generated by analyzing the datasets schema, and dictionaries:

Unique Values: The ratio of unique values based on the total occurrences.

Data Types: Such as float, integer, string, boolean, categorical, date, geolocation, a resource (i.e., a URI) or description (containing metadata information).

URI Prefix and Resource Name: The URI is parsed to obtain these features. In our running example, <ds:location> (in C_{UK}), is parsed as <ds> and <location>.

URI Resource Name Length: The total number of characters in a URI.

Additive Property: Identifies numerical type resources as additive or non-additive.

Data Classification. Each resource in D_I is enriched with the above-mentioned features and passed to the Classification module. This module contains the model that is trained with the union of all previous datasets (D_P), with user corrected labels. Enriched resources of D_I are given as input to the trained model that labels its new resources based on the rules identified in previous iterations. Table 1 shows some labeled resources obtained in our running example. The next steps will only be performed between resources with the same label. Note that resources labeled as metadata are excluded from the next steps.

Table 1. Sample of labelled URIs

Label	Dataset	URI
Measure	C_{EU}	<sdmx-measure:obsValue>
Measure	C_{UK}	<ds:carbon_emissions_kgco2e>
Measure	C_{US}	<ds:total_emissions_mt_co2e>
Dimension	C_{EU}	<geo rdf:resource="">
Dimension	C_{UK}	<ds:location>
Dimension	C_{US}	<ds:location_latitude_longitude>

Label Validation. The Classification module provides the new annotated resources from D_I as a input to the User Validation module. The user corrects the labels of D_I, which is then merged with dataset D_{P-1} obtaining dataset D_P. Using reinforcement learning, the dataset D_P is re-fed to the Classification module to retrain the DT for the next iteration (i.e., when starting the process for a new RDF dataset to be integrated with IMD). In our running example, the DT is initially trained with the (manually labeled) dataset QBOAirbase. We then used this DT to label the dataset C_{UK}. If so, it would not assign a label to <ds:period> and <ds:location>, as these resources do not fit any of the current rules. We corrected these labels, as these resources contain dimensional data, and the combination of the QBOAirbase dataset and the corrected C_{UK} dataset will conform the training dataset for the next iteration (i.e., D_P). We retrain the DT with the new D_P training dataset and then use it to label the dataset C_{US}. Table 2 shows the accuracy and recall of the model achieved to label a newcoming dataset in each iteration as new (correctly labeled) resources of the previous iteration are added to the training dataset (D_P). In our running example, our experiments show that the model accuracy and recall improves drastically with few new input resources per iteration.

As one may infer, the order in which the datasets are ingested in our system may impact the accuracy and recall of each iteration. We discuss more about our approach for automatic dimension and measure elicitation in Sect. 4.1.

Table 2. Model accuracy per iteration

Iteration	Model accuracy	Recall	Dataset labeled	New resources added
First	22.2%	15.8%	C_{UK}	-
Second	70%	83.3%	C_{US}	30
Third	81.3%	100%	C_{EU}	27

2.2 Entity Resolution

Entity resolution plays a significant role in our research. Once the resources are labeled we obtain an annotated dataset D_A. We perform ER on the dimensional and factual resources of D_A separately to identify links, with IMD, between the same resources. This step is required due to the usage of different vocabularies in the input RDF datasets and IMD.

ER is done by obtaining lexicons from resource names of D_A and indexing them for linkage purposes (referred to as Rule 1). Additionally, we enrich the resources of D_A with their synonym-map and hierarchy-map (that are obtained using external dictionaries) and index them to perform ER (referred to as Rule 2 and Rule 3). We introduce a final rule to consider equivalence of *two dimensional resources* by means of instances. We apply Rules 1–3 on instance level to identify if two instances of separate concepts are the same (referred to as Rule 4). A resource in D_A can be linked with a resource in IMD either through equivalence (by having the same name lexicons or synonyms) or subsumption (through taxonomies). The formal definition of rules are as follows:

Rule 1. Given two resources, d_1 and d_2, d_1 is the same resource as d_2 if there is an equivalence when considering the lemmas in the names of both resources:
$$\{d_1, d_2\} \mapsto \{d_2\} \ \textit{iff} \ d_1 \equiv d_2 \ \textit{where} \ d_1 \in D_A \wedge d_2 \in \textit{IMD}$$

Rule 2. Given two resources, d_1 and d_2, d_1 is the same resource as d_2 if there is an equivalence (i.e., same lemma) when considering the synonym map (S_d) of both resources:
$$\{d_1, d_2\} \mapsto \{d_2\} \ \textit{iff} \ d_3 \equiv d_4 \ \textit{where} \ d_1 \in D_A \wedge d_2 \in \textit{IMD} \wedge d_3 \in S_{d1} \wedge d_4 \in S_{d2}$$

Rule 3. Given two resources, d_1 and d_2, d_2 subsumes or is subsumed by d_1 if d_1 and d_2 (or their synonyms) participate in the same hierarchical-map (H_d) either directly or through their synonym map:
$$d_3 \equiv d_4 \wedge d_4 \sqsubseteq d_5 \ \textit{where} \ (d_3 \in (d_1 \cup S_{d1}) \wedge d_4 \in (d_2 \cup S_{d2}) \wedge d_5 \in H_{d2}) \vee (d_3 \in (d_2 \cup S_{d2}) \wedge d_4 \in (d_1 \cup S_{d1}) \wedge d_5 \in H_{d1})$$

Rule 4. Given two resources, d_1 and d_2, there is an equivalence if I_{d1}, the instance space of d_1, has an intersection with I_{d2}, the instance space of d_2, that is greater than an input parameter θ, which is the required level of equal instances in both resources:
$$d_1 \equiv d_2 \ \textit{iff} \ I_{d1} \cap I_{d2} > \theta \ \textit{where} \ I_{d1} = \text{instances of } d_1 \wedge I_{d2} = \text{instances of } d_2 \wedge \theta \in \mathbb{R}$$

If d_1 maps to d_2 by any of the rules given above, the other rules are not checked. As result, we link both concepts either by an equivalence relationships (Rules 1, 2 or 4) or by subsumption (Rule 3).

We now present some examples of each rule based on our running example. For example, there is an equivalence between dimensional URIs <ds:location>

(in C_{UK}) and <ds:location_latitude_longitude> (in C_{US}), as they provide information for the same lemma **location** (Rule 1). These URIs also contain instances on the same dimensional level. However, <geo rdf:resource=""> (in C_{EU}) is linked with the location URIs via subsumption (Rule 3) at the instance level. When we apply Rule 4, it shows that the instances of <geo rdf:resource=""> subsumes the instances of <ds:location_latitude_longitude> and <ds:location>, as C_{UK} and C_{US} provide location information at a finer granularity level than C_{EU} (at country level). Additionally, <ds:period> (in C_{UK}) is linked with <dcterms:date> (in C_{EU}), as **period** and **date** share the same synonym-map (Rule 2). The factual resources in the datasets, <ds:carbon_emissions_kgco2e> (in C_{UK}) and <ds:total_emissions_mt_co2e> (in C_{US}), are linked based on the same lemmas **emission** and **co2** (Rule 1).

Before finalizing this step, and provided that ER methods can hardly achieve a 100% recall, we let the user to manually fix any missing link we could not detect.

2.3   Dimensional Alignment

The resultant dataset D_R is linked to IMD by means of the relationships identified in the previous step: either through equivalence or through subsumption relations. This component identifies updates to be performed in the IMD dimensional data (both schema and instances) according to the links identified. As illustrated in Fig. 2, when the dataset D_R is given as input to the Dimensional Alignment component, the dimensional space of IMD can be updated by (i) adding new dimensions, (ii) adding new hierarchical levels to existing dimensions (schema level) or (iii) adding new instances to an existing dimension level. We formalize IMD with QB4OLAP notation and to add a new dimension, or dimensional level to IMD, we add the corresponding QB4OLAP triple.

Add Dimension. If a dimension in D_R is not linked to any dimensional resource from IMD (i.e., Rules 1–4 from Sect. 2.2 are not satisfied) it means that there is no correspondence in IMD for this concept. Thus, we need to create a new dimension in the IMD multidimensional schema. In our running example, we add a new dimension for location in IMD and define it using QB4OLAP annotation:

```
10   schema:locationDim rdf:type qb:DimensionProperty;
11    rdfs:label "Location class dimension"@en;
12    qb4o:hasHierarchy schema:locationHier.
13   schema:locationHier rdf:type qb4o:Hierarchy;
14    rdfs:label "Location Hierarchy"@en;
15    qb4o:inDimension schema:locationDim;
16    qb4o:hasLevel schema:propertyLevel, schema:stateLevel.
17   _:station_hl1 rdf:type qb4o:HierarchyStep;
18    qb4o:inHierarchy schema:locationHier;
19    qb4o:childLevel schema:propertyLevel;
20    qb4o:parentLevel schema:stateLevel;
21    qb4o:pcCardinality qb4o:ManyToOne;
22    qb4o:rollup schema:inState.
23   schema:inState rdf:type qb4o:RollupProperty.
```

At lines 10–12 we define the schema for the location dimension and add hierarchies at lines 13–16. The hierarchical level for "State" is defined at lines 17–22, and the roll-up property from "Property" to "State" is defined at line 23.

With regard to our running example, this excerpt of QB4OLAP would be added when the location dimension is introduced in IMD for first time (i.e., with C_{UK} or C_{US})

Add Dimensional Level. If Rule 3 identifies a subsumption relationship between dimensional concepts in D_R and IMD, we add a new dimensional level to the proper dimension in IMD. In our running example, as stated in Sect. 2.2, the URI <geo rdf:resource=""> (in C_{EU}) provides details on a higher taxonomy (country level). Therefore, we add a new lattice level to the schema of IMD. The following QB4OLAP-compliant definition states how we add the dimensional level for "Country" and update the schema of IMD:

```
24   schema:locationHier rdf:type qb4o:Hierarchy;
25     qb4o:hasLevel schema:propertyLevel, schema:stateLevel,
26     schema:countryLevel.
27   _:station_hl2 rdf:type qb4o:HierarchyStep;
28     qb4o:inHierarchy schema:locationHier;
29     qb4o:childLevel schema:stateLevel;
30     qb4o:parentLevel schema:countryLevel;
31     qb4o:pcCardinality qb4o:ManyToOne;
32     qb4o:rollup schema:inCountry.
33   schema:inCountry rdf:type qb4o:LevelProperty;
34     rdfs:label "type Level"@en;
35     qb4o:hasAttribute property:country.
36   property:country rdf:type qb4o:LevelAttribute;
37     rdfs:label "country Name"@en;
38     rdfs:range xsd:string.
39   schema:inCountry rdf:type qb4o:RollupProperty.
```

Notice that we update line 16 with lines 25–26 to add a new dimensional level. We define the new hierarchy in lines 27–38 and finally add the roll-up property from "State" to "Country" in line 39.

Add Instances. We add instances to IMD in two cases, either by identifying equivalence relationship between dimensional resources (defined by Rules 1, 2 and 4 in Sect. 2.2) or when a new dimensional level is added (defined by Rule 3 in Sect. 2.2). In the first case, we need to apply ER again, but this time between the instances of D_R and IMD. This step is required for identifying equivalences (Rules 1 and 2) between the instances of D_R and IMD, to avoid duplication of instances in IMD. In the second case, however, we simply add all the instance resources related to that dimensional resource on the newly created level.

For this second step we rely on previous work [6]. For example, in our running example, when a new dimensional level from C_{EU} is identified (country level), we add instances from URI <geo rdf:resource="">. The instances are added to populate "Country Name" property as specified in lines 36–38 of the QB4OLAP-compliant definition when a new dimensional level is identified. Each cuboid in the lattice is required to have the unit of measure associated with it (further explained in Sect. 2.4). Hence, when adding the instances for "Country" dimension, we also add the unit of measure for the dataset C_{EU}.

2.4   Factual Alignment

The last phase of our framework is to populate the IMD with additional factual instances. Using Rules 1, 2 and 4 in Sect. 2.2 we obtain resources that contain factual data and we transform instances of dataset D_R to guarantee uniformity. Additionally, for Factual Alignment component, we require the unit of measure. The unit can be identified by either being explicitly provided in the dataset or we ask the user to provide this information. We require these units so that we can transform the factual data in D_R to that of IMD, using conversion functions. The conversion functions are stored in our framework for each unit of measure.

The factual instances of D_I can either have an equivalent measure from IMD, or do not have any counterpart in IMD yet. As IMD is QB4OLAP-compliant, the factual instances are instantiated as cuboids at a certain granularity level. As can be seen in Fig. 1, the datasets sit at different cuboid levels based on the granularity detail of the factual instances provided in those datasets. Therefore, if there is an equivalent measure from IMD in D_I then we compare the units, apply the transformation on instances of D_I to make them compatible, and add the transformed factual instances to IMD. If there is no equivalent measure from IMD in D_I, we import the factual instances of D_I as they are and annotate the unit of measure at that cuboid level. Note that we annotate the unit and transformation between units as this information is elicited to automate these tasks in future steps.

In our running example, the URIs that contain data for carbon emission are <ds:carbon_emissions_kgco2e> (in C_{UK}), <ds:total_emissions_mt_co2e> (in C_{US}), and <sdmx-measure:obsValue> (in C_{EU}). Each dataset contains emission data in a different unit of measure: kg for C_{UK}, mt for C_{US}, and tonnes for C_{EU}. C_{US} and C_{UK} contain factual instances on the same cuboid (3-D) level. Yet, they have different units of factual measures. We select the unit of measure from C_{UK} (kg) as the base unit at that cuboid level and transform factual instances of C_{US} from mt to kg before combining them. C_{EU} contains factual instances on a higher cuboid (2-D) level and therefore we can import the new factual instances to that cuboid level but we should also transform their values from tonnes to kg (to guarantee correct Roll-up between both cuboids). Additionally, the unit of measure for carbon emission are available in C_{UK} and C_{US}, but C_{EU} does not implicitly contain the unit of measure. Therefore, we extract this information from the Eurostat website and annotate it. We add the new unit of measure in our schema:C with use of rdfs:comment since each measure can have it's own unit. For future work, we plan to incorporate more advanced state-of-the-art techniques to integrate measures into the IMD through either formal representation of the measures dependencies by means of specific predicates [4,13] or through the semantic representation of the calculation formulae of such measures [3].

3   Implementation

In this section, we first discuss the prototype created to implement our method and then discuss a set of experiments conducted to show its feasibility. We build a Java project to extract entities from input graphs using Apache Jena[6] and add features that are computed using the OpenLink Virtuoso[7] platform. Then, for building the DT model to label resources as dimensions, metadata or measures we opt for the KNIME Analytical Platform[8] and perform ER operations on the annotated datasets using LogMap [8] and Instance-based Unified Taxonomy [18] to finally merge the datasets. We use LogMap over other ontology alignments tools as it has low time complexity and high (successful) linkage discovery [1].

Next subsections elaborate on the implementation of the framework: Sect. 3.1 describes how we built our classification model. Sect. 3.2 describes the techniques used for performing ER operations and finally, Sect. 3.3 elaborates on how we extend the IMD schema. We use real datasets (C_{US}, C_{UK}, and C_{EU}) to showcase the feasibility of our method.

3.1 Model Building in KNIME

In KNIME, we use the *Excel Reader nodes* that read the features specified in Sect. 2.1 (Feature Extraction). We train the *Decision Tree Learning node* using the training input data and we test the *Decision Tree Predictor node* using the testing data. In Sect. 2.1 we specify that in each iteration we increment the training datasets with new (user validated) label resources. This incremented data is then used to retrain the DT model. In Sect. 4.1 we perform Leave One Out (LOO) cross-validation to show the accuracy of our approach. That is, how likely is the trained model able to identify measures and dimensions from previous examples.

3.2 Entity Resolution for Dimensions and Measures

Our focus is to integrate RDF datasets in a common multidimensional space and build a lattice, therefore, we use state-of-the-art techniques to perform ER. We use LogMap for ER operations defined using Rules 1–3 (see Sect. 2.2) as LogMap uses WordNet[9] to obtain lexicons from resource names, indexes resources for linkage purposes with their synonym-map, and builds hierarchies for each class. Additionally, for ER Rule 3, we use the mapping technique (on instance level) proposed in [6] along with matching of resources using rdfs:subClassOf property.

Furthermore, we use Instance-based Unified Taxonomy [18] for ER Rule 4 stated in Sect. 2.2, as it builds a unified taxonomy of classes and links instances through owl:sameAs property and exact matching. Sect. 4.2 states the performance results obtained when applying ER operations on our datasets.

---

[6] https://jena.apache.org/.
[7] https://virtuoso.openlinksw.com/.
[8] https://www.knime.com/.
[9] https://wordnet.princeton.edu/.

3.3 Schema Alignment

To add dimensions, dimensional hierarchies, and dimensional instances: we use Apache Jena to extract the annotated data from the input dataset, add QB4OLAP-compliant triples at new granularity level (if required), remove duplicate instances, and add it in the IMD dataset. To add factual instances, we extract the units of measure in the input dataset and compare it to the unit of measure of IMD, if the unit of measures differ, conduct transformations on the resource values. We also ask the user to provide conversion function, using the Java interface, where the unit of measure is not automatically convertible. We extract the dimensional hierarchy of measures from the input graph and compare them to the IMD dataset to either remove duplicate instances or to add a QB4OLAP-compliant triple at newly identified granularity level. Lastly, we add new instances to the IMD dataset.

4 Experiments and Results

This section discusses about the feasibility of our framework defined in Sect. 2 and implemented using the techniques provided in Sect. 3.

4.1 Model Testing in KNIME

Once we have (correctly) labeled all the resources in our running example, we use it as ground truth to perform LOO cross-validation to show the accuracy of our approach by picking a resource from the dataset, training the model on rest of the resources and check how well the model labels the one resource not used in training. This exercise is repeated for each resource, and the goal is to test the model's ability to see how our approach behaves with a resource that it has not seen before. We have 230 resources in total and therefore the LOO cross-validation is performed 230 times (once for each resource). After all 230 iterations are conducted, the model yielded an average error rate of 7.39% with a 6.84% variance between these error rates. Most of the errors produced were due to the resources not satisfying any of the rules obtained and hence the model labeled them as "?". Additionally, to check the validity of the model on a dataset from a different domain, we tested the DT (trained on *carbon* datasets) on a dataset from the *crime* domain[10]. This dataset contained 40 resources, of which the model correctly labeled 29 resources (i.e. obtained an accuracy of 74.36%). These experiments show that the domain is irrelevant when deciding the multidimensional role of a resource. In the future, we plan to validate this hypothesis with more thorough experiments.

Also, as one may expect, the order in which we take the resources has a big impact on the training. We will investigate how to make our method more stable in the future. Nevertheless, the big advantage of building a DT is that we obtain

---

[10] https://data.cityofchicago.org/Public-Safety/Crimes-2001-to-present-Dashboard/5cd6-ry5g.

rules to label each resource. Table 3 contains the most relevant rules obtained from our tests. Relevantly, after training our model with different orders, we eventually get very similar rules. Thus, the order has an impact during the reinforced learning stage.

Indeed, after identifying the rules in this experiment, we notice that these rules are aligned with the main rules provided by previous research works in the literature (e.g., [2,14] present thorough surveys). However, most of these approaches manually identify dimensions and measures, while our approach embeds a semi-automatic approach.

Table 3. Interesting rules identified for each label

Label	Rule
Measure	Resources that have *additive* feature marked as "Yes"
Metadata	Resources that contain W3C defined prefix in *URI Prefix*, such as foaf, owl, rdf, and rdfs
Dimension	Resources that have *Data Type* as "GeoLocation" or have ">99%" *Unique Values* and *Data Type* as "String"

4.2 Entity Resolution

We label the resources either as dimensions or measures to facilitate ER in our framework. The labeling of resources indicates that instead of performing an NxM comparison, only compare the resources that are equally labeled either as "dimension" or "measure". For example, when we compare C_{UK} with C_{US} with a generic ER framework, not considering dimensions or measures (i.e., labels), there would be 810 comparisons to be done. But, when we compare them with labels, the comparison drops to 139. Table 4 states the total number of resources in each dataset, and how many of them are dimensional and factual resources. Table 5 provides the number of comparisons made for matching these datasets, with and without labeling the resources. We observe that using labels aids in comparison as most of the resources in RDF datasets are metadata that only provide additional information. Also, when there are links identified in the previous ER iteration, the number of resources to be compared decrease in the next iteration. After labeling and comparing all three datasets (C_{UK}, C_{US}, and C_{EU}), as stated in Table 5, we reduce the number of comparisons by 88% and the runtime by 81%.

As defined in Sect. 2.2, whenever a resource d_1 in D_A maps onto a resource d_2 in IMD, we always select the URI of d_2 to define the newly identified resource. To avoid re-computation of synonym-maps and hierarchical-maps every time a new resource needs to be matched, we store them in IMD.

Table 4. Total & labeled resources

Dataset	Total	Dimension	Measures
C_{UK}	30	16	3
C_{US}	27	7	9
C_{EU}	16	2	1

Table 5. ER with & without labels

Using labels	Comparisons	Run-time (s)
Yes	201	31
No	1658	165

Missing Links. The framework missed some links between resources that had to be added manually. For example, <sdmx-measure:obsValue> in C_{EU} contains carbon emission, yet it was not linked with either <ds:carbon_emissions_kgco2e> in C_{UK} or with <ds:total_emissions_mt_co2e> in C_{US}. Similarly, <ds:sub_sector> from C_{US} and <ds:category> from C_{UK} contain the type of areas (e.g. primary schools area, chemical production area, glass production area, etc.) where the carbon emission value is collected, yet these two resources were not linked. Based on this the recall of our approach, for identifying linkage, is 71.4%.

5   Related Work

In this section we mention some frameworks that provide entity resolution component for integration of RDF datasets.

The user can obtain a clean and consistent local target vocabulary using [16] while tracking data provenance. It includes an identity resolution component for discovering URI aliases in data and maps them to single target URI based on user-provided matching heuristics using SILK [7].

The authors of [12] provide a mapping technique that achieves integration automatically using SPARQL CONSTRUCT mappings. The mappings are translated to the new consolidated, well specified and homogenous target ontology model, thereby, facilitating data integration from different academic RDF datasets by providing a bridge between heterogeneities of RDF datasets.

In [11], the authors define a multidimensional model based on the QB vocabulary and present OLAP algebra to SPARQL mappings. They use graph-based RDF data model along with QB vocabulary for querying and storing of Data cubes. They also define common OLAP operations on the Data cubes by re-using QB. The drawback in their approach is that it does not allow optimized OLAP queries to RDF and the entire ETL process needs to be repeated if new statistics are defined.

OLAP4LD [10] provides a platform for developers of applications using Linked Data sources reusing the QB vocabulary and explore the Eurostat data via Linked Data Cube Explorers. The authors focus on translating analytical operations to queries over LD sources and pre-processing to integrate heterogeneously modeled data.

A Federated Datawarehouse (F-DW) [9] is the integration of heterogeneous business intelligence systems set to provide analytical capabilities across the different function of an organization. Reference data is made common across various data warehouses for data consistency and integrity, and identical data value for a confirmed fact will be ensured. The prime design objective of a F-DW is to achieve a "single version of truth" and the authors stress that there should be defined, documented, and integrated business rules used across the component data warehouses in the whole architecture.

Our approach is different in the sense that our inputs are datasets and not multidimensional (MD) schemas. We aim towards a generic approach that generates a MD interpretation of RDF datasets. RDF datasets can embed a MD interpretation by using QB4OLAP but, unfortunately, publishers (such as Eurostat) are not yet publishing native MD datasets. Therefore, we first identify and assign a MD interpretation to online RDF datasets. Then, we perform ER but over the identified MD concepts of the same class. This way, we drastically reduce the problem complexity (typical ER tools are quadratic on the dataset size). In our case, many triples are disregarded (RDF metadata without MD meaning and not useful for data analysis) and create two big groups of concepts: facts and dimensions. In our tests, 48% of the resources were disregarded for data analysis, 34% were identified as dimensions and 18% as facts. As a consequence, we simplified the complexity of the entity resolution process considerably.

6 Conclusion

In this paper, we present a technique to semi-automatically combine RDF datasets into the same multidimensional space and identify each dataset as a cuboid of its lattice. Unlike other frameworks, we perform integration based on a multidimensional interpretation of arbitrary datasets. We showcase our method choosing RDF data as a use case, but the proposed methods could be generalized for other data models. Specifically, we have defined a set of steps to generate a single data cube out of independent, yet overlapping, RDF datasets. We use QB4OLAP to annotate the integrated data and define, for each input dataset, a cuboid at the right aggregation level of the IMD. As a consequence, the schema of the input RDF datasets is integrated, and we can use state-of-the-art tools to analyze QB4OLAP data to perform OLAP operations on the defined lattice. This novel method is a significative step towards automating effective multidimensional integration of related RDF datasets that were created independently. Indeed, our approach could be generalized to other data formats. Importantly, providing a MD interpretation of the datasets considerably reduce the complexity of ER when integrating the dataset schemas.

For our next work, we will propose a mechanism to query the source RDF datasets (example C_{UK}, C_{US}, and C_{EU} datasets used in this paper) via federated SPARQL queries and therefore avoiding to materialize the IMD.

Acknowledgements. This research has been funded by the European Commission through the Erasmus Mundus Joint Doctorate Information Technologies for Business Intelligence-Doctoral College (IT4BI-DC) program.

References

1. Achichi, M., et al.: Results of the ontology alignment evaluation initiative 2017. In: Proceedings of the 12th International Workshop on Ontology Matching Co-Located with the 16th International Semantic Web Conference, vol. 2032, pp. 61–113. CEUR-WS, October 2017
2. Cravero, A., Sepúlveda, S.: Multidimensional design paradigms for data warehouses: a systematic mapping study. J. Softw. Eng. Appl. **7**(1), 53–61 (2014)
3. Diamantini, C., Potena, D., Storti, E.: Multidimensional query reformulation with measure decomposition. Inf. Syst. **78**, 23–39 (2018)
4. Estrada-Torres, B., et al.: Measuring performance in knowledge-intensive processes. ACM Trans. Internet Technol. **19**(1), 151–1526 (2019)
5. Etcheverry, L., Vaisman, A.A.: QB4OLAP: a new vocabulary for OLAP cubes on the semantic web. In: Proceedings of the 3rd International Conference on Consuming Linked Data, vol. 905, pp. 27–38. CEUR-WS.org, November 2012
6. Gallinucci, E., Golfarelli, M., Rizzi, S., Abelló, A., Romero, O.: Interactive multidimensional modeling of linked data for exploratory OLAP. Inf. Syst. **77**, 86–104 (2018)
7. Isele, R., Jentzsch, A., Bizer, C.: Silk server - adding missing links while consuming linked data. In: Proceedings of the 1st International Conference on Consuming Linked Data, vol. 665, pp. 85–96. CEUR-WS.org, November 2010
8. Jiménez-Ruiz, E., Cuenca Grau, B.: LogMap: logic-based and scalable ontology matching. In: Aroyo, L., Welty, C., Alani, H., Taylor, J., Bernstein, A., Kagal, L., Noy, N., Blomqvist, E. (eds.) ISWC 2011. LNCS, vol. 7031, pp. 273–288. Springer, Heidelberg (2011). https://doi.org/10.1007/978-3-642-25073-6_18
9. Jindal, R., Acharya, A.: Federated data warehouse architecture. Wipro Technologies White Paper (2004)
10. Kämpgen, B., Harth, A.: OLAP4LD – a framework for building analysis applications over governmental statistics. In: Presutti, V., Blomqvist, E., Troncy, R., Sack, H., Papadakis, I., Tordai, A. (eds.) ESWC 2014. LNCS, vol. 8798, pp. 389–394. Springer, Cham (2014). https://doi.org/10.1007/978-3-319-11955-7_54
11. Kämpgen, B., O'Riain, S., Harth, A.: Interacting with statistical linked data via OLAP operations. In: Simperl, E., Norton, B., Mladenic, D., Della Valle, E., Fundulaki, I., Passant, A., Troncy, R. (eds.) ESWC 2012. LNCS, vol. 7540, pp. 87–101. Springer, Heidelberg (2015). https://doi.org/10.1007/978-3-662-46641-4_7
12. Moaawad, M.R., Mokhtar, H.M.O., Al Feel, H.T.: On-the-fly academic linked data integration. In: Proceedings of the International Conference on Compute and Data Analysis, pp. 114–122. ACM, May 2017
13. Popova, V., Sharpanskykh, A.: Formal modelling of organisational goals based on performance indicators. Data Knowl. Eng. **70**(4), 335–364 (2011)
14. Romero, O., Abelló, A.: A survey of multidimensional modeling methodologies. Int. J. Data Warehous. Min. **5**(2), 1–23 (2009)
15. Schmachtenberg, M., Bizer, C., Paulheim, H.: Adoption of the linked data best practices in different topical domains. In: Mika, P., et al. (eds.) ISWC 2014, Part I. LNCS, vol. 8796, pp. 245–260. Springer, Cham (2014). https://doi.org/10.1007/978-3-319-11964-9_16

16. Schultz, A., Matteini, A., Isele, R., Bizer, C., Becker, C.: LDIF - linked data integration framework. In: Proceedings of the 2nd International Conference on Consuming Linked Data, vol. 782, pp. 125–130. CEUR-WS.org, October 2011
17. Suchanek, F.M., Abiteboul, S., Senellart, P.: Paris: probabilistic alignment of relations, instances, and schema. Proc. VLDB Endow. **5**(3), 157–168 (2011)
18. Zong, N.: Instance-based hierarchical schema alignment in linked data. Ph.D. thesis, Seoul National University Graduate School, Seoul, South Korea (2015)

RDFPartSuite: Bridging Physical and Logical RDF Partitioning

Jorge Galicia[1(✉)], Amin Mesmoudi[1,2], and Ladjel Bellatreche[1]

[1] LIAS/ISAE-ENSMA, Chasseneuil-du-Poitou, France
{jorge.galicia,bellatreche}@ensma.fr
[2] Université de Poitiers, Poitiers, France
amin.mesmoudi@univ-poitiers.fr

Abstract. The Resource Description Framework (RDF) has become a very popular graph-based standard initially designed to represent information on the Web. Its flexibility motivated the use of this standard in other domains and today RDF datasets are big sources of information. In this line, the research on scalable distributed and parallel RDF processing systems has gained momentum. Most of these systems apply partitioning algorithms that use the triple, the finest logical data structure in RDF, as a distribution unit. This merely physical strategy implies losing the graph structure of the model causing performance degradation. We believe that gathering the triples storing the same logical entities first contributes not only to avoid scanning irrelevant data but also to create RDF partitions with an actual logical meaning. Besides, this logical representation allows defining partitions with a declarative language leaving aside implementation details. In this study, we give the formal definition and detail the algorithms to gather the logical entities, which we name graph fragments $(\mathcal{G}f)$, used as distribution units for RDF datasets. The logical entities proposed, harmonize with the notion of partitions by instances (horizontal) and by attributes (vertical) in the relational model. We propose allocation strategies for these fragments, considering the case when replication is available and in which both fragments by instances and by attributes are considered. We also discuss how to incorporate our declarative partitioning definition language to the existing state of the art systems. Our experiments in synthetic and real datasets show that graph fragments avert data skewness. In addition, we show that this type of data organization exhibits quantitative promise in certain types of queries. All of the above techniques are integrated into the same framework that we called *RDFPartSuite*.

Keywords: RDF · Partitioning · Distributed computing

1 Introduction

The Resource Description Framework (RDF) has been widely accepted as the standard data model for data interchange in the Web. The model uses triples

© Springer Nature Switzerland AG 2019
C. Ordonez et al. (Eds.): DaWaK 2019, LNCS 11708, pp. 136–150, 2019.
https://doi.org/10.1007/978-3-030-27520-4_10

consisting of a subject, a predicate and an object $<s, p, o>$ as the core abstract data structure to represent information. Formerly, RDF intended to model exclusively information in the Web. Moreover, its flexibility allowed the standard to be used in other domains (e.g. genetics, biology). Currently, collections of RDF triples (known as knowledge bases) are extensive sources of information popularly queried and aggregated. Its size has considerably augmented in the last years and likewise, the development of scalable distributed RDF processing systems[1]. The data in most of these systems are partitioned for scaling purposes with hardly customizable strategies. Therefore, it is necessary to establish a general model encompassing the partitioning approaches of existent distributed RDF systems. Such model must be based on the manipulation of higher-level entities at a *logical* level allowing its incorporation into existing systems.

The RDF model uses the triple as its key data structure giving it more flexibility but at the price of data dispersion. Raw RDF files store lots of triples (sometimes billions) in which the facts of the same high-level entity are scattered through the batch of data. The partitioning strategies used by several systems take the triples as distribution units and apply, for example, hashing or graph-partitioning functions sharding the triples at its finest granularity. These strategies do not allow to keep the graph structure of the model at query runtime impacting greatly the system's performance.

Our work introduces a logical dimension to the partitioning process of RDF graphs. We define two types of high-level entities, which we name forward and backward graph fragments $(\mathcal{G}f)$. These fragments harmonize with the notion of partitions by instances (horizontal) and by attributes (vertical) in the relational model. We detail an identification algorithm for both types of entities with a motivating example. Next, the graph fragments are distributed among the nodes of a distributed system. We then propose a declarative partitioning definition language for RDF stating the higher-level entities and partitions. Finally, we integrate all the proposed techniques into the same framework named *RDFPartSuite*.

The contributions of this paper are:

1. The formalization of a logical dimension generalizing RDF partitions by the identification of high-level entities.
2. An analysis of allocation methods integrating both partitioning by attributes and by instances in RDF systems.
3. The proposition of a declarative definition language to state and allocate the high-level entities (graph fragments) in an RDF graph.

The rest of the paper is organized as follows. We start in Sect. 2 giving an overview of our work. We present briefly the algorithms to create and allocate the forward and backward graph fragments by the means of a motivating example. In Sect. 3, we formalize our approach and define the graph fragment's allocation problem. Next, Sect. 4 presents our experimental study and Sect. 5 discusses the

---

[1] We use the term *distributed RDF system* indistinctly to denote both parallel and distributed RDF systems.

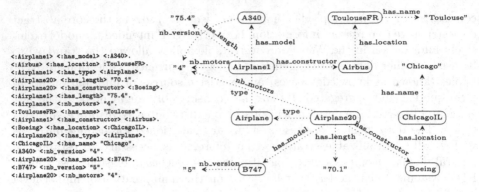

```
<:Airplane1> <:has_model> <:A340>.
<:Airbus> <:has_location> :ToulouseFR>.
<:Airplane1> <:has_type> <:Airplane>.
<:Airplane20> <:has_length> "70.1".
<:Airplane20> <:has_constructor> <:Boeing>.
<:Airplane1> <:has_length> "75.4".
<:Airplane1> <:nb_motors> "4".
<:ToulouseFR> <:has_name> "Toulouse".
<:Airplane1> <:has_constructor> <:Airbus>.
<:Boeing> <:has_location> <:ChicagoIL>.
<:Airplane20> <:has_type> <:Airplane>.
<:ChicagoIL> <:has_name> "Chicago".
<:A340> <:nb_version> "4".
<:Airplane20> <:has_model> <:B747>.
<:B747> <:nb_version> "5".
<:Airplane20> <:nb_motors> "4".
```

Fig. 1. RDF graph G

related work. Finally, in Sect. 6 we conclude and give insights on our future research.

2   Logic Partitioning Overview: Motivating Example

In most of RDF systems, data partitioning strategies are directly applied to triples (e.g. a hashing function on the subject). Partitions are thus created with purely physical techniques. We aim to create partitions with a more logical representation whose implementation is independent of the execution layer of any system. To achieve this, triples must be clustered using structures with a higher abstraction level than the one offered by triples.

We propose to gather the data similarly as it is done in the relational model, clustering the triples in higher level entities first. Then, these entities are partitioned at the instance or attribute levels. However, in RDF there is no conceptual design stage stating high-level entities like in relational databases. This feature gives RDF flexibility and makes it appropriate to model the varying data on the Web. Our proposal allows to detect high-level logical entities from existing data sources based on the concept of characteristic sets introduced in [1].

Furthermore, clustering the triples by instances or by its attributes contributes at query runtime. Indeed, horizontal and vertical partitions in the relational model let to retrieve only relation's subsets avoiding full table scans. Similarly, building meaningful clusters of triples avoids traversing the entire graph when looking for specific query matches. In RDF, a query language (e.g. SPARQL in Fig. 5) expresses graph patterns and its result corresponds to entity sets or RDF sub-graphs. From a graph-based processing perspective, the query matches are found traversing the graph starting from a node and moving through its forward or backward edges. In a partitioned graph, the execution system checks firstly for matches using the triple's clusters averting to traverse the entire graph. These clusters are used as fragments in the allocation strategies that we detail in this section. Also, we describe a partitioning definition language that simplifies the creation of partitions hiding implementation details to the final user.

The following example gives an overview of our approach. Let us consider the RDF graph G shown in Fig. 1. We simplified the N-Triples syntax to encode the dataset by adding a ":" representing the complete IRI (Internationalized Resource Identifier).

2.1 Creation of RDF Groups of Instances

To gather the triples of the same high-level entity we group in the first place the triples by its subjects. We obtain 8 groups of triples. We name each of these groups a *forward graph star* $\overrightarrow{\mathcal{G}s}$ representing instances of higher-level entities (e.g. an Airplane, Constructor, Airplane Model). We observe that two instances of the same high-level entity share the same (or almost the same) set of predicates. In general, the forward graph fragments can be uniquely identified by the set of its emitting edges (i.e. predicates). This concept is quite similar to the characteristics sets mentioned in [1]. In the example, the forward entities Airplane1 and Airplane20 share the same predicates. Airplane1 and Airplane20 belong to the same group that we name a *forward graph fragment* $\overrightarrow{\mathcal{G}f_i}$ of G.

```
SELECT ?y ?z
WHERE {
?x :has_model :B747 .
?x :has_length ?y .
?x :has_constructor ?z   . }
```

```
SELECT COUNT(?x)
WHERE {
?x :has_motors "4"   . }
```

```
SELECT ?y ?z
WHERE {
?x :has_model :A340 .       #1
?x :has_length ?y .         #2
?x :has_constructor ?w .    #3
?w :has_location ?z .       #4
?z :has_name "Toulouse" .   #5 }
```

(a) Example query 1   (b) Example query 2   (c) Example query 3

Fig. 2. Example SPARQL queries

A forward graph fragment represents a high-level entity on the graph G. In Sect. 3 we prove that the set of forward graph fragments is a partition set of the RDF dataset. The set of forward graph fragments is represented as $\overrightarrow{\mathcal{G}f} = \{\overrightarrow{\mathcal{G}f_1}, ..., \overrightarrow{\mathcal{G}f_m}\}$. In the example, there are four forward graph fragments representing the following logical entities: an airplane containing 10 triples, a location with 2 triples, an airplane constructor with 2 triples and an airplane model with 2 triples.

The organization of the RDF triples in forward graph fragments is ideal when solving star-pattern BGP queries like the one shown in Fig. 2a. In order to solve it, an index of the characteristic sets as the one illustrated on Fig. 3a could be built to scan only the relevant forward graph fragments.

Still, an organization of the data in forward graph fragments is not optimal when solving queries with a very reduced number of predicates like the one shown in Fig. 2b. To solve the query, all the triples of the forward fragment must be read, even the triples with predicates other than **nb_motors**. When exploring the graph for matches, 10 triples in the airplane forward graph fragment must be scanned when only 2 triples are relevant. The problem is quite similar to the one that motivated vertical partitions in relational databases in which only the

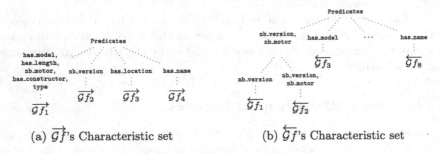

(a) $\overrightarrow{\mathcal{G}f}$'s Characteristic set   (b) $\overleftarrow{\mathcal{G}f}$'s Characteristic set

Fig. 3. Characteristic sets index

pertinent attributes of a table are accessed. Partitioning a table by attributes is very performant when the query does not access many attributes of the relation. This inspired us to propose another organization model for the RDF data that we name *backward graph fragment* described in the following section.

2.2   Creation of RDF Groups of Attributes

The vertical partitions for RDF data are obtained gathering first the triples by its incoming edges. In other words, we group the triples by its object. We obtain 13 groups for the example of Fig. 1. We name these groups *backward graph stars* $\overleftarrow{\mathcal{G}s}$. Similarly to what was done for the forward entities, we use a characteristic set to identify the backward graph fragments and index them as shown in Fig. 3b. We name a *backward graph fragment* $\overleftarrow{\mathcal{G}f_i}$ of G to the groups of backward stars with similar characteristic sets. As it is proven in Sect. 3, the set of backward graph fragments form a partition set of the RDF graph.

For the example graph of Fig. 1 we obtain 8 backward graph fragments: the group of names with the has_name predicate (2 triples), the group of constructors (2 triples), lengths (2 triples), locations (2 triples), models (2 triples) and types (2 triples). There is a group gathering the triples having only the nb_version predicate (1 triple), and another group gathering the triples with the predicates nb_version and nb_motor (3 triples). The predicates could be organized with a simple index structure like the one shown in Fig. 3b.

The solution to the query of Fig. 2b is found scanning only the backward graph fragment storing triples with the predicate nb_motors. The execution engine checks the characteristic set's index of Fig. 3b and scans only the backward graph fragment $\overleftarrow{\mathcal{G}f_2}$ that gathers 3 triples. Comparing to the result obtained when the data are organized as *forward graph fragments* and assuming that the cost to explore the predicate's index is the same, the execution process on the backward fragments is more efficient. The process saves, for this case, the resources used to scan 7 triples. However, as it is the case for vertical partitions in the relational model, the performance can be degraded in SPARQL queries joining many single patterns.

2.3   Integration of the Partitions by Attributes and by Instances

The integration of horizontal and vertical partitions in the relational model has been explored by many researchers in the past. For example, the advisors of a database physical design described in [2] or the Fractured Mirrors storage shown in [3]. Meanwhile, many massive processing systems propose replication strategies not only to recover and support fault tolerance but to improve the response time of queries. We consider a system storing two copies of the data. One copy organizes the data as forward graph fragments and another as backward graph fragments. This configuration is useful especially when the workload is unknown in the initial partitioning stage. We assume that the execution engine of the system is able to consider simultaneously forward and backward graph fragments. Let us consider the query of Fig. 2c that contains 5 single patterns. The single patterns 1 to 3 form a star-query pattern, the patterns 3–4 and 4–5 form two path-query patterns. The execution engine proposes a plan in which, for instance, the matches of the query patterns 1 to 3 are found using the forward graph fragment $\overrightarrow{\mathcal{G}f_1}$. Then to find matches for pattern 4 the intermediate results are with the forward fragment $\overrightarrow{\mathcal{G}f_3}$. Finally, to find matches from patterns 1 to 5, the results could be joined with the backward segment $\overleftarrow{\mathcal{G}f_8}$.

Forward and backward graph fragments could be used as *distribution units* to be allocated in a distributed system. The allocation strategies are discussed in the following section.

2.4   Allocation of Graph Fragments

The allocation of fragments (i.e. $\overrightarrow{\mathcal{G}f}$ or $\overleftarrow{\mathcal{G}f}$) is a mandatory stride in distributed systems. We assume that the size of a fragment is smaller than the maximum capacity of a site. We treat the allocation of big fragments later in the paper. Using a straightforward strategy like a round-robin for instance, may not produce an optimal performance. Firstly because the size of the segments is not uniform, therefore the data may be unequally distributed causing data skewness in some nodes. Secondly, queries like the one presented on Fig. 2c may be affected by the cost to send intermediate results through the network. The network costs are the bottleneck of distributed systems and should be minimized to improve the performance. We consider an allocation strategy in which the fragments connected with a path pattern should be allocated in the same site. For example in the execution plan detailed previously, the segments $\overrightarrow{\mathcal{G}f_1}, \overrightarrow{\mathcal{G}f_3}$ and $\overleftarrow{\mathcal{G}f_8}$ should be located in the same machine to avoid unnecessary network costs. The allocation problem is formalized in [4].

Allocation Heuristics: Graph Partitioning. The fragments could be represented in a weighted directed graph $\mathcal{G}$ as it is illustrated in Fig. 4b. Graph partitioning heuristics (e.g. METIS [5]) are then used to find good approximate solutions to the distribution problem. In the graph of Fig. 4b, the colored nodes correspond to the forward graph fragments. The red dashed line separates the

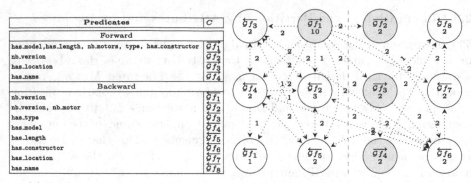

(a) Predicates by graph fragment   (b) Extract of graph of fragments $\mathcal{G}$

Fig. 4. Allocation example of graph G

graph into two hypothetical partitions. The node's weights represent the total number of triples on each segment. The weights are represented inside the nodes of the graph of Fig. 4b. The predicates on each segment are shown on the table of Fig. 4a. The edge's weights represent the number of triples that should be transferred between two fragments when they are joined. The weights are calculated according to the cases shown on Table 1. Not all the edges between fragments are represented in the graph of Fig. 4b for readability. Let us consider for example the edge's weight between the forward graph fragments $\overrightarrow{\mathcal{G}f_1}$ and $\overrightarrow{\mathcal{G}f_2}$. The weight in this case is 2 since according to Table 1, the objects (A340 and B747) of two triples in $\overrightarrow{\mathcal{G}f_1}$ are subjects of two triples in $\overrightarrow{\mathcal{G}f_2}$.

Table 1. Edge's weights in fragment graph $\mathcal{G}$

Edge type	$W(edge) = \#$ triples t_i such that $t_i \in G_i, t_j \in G_j$ and:
$(\overrightarrow{\mathcal{G}f_i}, \overrightarrow{\mathcal{G}f_j})$	$object(t_i) = subject(t_j)$
$(\overleftarrow{\mathcal{G}f_i}, \overrightarrow{\mathcal{G}f_j})$	$subject(t_i) = subject(t_j)$
$(\overrightarrow{\mathcal{G}f_i}, \overleftarrow{\mathcal{G}f_j})$	$object(t_i) = subject(t_j)$
$(\overrightarrow{\mathcal{G}f_i}, \overleftarrow{\mathcal{G}f_j})$	$subject(t_i) = subject(t_j)$
$(\overrightarrow{\mathcal{G}f_i}, \overleftarrow{\mathcal{G}f_j})$	$object(t_i) = subject(t_j)$

Allocation of Big Fragments. When the size of a forward or backward graph fragment is bigger than the maximum available space for a site, the entity needs to be repartitioned. To partition within a fragment, we apply a function that sub-partitions the segments in such a way that all the triples of a backward or forward graph stars belong to the same partition. In other words, the repartitioning function should not subdivide the forward or backward graph stars. The repartitioning function selects the forward graph fragments using a set of min-term predicates (defined in [6]).

2.5   Use of a Declarative Language to Describe a Partition

The creation of forward and backward graph fragments before partitioning a raw RDF file contributes to integrate a logical dimension to the purely physical partitioning process. The forward and backward fragments are described by its characteristic sets. A declarative language could be then used to describe the creation and partitioning processes for both types of segments. The declarations are done similarly to the declarations of tables with the DDL (data definition language) in SQL.

Declaration. Graph fragments are declared as follows:

```
CREATE {FORWARD,BACKWARD,BOTH}_FRAGMENTS FROM [raw rdf file path]
      WITH [max_size]
```

The CREATE command is taken from the SQL DDL. The instruction {FORWARD, BACKWARD}_FRAGMENTS indicate the creation of only forward or backward fragments respectively. To create both types of fragments the instruction: BOTH _FRAGMENTS must be used. The [raw rdf file path] is the path to the raw file to be loaded into the processing system. Finally the parameter WITH [max_size] states the maximum allowed size for a graph fragment. At the end of both instructions, the forward and backward graph fragments should be created and a list of the fragments, its predicates and statistics should be available to the user. This file creates an id for each fragment used mainly to repartition. At the end of the instruction the forward and backward graph fragments that need to be repartitioned, because their size is bigger than the max_size parameter, should be also indicated.

Repartition. To repartition a graph fragment, the following syntax is used:

```
CREATE {FORWARD,BACKWARD}_FRAGMENT [id] PARTITION BY ([predicate function])
```

This instruction overwrites a graph fragment (identified with id) and creates the partitions according to the predicate's function. The predicate's function model is inspired as well from the DDL of the relational model:

```
(PARTITION [partition_name_1] [predicate] {LESS,MORE} THAN [VALUE],...,
 PARTITION [partition_name_m] [predicate] {LESS,MORE} THAN [VALUE]. )
```

The following expression declares the repartition of the forward graph fragment Airplane (we assumed that its ID is 1) used as example int the previous section:

```
CREATE FORWARD_FRAGMENT 1
PARTITION BY (PARTITION part1 has_length LESS THAN   (73),
             PARTITION part2 has_length VALUES LESS THAN <(maxvalue))
```

The maxvalue is a reserved word indicating the maximum value of the object with the predicate has_length.

Allocation. To allocate the graph fragments to the processing sites, the declaration is done with the following syntax:

```
ALLOCATE {FORWARD,BACKWARD,BOTH}_FRAGMENTS BY [partition function]
         WITH [nb_partitions]
```

As it was done at the creation stage of fragments, we can allocate individually forward and backward edges with the instruction ALLOCATE {FORWARD,BACKWARD} _FRAGMENTS. To allocate both segments the keyword BOTH_FRAGMENTS should be used. The partitioning strategy to be used by the system is indicated in the [partition function] instruction. The available partitioning strategies are round-robin and METIS.

The nb_partitions parameter indicates the number of sites in which the data should be distributed. If one segment must be moved to another partition by implementation constraints, the declaration is:

```
ALLOCATE {FORWARD,BACKWARD}_FRAGMENT [id] TO SITE ([site id])
```

Integration of the Language to Other Systems. In this section we show that the proposed language is able to express the partitioning strategies used by most of distributed RDF systems. For example, HadoopRDF [7] applying a hashing function on the subject is mimicked creating forward star graphs fragments that are allocated using a hashing function. If the system uses vertical partitions (e.g. SW-Store [8]), backward graph fragments are then declared. The declaration of fragments with maximum partition size of 1000 is for example:

```
CREATE {FORWARD,BACKWARD}_FRAGMENTS FROM [raw_rdf];
ALLOCATE FORWARD_FRAGMENTS BY HASH WITH 1000
```

The gStoreD [9] system creates the forward entities and leaves to the user the freedom to choose the allocation strategy. The declaration of fragments allocated with the METIS heuristic is for instance:

```
CREATE FORWARD_FRAGMENTS FROM [raw_rdf];
ALLOCATE FORWARD_FRAGMENTS BY {HASH,METIS} WITH 1000
```

The prototype system QDAG [10] uses both, forward and backward segments, and allocate them with a graph partitioning heuristic declared as:

```
CREATE BOTH_FRAGMENTS FROM [raw_rdf];
ALLOCATE FORWARD_FRAGMENTS BY METIS WITH 1000
```

In the following section we give the formal definitions for the graph stars, graph fragments and partitioning algorithms applied throughout the example developed on this section.

3   Formal Definitions

In this section we define the forward and backward graph fragments for an input RDF dataset that we represent as G. We define as well the allocation problem when the forward and backward fragments are considered as allocation units. Finally, we show the formal mapping of the allocation problem to a graph partitioning problem.

3.1 Preliminaries

Definition 1. *Triple*. *In our work, data are described using triples of the form* $t = (s, p, o)$, *where s is the subject, p is the predicate and o is the object.*

The functions $f_s(t) \rightarrow s$ and $f_o(t) \rightarrow o$ return the subject and object of an RDF triple respectively. Both functions are applied in some of the definitions of this section.

3.2 Forward Graph Fragments Construction

Definition 2. *Forward graph star*. *A forward graph star denoted as $\overrightarrow{\mathcal{G}s}$ is a subset of G in which the triples share the same subject. Formally $\overrightarrow{\mathcal{G}s} \subseteq G$, such that $\overrightarrow{\mathcal{G}s} = \{t | \forall_{i \neq j}(f_s(t_i) = f_s(t_j))\}$.*

Theorem 1. *A triple belongs to one and only one forward graph star. Let t be a triple $t \in G$, if $t \in \overrightarrow{\mathcal{G}s}_i \Rightarrow \forall_{i \neq j}(t \notin \overrightarrow{\mathcal{G}s}_j)$.*

Definition 3. *Characteristic set*. *The characteristic set is the set of distinct predicates in the forward or backward graph fragment. The functions $p(\overrightarrow{\mathcal{G}s}_i), p(\overleftarrow{\mathcal{G}s}_i)$ return the fragment's characteristic set.*

The similarity function $Sim(p(\overrightarrow{\mathcal{G}s}_i), p(\overrightarrow{\mathcal{G}s}_j))$ returns the similarity between two characteristic sets. This function could be, for example, a Jaccard similarity between both sets. More details are found in [4].

Definition 4. *Forward graph fragment*. *A forward graph fragment is a set of forward graph stars $\overrightarrow{\mathcal{G}s}$ that have the similar characteristic sets according to a threshold τ). A forward graph star $\overrightarrow{\mathcal{G}s}_i$ belongs to one and only one forward graph fragment $\overrightarrow{\mathcal{G}f}_i$. Formally $\overrightarrow{\mathcal{G}f} = \{\overrightarrow{\mathcal{G}s} | \forall_{i \neq j}(Sim(p(\overrightarrow{\mathcal{G}s}_i), p(\overrightarrow{\mathcal{G}s}_j)) \geq \tau)\}$.*

Theorem 2. *The set $\overrightarrow{\mathcal{G}f} = \{\overrightarrow{\mathcal{G}f}_1, ..., \overrightarrow{\mathcal{G}f}_l\}$ of all forward graph fragments for the graph G is a correct[2] partition set of the graph G.*

3.3 Backward Graph Fragments Construction

Definition 5. *Backward graph star*. *A backward graph star denoted as $\overleftarrow{\mathcal{G}s}$ is a subset of the original RDF graph G in which the triples share the same object. Formally $\overleftarrow{\mathcal{G}s} \subseteq G$ such that $\overleftarrow{\mathcal{G}s} = \{t | \forall_{i \neq j}(f_o(t_i) = f_o(t_j))\}$.*

Theorem 3. *A triple belongs to one and only one backward graph star. Let t be a triple $t \in G$, if $t \in \overleftarrow{\mathcal{G}s}_i \Rightarrow \forall_{i \neq j}(t \notin \overleftarrow{\mathcal{G}s}_j)$.*

Definition 6. *Backward graph fragment*. *A backward segment $\overleftarrow{\mathcal{G}f}$ is a set of backward graph stars $\overleftarrow{\mathcal{G}s}$ with similar characteristic sets according to a threshold τ. Formally $\overleftarrow{\mathcal{G}f} = \{\overleftarrow{\mathcal{G}s} | Sim(p(\overleftarrow{\mathcal{G}s}_i), p(\overleftarrow{\mathcal{G}s}_j)) \geq \tau\}$.*

---

[2] According to the correctness fragmentation rules in [6].

Theorem 4. *The set $\overleftarrow{\mathcal{G}f} = \{\overleftarrow{\mathcal{G}f}_1, ..., \overleftarrow{\mathcal{G}f}_m\}$ of all backward segments for the graph G is a correct partition set of the graph G.*

The proofs and more details of the theorems 1 to 4 stated in this section are skipped due to page limit and can be found in our technical report [4]. We have shown that the sets of forward and backward segments $(\overrightarrow{\mathcal{G}f}, \overleftarrow{\mathcal{G}f}_l\})$ induce *correct* partitions (i.e. complete, disjoint and rebuildable) of the original RDF dataset G. The elements on each set correspond to *fragments* of G that are adopted as distribution units during the *allocation* step. The allocation problem is formalized in our technical report [4].

4   Experimental Evaluation

We evaluated the different parts of RDFPartSuite in three phases: pre-processing assessed in terms of execution time and the cardinality of forward and backward fragments, theoretical allocation evaluated in terms of a precision measurement defined in Sect. 4.3 and query performance.

4.1   Experimental Setup

Hardware: We conducted the partition experiments of Sect. 4.2 on a Dell Tower Precision 3620 running Windows 10. This computer features an Intel(R) Core(TM) i7-7700 CPU @ 3.60 GHz processor, 16 GB of main memory, 2 TB of hard disk and a solid state disk SSD of 250 GB.

Software: We used Scala 2.11.8 as programming language to gather the triples in forward and backward graph fragments.

Datasets: we tested on real and synthetic datasets (described in Table 2) of variable sizes. We utilize the popular LUBM and Watdiv benchmarks as synthetic datasets and a Yago2 and DBLP as real datasets.

Table 2. Datasets pre-processing M: millions, #S #O: number of distinct subjects and objects

Dataset	Triples (M)	#S (M)	#P	#O (M)	Size (GB)	Time (min)	$|\overrightarrow{\mathcal{G}f}|$	$|\overleftarrow{\mathcal{G}f}|$
WatDiv100M	108.99	5.21	86	9.76	15	70.46	39,855	1,181
Yago2	284.4	10.12	98	52.37	42	815.40	25,511	1,216
LUBM100M	100.07	16.27	17	12.10	17	55.14	11	13
DBLP	207.21	6.84	27	35.52	32	126.61	247	26

4.2   Pre-processing Evaluation

Our first experiment measures the time to build the forward and backward graph fragments for each dataset. The results are presented in Table 2. The loading times are comparable to the ones obtained by other systems with a lightweight partitioning strategy detailed in [11]. The forward and backward entities are obtained using $\tau = 1$ to show the maximum number of created segments. For all datasets the number of backward graph fragments is greater than the number of backward graph fragments due to the cardinality of the characteristic sets which is smaller in forward graph fragments.

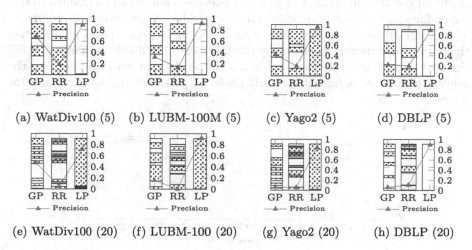

(a) WatDiv100 (5)   (b) LUBM-100M (5)   (c) Yago2 (5)   (d) DBLP (5)

(e) WatDiv100 (20)   (f) LUBM-100 (20)   (g) Yago2 (20)   (h) DBLP (20)

Fig. 5. Triple distribution by allocation method GP: graph partitioning, RR: rond-robin, LP: linear programming

4.3   Allocation Evaluation

First, we compare the round-robin, linear programming and min-cut allocation strategies in term of data skewness. In Fig. 5 we show the distribution of data when 5 and 20 partitions are created. We validate that the round-robin distribution strategy leads to data skew problems. The linear programming solution, which is the solution to the optimization problem of [4], was implemented used the Gurobi [12] optimization library. To obtain a solution with a reasonable time, we configured the tolerance $\epsilon = 500{,}000$. With this big tolerance, the majority of the data was allocated to one partition.

The precision of the partitioning strategy is calculated as: $\frac{w(e_c)}{w(e)}$, in which $w(e_c)$ and $w(e)$ are respectively the sum of the weights of the cut edges and the total number of edge's weights. The results are shown red in Fig. 5. As expected the higher precision is obtained by the linear program, however the scalability on this program is very low. The min-cut partitioning algorithm, used by systems like EAGRE [13], gets better results than the round-robin strategy, not only in terms of data distribution but also on the precision.

4.4   Query Performance

We evaluated the performance of queries when the data are organized as forward and backward graph fragments but not simultaneously. For this, we configured gStoreD [9] to store an adjacency list organized on subjects, and another organized on objects. gStoreD was deployed on a 5 machine cluster connected by a 10 Gbps Ethernet switch. The cluster runs a 64-bit Linux and each site has a 8 GB RAM, a processor Intel(R) Xeon(R) Gold 5118 CPU @ 2.30 GHz and 100 GB of hard disk. We performed the evaluation with the four query types (linear, star, snowflake and complex queries) of the benchmark for 10 million triples of the WatDiv dataset. The list of queries is not found mentioned in this paper for space reasons, but it is in the appendix of our technical report [4]. The results are shown in Fig. 6. The system has much better performance when the data is organized as backward graph fragments when solving linear queries. For snowflake shaped queries, there is not a strategy that works better than the other. The forward organization is still preferred when solving complex and star queries.

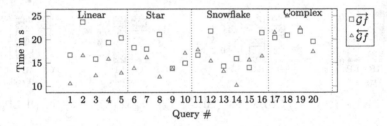

Fig. 6. Query performance of forward and backward graph fragments

5   Related Work

In this section we describe briefly the principal storage models used by RDF systems. Then we classify the RDF partitioning approaches and we cite the studies comparing the performance between partitioning strategies.

5.1   RDF Data Storage Models

The data storage model of several RDF systems is based on the relational model. The triples of these systems are stored on disk using one of the following strategies: (i) *Triple table*: a single table with three columns (subject, predicate, object) is created. The most popular system using this strategy is RDF-3X [14], (ii) *Property table:* storing the data in a single table whose dimensions are determined by the number of subjects and distinct predicates. This strategy is applied by the Jena2 [15] system. (iii) *Vertical partition:* this strategy creates a two-column table (subject, object) per predicate. The strategy is applied by SW-Store [8] and

HadoopRDF [7] for example, *(iv) More sophisticated storage models* are used by other systems, like G-Store [16], Trinity.RDF [17] and H2RDF [18] in which the data are stored using graph-representation models like adjacency lists.

5.2 Partitioning RDF Techniques

Distributed RDF systems are classified in two major groups: federated and clustered storage systems. In this section we describe the partitioning strategies for clustered storage systems that distribute the data among different data nodes being part of a single RDF storage solution. Federated systems are out of the scope of our study. We classified distributed RDF systems according to their distribution strategy. These strategies are: *(i) Hash:* the allocation of triples is performed according to a hash value computed on the subject, predicate or object of a triple modulo the number of computer nodes (e.g. Trinity.RDF [17], HadoopRDF [7]), *(ii) Hierarchical hash:* this strategy builds a path hierarchy based on the subject's IRIs (e.g. SHAPE [19]), *(iii) Minimal edge-cut:* it solves the partitioning problem as a k-way graph partitioning using heuristic packages like METIS [5] (e.g. EAGRE [13], WARP [20]), *(iv) Round robin:* this technique assigns the triples with a round-robin approach (e.g. [21]), and *(v) Other approaches:* these works are built on top of distributed platforms. In these systems, the storage back-end is in charge of distributing the triples (e.g. H2RDF [18]). Recently, the study [22] evaluated and compared the impact of certain data placement strategies in distributed RDF stores.

6 Conclusion

In this paper, we introduce a logical dimension to the partitioning process of RDF graphs. We formalize and give an overview of the algorithms used to create and allocate the logical entities that we named graph fragments $\mathcal{G}f$. We propose a declarative language to facilitate the separation of the distribution design from the execution layer hiding implementation details to the final user. The previous techniques are incorporated into the same framework named *RDFPartSuite*. Our experiments confirmed that creating logical fragments is not more costly than creating partitions with purely physical partitioning methods. We used synthetic and real datasets to show that graph fragments avert data skewness. Also, we showed the trade-offs between forward and backward graph fragments at query runtime. Our on-going works include the consideration of the query workload in the declaration of partitions and the consideration of dynamic datasets (updates).

References

1. Neumann, T., Moerkotte, G.: Characteristic sets: accurate cardinality estimation for RDF queries with multiple joins. In: Proceedings of the 27th ICDE, Hannover, Germany, 11–16 April, pp. 984–994 (2011)

2. Agrawal, S., Narasayya, V.R., Yang, B.: Integrating vertical and horizontal partitioning into automated physical database design. In: Proceedings of SIGMOD, Paris, France, 13–18 June, pp. 359–370 (2004)
3. Ramamurthy, R., DeWitt, D.J., Su, Q.: A case for fractured mirrors. In: Proceedings of 28th VLDB 2002, Hong Kong, China, 20–23 August, pp. 430–441 (2002)
4. Galicia, J., Mesmoudi, A., Bellatreche, L.: A logic dimension on RDF partitioning, Technical report (2019)
5. Karypis, G., Kumar, V.: A fast and high quality multilevel scheme for partitioning irregular graphs. SIAM J. Sci. Comput. **20**(1), 359–392 (1998)
6. Özsu, M.T., Valduriez, P.: Principles of Distributed Database Systems. Springer, New York (2011). https://doi.org/10.1007/978-1-4419-8834-8
7. Du, J.-H., Wang, H.-F., Ni, Y., Yu, Y.: HadoopRDF: a scalable semantic data analytical engine. In: Huang, D.-S., Ma, J., Jo, K.-H., Gromiha, M.M. (eds.) ICIC 2012. LNCS (LNAI), vol. 7390, pp. 633–641. Springer, Heidelberg (2012). https://doi.org/10.1007/978-3-642-31576-3_80
8. Abadi, D.J., Marcus, A., Madden, S., Hollenbach, K.: SW-store: a vertically partitioned DBMS for semantic web data management. VLDB J. **18**(2), 385–406 (2009)
9. Zeng, L., Zou, L.: Redesign of the gstore system. Front. Comput. Sci. **12**(4), 623–641 (2018)
10. Mesmoudi, A.: Declarative parallel query processing on large scale astronomical databases. Ph.D. thesis, Doctoral School in Computer Science and Mathematics, Lyon, France (2015)
11. Abdelaziz, I., Harbi, R., Khayyat, Z., Kalnis, P.: A survey and experimental comparison of distributed SPARQL engines for very large RDF data. PVLDB **10**(13), 2049–2060 (2017)
12. Gurobi Optimization, LLC: Gurobi optimizer reference manual (2018)
13. Zhang, X., Chen, L., Tong, Y., Wang, M.: EAGRE: towards scalable I/O efficient SPARQL query evaluation on the cloud. In: 29th IEEE ICDE Brisbane, Australia, 8–12 April, pp. 565–576 (2013)
14. Neumann, T., Weikum, G.: RDF-3X: a risc-style engine for RDF. PVLDB **1**(1), 647–659 (2008)
15. Wilkinson, K.: Jena property table implementation (2006)
16. Zou, L., Özsu, M.T., Chen, L., Shen, X., Huang, R., Zhao, D.: gStore: a graph-based SPARQL query engine. VLDB J. **23**(4), 565–590 (2014)
17. Zeng, K., Yang, J., Wang, H., Shao, B., Wang, Z.: A distributed graph engine for web scale RDF data. PVLDB **6**(4), 265–276 (2013)
18. Papailiou, N., Konstantinou, I., Tsoumakos, D., Karras, P., Koziris, N.: H2RDF+: high-performance distributed joins over large-scale RDF graphs. In: Proceedings of the 2013 IEEE International Conference on Big Data, Santa Clara, CA, USA, 6–9 October, pp. 255–263 (2013)
19. Lee, K., Liu, L.: Scaling queries over big RDF graphs with semantic hash partitioning. PVLDB **6**(14), 1894–1905 (2013)
20. Hose, K., Schenkel, R.: WARP: workload-aware replication and partitioning for RDF. In: Workshops Proceedings of the 29th IEEE ICDE 2013, Brisbane, Australia, 8–12 April, pp. 1–6 (2013)
21. Saleem, M., Khan, Y., Hasnain, A., Ermilov, I., Ngomo, A.N.: A fine-grained evaluation of SPARQL endpoint federation systems. Semant. Web **7**(5), 493–518 (2016)
22. Janke, D., Staab, S., Thimm, M.: Impact analysis of data placement strategies on query efforts in distributed RDF stores. J. Web Semant. **50**, 21–48 (2018)

Mining Quantitative Temporal Dependencies Between Interval-Based Streams

Amine El Ouassouli[1,2](✉), Lionel Robinault[2], and Vasile-Marian Scuturici[1]

[1] Univ Lyon, INSA Lyon, LIRIS (UMR 5205 CNRS), 69621 Villeurbanne, France
{amine.el-ouassouli,marian.scuturici}@insa-lyon.fr
[2] Foxstream, 69120 Vaulx-en-Velin, France
l.robinault@foxstream.fr

Abstract. Data streams gathered from sensor systems can contain a significant amount of noise and are challenging to sequential pattern mining algorithms. A majority of existing approaches deals with such data as time point events to find *before/after* relations that induces loss of information when dealing with events lasting in time, i.e intervals. Other interval-based approaches focus on qualitative patterns and are sensitive to temporal variability. We consider that quantitative patterns maintaining lag information permit to deal better with this problem. In this work, we propose an efficient algorithm devised to extract quantitative patterns from interval streams: Interval Time Lag Discovery (ITLD). It is based on an intersection-based confidence that is automatically assessed with a statistical χ^2 test. Experimental results, on both synthetic and real-life data, show that our method is more suitable for sensor interval streams and provides more precise information in comparison with existing approaches.

1 Introduction

Extensive usage of sensors makes it possible to obtain a precise monitoring of temporal activities. In many application domains, the data produced by the monitoring raises the need to extract patterns describing hidden temporal phenomena or behaviors. As a motivation example, we describe in Figs. 1 and 2 a sensor system composed of 4 outdoor cameras situated in an office area. These cameras are capturing motion using real time video processing. Starting from images taken by these cameras, we defined 10 "virtual" motion sensors (displayed with red polygons) corresponding to physical regions and labeled as 1-1, 1-2, ..., 4-3. Each virtual motion sensor produces a sequence of time point events of two types: *"B: Motion Begin"* and *"E: Motion End"*. We can associate these events with a change in the state of the data source. Such a sequence can be transformed to an interval-based state sequence composed of intervals $[t_b, t_e)$ signaling the activation of an environment state *"Motion in area x"*, noted Mx for concision. For example, Fig. 3 describes an initial sequence of raw time point events produced by *4-1* and *4-3* and Fig. 4 shows the corresponding interval sequences.

© Springer Nature Switzerland AG 2019
C. Ordonez et al. (Eds.): DaWaK 2019, LNCS 11708, pp. 151–165, 2019.
https://doi.org/10.1007/978-3-030-27520-4_11

Fig. 1. Four outdoor camera views. Red contours describe motion analysis areas. (Color figure online)

Fig. 2. Position of motion analysis areas (aerial view)

We consider that an interval based representation provides more insight about the studied phenomenon. Our objective is to identify temporal correlations between these state streams in order to describe quantitatively typical trajectories in this environment. For example, the temporal relation "M4-1 → M4-3$^{(3,2)}$" means that the activation of the state *M4-3* appears 3 seconds after the activation of *M4-1*, and the state is deactivated 2 seconds after the end of *M4-1*. In Fig. 2 this relation is not always true, and we evaluate the confidence in this relation at 0.7.

Sequential and temporal pattern mining is a well studied problem. Most of the existing algorithms (e.g [8,10]) focus on point-based data (events without duration) and are designed to extract qualitative patterns describing *before/after* relations. In the case of interval-based data (events with duration), using such approaches implies loss of information as they do not take into account event duration. As discussed in [1], richer relations can be expressed using interval-based models. Most of existing quantitative algorithms utilizes pattern models based on Allen' logic. This makes them sensitive to variability of data as slight changes in interval endpoints can lead to different qualitative relations. More expressive quantitative patterns, including time lag information, permit to deal better with this problem. Among the existing approaches, the majority [4,9,12] do not allow to process straightforwardly interval streams. Input data for these algorithms is assumed to be a transaction database of sequences identified by an ID: each transaction corresponds to a time-stamped activity.

Our problem is different: sensor data are unbounded sequences with no separation between activities. Two existing algorithms, PIVOTMiner [5] and TEDDY [11] permit to process interval streams. PIVOTMiner is a support-based

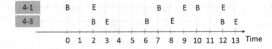

Fig. 3. Raw event streams produced by sensors *4-1* and *4-3* in Fig. 2.

Fig. 4. State (interval) streams built from Fig. 3 raw data

algorithm while TEDDY considers a novel confidence measure based on intervals intersection length. In the case of sensor data the usage of the intersection between intervals allows a better evaluation of certain temporal relations. Given the example described in Fig. 2, if several pedestrians cross successively *4-1* then *4-3*, a single long interval for each analysis zone will be generated: a support-based assessment will count 1 occurrence while intersection will be impacted according to the intersection length. To validate the detected temporal relations, PIVOTMiner uses a given minimum support as a significance criterion and TEDDY considers automatic statistical assessment of intersection length.

This work focuses on the problem of finding pairwise relations describing significant time lags between interval-based state streams. More precisely, we extend the intersection-based pattern model introduced in [11] and propose an efficient and precise time lag discovery algorithm, named Interval Time Lag Discovery (ITLD). It is based on the statistical assessment of the confidence variation. We also report on experiments on both synthetic and real life data showing that ITLD is capable of discovering efficiently and precisely significant time lags between pair of state streams in comparison with the existing approaches.

The rest of this paper is organized as follows: Sect. 2 surveys temporal pattern mining approaches; Sect. 3 introduces necessary definitions; Sect. 4 introduces the ITLD algorithm; Sect. 5 reports on experiments on ILTD and provides comparison with existing algorithms and finally Sect. 6 concludes this paper and discusses some perspectives.

2 Related Work

Sequential and temporal pattern mining is a well studied research area designed to discover regularities among temporally ordered data. Contributions in this field can be categorized following three main criteria: data format, time models and temporal description.

The pattern mining problem differs following the input data format. On the one hand, this task consists of finding regularities between transactions within a database of temporally ordered sequences. Each sequence describes an activity/behavior associated with an unique ID referring to a subject (e.g market basket data or medical records). On the other hand, pattern mining is designed to discover frequent relations within a unique sequence/stream of items without separation between activities. Our work belongs to this category. For both cases, existing contributions use a user-given minimum support referring to the part of data where a relation stands. The support assessment can be based on number of transaction for sequence databases, time windows [8] or number of items [13].

Most of the contributions dealing with streaming data or single sequences focus on time points with qualitative patterns reporting on *before/after/co-occurring* relations (e.g [8]). Quantitative patterns do not permit a time delay discrimination which is rather useful for many application domains. While some contributions integrate temporal constraints (time windows [8], gap constraint [2]), several approaches tackled quantifying time delays [3,7,13,14]. Another category of contributions deals with interval data. As discussed in [1] this time model permits to express more complex relations (13 relations in Allen' logic). This form of patterns may suffer from expressiveness limitations as ambiguity (a pattern may lead to different relations) or completeness (not allowing to express all possible relations). We refer to [6] for a detailed analysis of expressiveness. Besides, qualitative patterns are sensitive to temporal variability as slight variation in intervals endpoints that may lead to different qualitative relations.

Quantitative patterns permit to deal with these problems as they include temporal information permitting to infer Allen' logic. Among the existing contributions [4,9,12] only two recent approaches are designed or can be easily adapted to interval streams [5,11]. In [5], authors propose PIVOTMiner that uses a geometric approach consisting on the projection of each interval $[t_b, t_e)$ into a bi-dimensional plane (begin, end) where the temporal relation between two intervals is considered as a vector. This allows to perform a DBSCAN clustering after an origin transformation stage: to mine $A \rightarrow B$, all vector with A as source and B as target are moved such as all sources coincides with the space origin. Cluster centroides provides the time lag information. While designed for databases of sequences, this approach can be easily adapted to data streams as it is not endpoint sensitive (as in [4,12]). Finally, in [11] the authors proposed a novel form of temporal relations made possible with usage of intervals. It is based on intersection of streams. In that work, a data stream is formatted as several sub-streams containing each a unique label type. The main idea is to use a χ^2 test of independence to compare the observed intersection length with the expected one under the independence hypothesis. The latter, uses the assumption of uniform distribution of B's length. The authors propose an algorithm, TEDDY, devised to discover such dependencies given a temporal constraint Δ on temporal transformation and by extension time lags. It explores the search space as a semi-lattice defined by transformation inclusion (i.e if an (α, β)-transformation sequence contains all intervals of (α', β')-transformation then (α, β) includes (α', β')) using a breadth-first strategy. As the used confidence measure is monotonic, multiple temporal dependencies can express the same temporal relation. Therefore, TEDDY integrates pruning criteria based on a dominance relationship permitting to select the most specific dependencies by refining temporal transformation while controlling the loss in the intersection-based confidence measure.

3   Definitions and Problem Statement

Data Streams. We define a data stream D as a sequence of timestamped data produced by a source d: $D = \{(t, v) \mid t \in \mathcal{T}, v \in \mathcal{V}\}$. $\mathcal{T}$ is an infinite set of discrete timestamps and $\mathcal{V}$ is the set of possible values given by D. We assume that a data source cannot produce more than one value at a time. In the experiment described higher, $\mathcal{V}$ is the set of possible event types $\mathcal{V} = \{B, E\}$. In Fig. 3, the data stream corresponding to data source 4-1 is $\{(B, 0), (E, 2), (B, 7)...(E, 12)\}$

States and State Streams. A state corresponds to a data configuration of interest for the application domain. It is defined as a predicate on data produced by a data source. For example, the state *Motion in area 4-1*, noted $M4$-1 in Fig. 4, is defined with $M4\text{-}1(t) = (Last(4\text{-}1, t) == B)$ denoting that environment is at state *Motion in 4-1* at t if the previous value from data source 4-1 at the timestamp t is *Motion begin*. A state stream A related to state a corresponds to intervals of time where the predicate $a(t)$ is valid. It is formally defined with $A = (a, \langle [t_{b_i}, t_{e_i}) \rangle)$ where a is a state label and the active intervals $\langle [t_{b_i}, t_{e_i}) \rangle$ a temporally ordered sequence of intervals such that $\forall i, t_{b_i}, t_{e_i} \in \mathcal{T}, t_{b_i} < t_{e_i} < t_{b_{i+1}}$ and $\forall t \in [t_{b_i}, t_{e_i})$, $a(t)$ is verified. The length of A, $\mathbf{len}(A)$, is the sum of its active interval lengths. The size of A, noted $\#A$, corresponds to its intervals number. In Fig. 5 $\mathbf{len}(A) = 2 + 2 + 2 = 6$ and $\#A = 3$.

Operations on State Streams. State streams support several useful operations. The intersection of two state streams A and B, noted $A \cap B$ is a state stream containing intervals where both A and B are active (Fig. 5). Formally, $A \cap B = (a \wedge b, \langle [t_{b_i}, t_{e_i}) \rangle$ such that $\forall t \in [t_{b_i}, t_{e_i}), \exists [t_{b_j}, t_{e_j}) \in A, \exists [t_{b_k}, t_{e_k}) \in B$ such that $t \in [t_{b_j}, t_{e_j})$ and $t \in [t_{b_k}, t_{e_k})$. The algorithm computing the intersection has a temporal complexity of $\Theta(Max(\#A, \#B))$. We define a temporal transformation (shifting) on a state stream B as an (α, β)-transformation. This operation produces the state stream $B^{(\alpha,\beta)} = \langle [t_{b_i} - \alpha, t_{e_i} - \beta) \mid [t_{b_i}, t_{e_i}) \in B \rangle$. Hereafter, α is called expansion and β reduction. $B^{(2,1)}$ (Fig. 5). The temporal complexity of an algorithm computing a temporal transformation is $\Theta(\#B)$.

Temporal Dependencies. A temporal dependency between two state streams $A \to B$ notifies that A *occurs simultaneously with* B. We call hereafter A the premise and B the conclusion. It is assessed via the length of $A \cap B$, and the corresponding confidence measure [11] is: $\mathbf{conf}(A \to B) = \dfrac{\mathbf{len}(A \cap B)}{\mathbf{len}(A)}$ It is to notice that the confidence is equal to 1 if all active intervals of A are included in the active intervals of B, i.e. A occurs always simultaneously with B. For example, in Fig. 5 $\mathbf{conf}(A \to B) = 3/6$. To avoid the utilization of user-given thresholds, this confidence measure is assessed with a Pearson χ^2 test of independence. The independence hypothesis states that A and B are statistically independent and

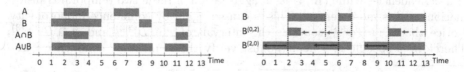

Fig. 5. Examples of intersection (left) and (α, β)-transformation (right) for state streams A and B

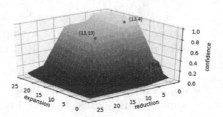

Fig. 6. Search space for temporal dependencies between 2 state streams

relies on the assumption that if the length of B is uniformly distributed in $\mathcal{T}$, there is no significant correlation between A and B. The given validity threshold is noted $th(len(A), len(B))$[1] and obtained for a significance level of 0.05. Simultaneity do not permit to express dependencies when it happens that B is time-delayed regarding A. To capture such relations, the statistical assessment is performed after applying (α, β)-transformations on the conclusion streams.

Problem Statement. Given a temporal constraint $\Delta = [min, max]$ on accepted time delays and a set of state streams $\mathcal{S}$, detect all pairwise statistically significant temporal dependencies $A \rightarrow B^{(\alpha,\beta)}$ with $A, B \in \mathcal{S}$ and $\alpha, \beta \in \Delta$. The search space for a pair of streams have a size of $|\Delta|^2$ and its brute force exploration is done in $\Theta(|\Delta|^2 * Max(\#A, \#B))$. Figure 6 describes the search space for a pair of state streams with two temporal dependencies and $\Delta = [0, 25]$.

Specific Dependencies Selection. Due to confidence monotonicity w.r.t expansions and reductions, numerous temporal transformations can lead to valid dependencies describing the same relation. For example in Fig. 7, while $B^{(2,2)}$ and $B^{(3,1)}$ include all intervals of A and obtain a maximal confidence for a temporal dependency of type $A \rightarrow B^{(\alpha,\beta)}$, we consider that $B^{(2,2)}$ reports better the phenomenon behind this relation. To reduce the redundancy problem, one must select temporal transformation of B such that intervals involved in a relations intersect the most tightly their corresponding intervals in A. These transformation leads to specific temporal dependencies. One property of such a transformation is that for any of B's intervals expansion by adding (resp. subtracting) a time unit does not lead to a significant confidence gain. The opposite is true for intervals reduction (Fig. 7). As for intersection validity, we evaluate confidence variation statistically using $th(len(A \cap B^{(\alpha,\beta)}), \#B)$. However, this definition of specificity is not sufficient when several temporal relations coexist. In Fig. 6 we describe 3 transformations of interest. $(13, 13)$ and $(4, 4)$ reports on real dependencies while $(13, 4)$ is an aggregation of these two temporal transformations and is more general. This is caused again by confidence monotonicity. Notice that $A \cap B^{(13,4)}$ contains the intervals of $A \cap B^{(4,4)}$ and $A \cap B^{(13,13)}$. Therefore, a specific dependency must verify the elementary confidence varia-

---

[1] $th(lp, lc) = \dfrac{lp * lc + \sqrt{\dfrac{3.84}{T} lp * lc(T_{obs} - lc)(T_{obs} - lp)}}{T_{obs} * lp}.$

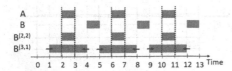
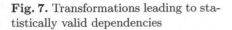

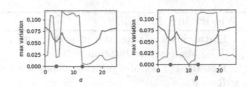

Fig. 7. Transformations leading to statistically valid dependencies

Fig. 8. Maximal gains (left) and losses (right) for Fig. 6 and corresponding thresholds (blue) (Color figure online)

tion conditions (significant losses, insignificant gains) and not be included in a more general dependency.

Maximal Confidence Elementary Variation. Specific expansion and reduction values can be extracted from different temporal relations. For example, in Fig. 6, the specific expansion $\alpha = 13$ can be found for $(13, \beta)$ with $\beta \leq 20$ and specific reduction $\beta = 4$ for $(\alpha, 4)$ with $\alpha \geq 2$. We also noticed that confidence gain for (α, β) is upper-bounded by the gain obtained for (α, α). The same property stands for all elementary confidence variations: shifts (i.e (α, α)-transformations) cannot induce overlaps and removal of intervals. As a consequence, all original transitions between active and inactive stream's portions are preserved. This is not guaranteed with other transformation. Figure 8 shows maximal variations corresponding to the example described in Fig. 6. Using partially the confidence variation conditions, specific values of expansions and reduction can be directly detected (red points) from maximal variations. Specific dependencies can be built using temporal order $((4,4),(13,13))$ and statistically assessed via the validity test. This heuristic permits to reduce the exploration in quadratic time of the lag search space to linear time w.r.t Δ.

4   Interval Time Lag Discovery (ITLD)

This section introduces the ITLD algorithm. Given a temporal constraint $\Delta = [min, max]$, it uses the maximal confidence elementary variation property to discover specific dependencies from a set of state streams. For each couple of state streams A and B, ITLD (lines 7 to 13) computes and stores maximal confidence elementary variations and thresholds for temporal transformations (α, α). Since elementary transformations (i.e adding/removing a unit to transformation values) do not cause interval overlaps, we infer the following equality:

$$\mathbf{conf}(A \rightarrow B^{(\alpha+1,\alpha)}) - \mathbf{conf}(A \rightarrow B^{(\alpha,\alpha)})$$
$$= \mathbf{conf}(A \rightarrow B^{(\alpha+1,\alpha+1)}) - \mathbf{conf}(A \rightarrow B^{(\alpha,\alpha+1)})$$

This relation states that adding a unit to expansion for the transformation (α, α) induce a confidence gain equal to the loss observed with subtracting a unit from expansion for $(\alpha + 1, \alpha + 1)$. This permit us to use the value of expansion gains

Algorithm 1. ITLD

Data: S: set of state streams, $\Delta = [min, max]$
T_{obs}: observation duration
Result: R : set of significant dependencies

```
 1  R ← ∅
 2  for p in S do
 3  │   for c in S do
 4  │   │   if p.label != c.label then
 5  │   │   │   G, L, TH ← ⟨⟩
 6  │   │   │   reference = Conf(p, c, (min, min))
 7  │   │   │   for α = min → max do
 8  │   │   │   │   gain ← Conf(p, c,(α + 1, α)) - reference
 9  │   │   │   │   loss ← reference - Conf(p,c,(α, α + 1))
10  │   │   │   │   Add gain to G
11  │   │   │   │   Add loss to L
12  │   │   │   │   Add th(len(p) * reference, #c) to TH
13  │   │   │   └   reference → reference + gain - loss
14  │   │   │   sigExp, sigRed ← GetSigVal(G, L, TH)
15  │   │   └   R ← R ∪ GetSpecDep(sigExp, sigRed)
16  return R
```

(resp. reduction losses) to assess losses (resp. reduction gains) and vice versa. Therefore, iteration on α (lines 7–13) computes only $gain$ with $(\alpha + 1, \alpha)$ and $loss$ with $(\alpha, \alpha + 1)$. For the same considerations, the $(\alpha + 1, \alpha + 1)$ confidence is obtained directly with: $\mathbf{conf}(A \to B^{(\alpha+1,\alpha+1)}) = \mathbf{conf}(A \to B^{(\alpha,\alpha)}) + gain - loss$. Thus, the reference confidence needs one computation (line 6) for $\alpha = min$ and is updated on line 13 for the rest of the values. The gain/loss statistical threshold $th(len(p \cap c^{(\alpha,\alpha)}), \#c)$ is computed in line 12. $len(p \cap c^{(\alpha,\alpha)})$ is calculated with $\mathbf{len}(p) * \mathbf{conf}(p \to c^{(\alpha,\alpha)})$ as $reference = \mathbf{conf}(p \to c^{(\alpha,\alpha)})$.

The second step of ITLD is the extraction of significant expansion and reduction values using the procedure *GetSpecVal* (line 14). This operation, done in $\Theta(|\Delta|)$, consists in verifying for each value of gain (G) and loss (L) the specificity conditions w.r.t their corresponding thresholds:

- For each element $i \in G$: $G[i] < TH[i]$ and $G[i-1] > TH[i-1]$
- For each element $i \in L$: $L[i] < TH[i]$ and $L[i-1] > TH[i-1]$

If the gain (res. loss) value at i is specific, $min + i$ is added to $sigExp$ (res. $sigRed$). The first value in L and last in G are tested with a relaxed version of the higher conditions to capture dependencies with potential specific values that are out of Δ:

- $L[0] > TH[0]$ means that the specific loss $l \leq min \Rightarrow min$ is added to $sigExp$
- $G[len(G)-1] > TH[0]$ means that the specific gain $l \geq max \Rightarrow max$ is added to $sigExp$

Procedure 2. GetSpecDep(sigExp, sigRed)

Data: sigExp, sigRed: temporally ordered specific values of expansions and reductions
Result: specDep : set of specific dependencies
1 specDep ← {}
2 **if** $|sigExp| == |sigRed|$ **then**
3 | $\quad i \leftarrow 0$
4 | **while** $i < |sigExp|$ **do**
5 | $\quad$ | $(\alpha, \beta) \leftarrow (sigExp[i], sigRed[i])$
6 | $\quad$ | **if** $Conf(A \rightarrow B(\alpha, \beta)) > th(len(A), len(B(\alpha, \beta))$ **then**
7 | $\quad$ | $\quad$ ⌊ **Add** $A \rightarrow B(\alpha, \beta)$ **to** $specDep$
8 | $\quad$ ⌊ $i++$
9 **else**
10 | $\quad (\alpha, \beta) \leftarrow (sigExp[0], sigRed[|sigRed| - 1])$
11 | **if** $Conf(A \rightarrow B(\alpha, \beta) > th(len(A, len(B(\alpha, \beta))$ **then**
12 | $\quad$ ⌊ **Add** $A \rightarrow B(\alpha, \beta)$ **to** $specDep$
13 **return** $specDep$

For example, in Fig. 8a with $\alpha \in \Delta = [0, 10]$, the real specific value 14 is out of bounds of Δ but statistically significant confidence variations are detected: max is considered as a specific gain to maintain this information.

The last step $GetSpecDep$ (Procedure 2) builds specific dependencies candidates using temporal order (first specific expansion with first specific reduction etc.) and tests if they are statistically valid (line 2 to 8). If an unequal number of specific expansion and reductions is found, the most general dependency (with the greatest expansion and lowest reduction, line 10 to 12) is considered and tested permitting to avoid losing temporal information.

5 Experiments

In the section, we describe experiments on both simulated and real world data. We also compare qualitatively and quantitatively ITLD with the existing similar approaches TEDDY [11] and PIVOTMiner [5] that we adapted to state streams. Algorithms were implemented in Python[2] (with DBSCAN from scikit-learn for PIVOTMiner) and tested on a Core i7 2.1 GHz with 8 GB memory running Windows 10.

5.1 Results on Synthetic Data

We developed a motion simulation tool[4] in order to obtain data sets using custom scenarios (trajectories, activity density, speed ...). This controlled testbed generates data sets with a corresponding ground truth permitting to evaluate the accuracy of discovered temporal dependencies. In the following, we use the F1-Score,

---

[2] https://github.com/AElOuassouli/Quantitaive-Interval-Stream-Mining.

Table 1. 11 simulated data sets with increasing occurrences number (Occ). $\#Int$: average number of intervals per stream, Den: average density % of T_{obs}

Occ	#Int	Den	Occ	#Int	Den	Occ	#Int	Den	Occ	#Int	Den
100	95	1.7	2000	1275	30.3	5000	1616	58.6	8000	1386	74.6
500	452	8.7	3000	1509	41.5	6000	1570	64.7	9000	1263	78.5
1000	798	16.4	4000	1606	50.5	7000	1493	69.9			

i.e the harmonic mean of precision and recall. Several matching condition types are used. Given two dependencies $D_1 = A_1 \rightarrow B_1^{(\alpha_1,\beta_1)}$ and $D_2 = A_2 \rightarrow B_2^{(\alpha_2,\beta_2)}$:

- Qualitative matching: D_1 and D_2 are equivalent if $A_1 = A_2$ and $B_1 = B_2$.
- Exact matching: D_1 and D_2 are equivalent if they match qualitatively and $(\alpha_1, \beta_1) = (\alpha_2, \beta_2)$
- Relaxed θ Matching: for $\theta \geq 0$, D_1 and D_2 are equivalent if they match qualitatively and $|\alpha_1 - \alpha_2| \leq \theta$ and $|\beta_1 - \beta_2| \leq \theta$.

In order to keep this comparison fair, we included the Δ constraint to PIVOT-Miner by simply keeping interval vectors with time gaps in Δ for the clustering step. In the case of a pair of interval streams with several temporal dependencies (e.g Fig. 6), the results given by TEDDY corresponds to the general temporal relation ($(13, 4)$ in Fig. 6) that matches qualitatively but not with respect to the temporal information. Moreover, as TEDDY considers temporal transformation (α, β) with $\alpha \geq \beta$, it outputs multiple dependencies if $\alpha < \beta$. For example, for $A \rightarrow B^{(9,10)}$, two results with $(9,9)$ and $(10,10)$ are given. In this case, we consider one dependency as a true positive and others as false positives.

Robustness to Density. Activity within a sensed environment can be more or less intense. In order to be efficient, a pattern mining approach must be able to give precise results for both sparse and dense data. We defined a linear trajectory with equidistant sensors and ran 11 simulations varying the number of occurrences for the same duration T_{obs} and object speed (cf. Table 1). The intervals number increases with the number of event occurrences for sparse data (100 to 5000 occurrences) and decreases for high density event occurrences due to intervals overlap. The streams active length always increases when the event occurrences increases and have a density ($\frac{len(stream)}{T_{obs}}$) ranging from 1.7% to 78.5%.

We report in Fig. 10 results for ITLD, TEDDY and PIVOTMiner ($\epsilon = 1$, $min\text{-}sup = 2/3$) with $\Delta = [0, 14]$, $T_{obs} = 10000$. Figure 10a shows that ITLD runs faster in comparison with PIVOTMiner and TEDDY. TEDDY benefits from its early pruning criteria for non-significant dependencies but requires more processing time for significant dependencies. PIVOTMiner suffers from its quadratic build of delta vectors linking each interval with all others. Therefore, its running time increases dramatically with intervals number. Notice that PIVOTMiner can be improved significantly using a sliding window technique.

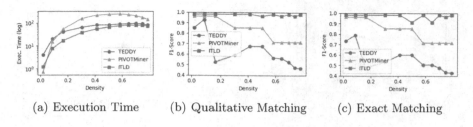

(a) Execution Time   (b) Qualitative Matching   (c) Exact Matching

Fig. 9. Execution time and F1-Scores w.r.t to density of streams

F1-Scores reported in Fig. 9b and c shows that ITLD is robust to density even for exact matching, in comparison with PIVOTMiner and TEDDY. The latter is able to detect significant dependencies (high recalls) but outputs a significant amount of false positives impacting precision. One observation is that these false positives often have large temporal transformations causing a significant increase in intervals lengths and, by extension, intersection. Large temporal transformations deforms, in some extent, the original temporal information and can lead to false statistically valid dependencies. This problem is tackled by ITLD as it considers confidence variations for non-deformed intervals. With the specified parameters, PIVOTMiner behaves well for sparse data and gives similar results to ITLD but lower recalls are obtained for dense data. This is explained by the fact that overlaps caused by great densities causes a fall in dependencies counts and then requires to lower the minimum support which may cause a pattern flooding problem. We conclude that the usage of intersection is more robust to variations of densities: both TEDDY and ITLD obtain good recalls. For a fair comparison, we executed PIVOTMiner varying ϵ and minimum support. The results are reported in Fig. 10. Even for the best ϵ value ($\epsilon = 2$), PIVOTMiner is sensitive to density. Low minimum supports causes pattern flooding (10%, 20%) and the best qualitative results are given for 30%. These results shows that using automatic statistical tests permits to adapt the significance threshold to activity intensity, contrary to user-given significance thresholds approaches requiring continuous parameter tuning.

Influence of $|\Delta|$. The empirical study of the influence of Δ was done on a data set generated using a linear trajectory with 10 equidistant steps. Its average density is 16% (sparse data) and contains 1000 occurrences in a $T_{obs} = 10000$. The

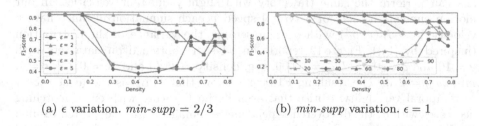

(a) ϵ variation. $min\text{-}supp = 2/3$ (b) $min\text{-}supp$ variation. $\epsilon = 1$

Fig. 10. F1 scores (qualitative matching) for PIVOTMiner

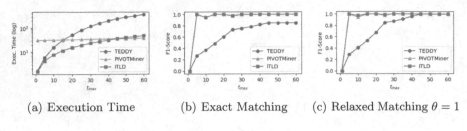

(a) Execution Time   (b) Exact Matching   (c) Relaxed Matching $\theta = 1$

Fig. 11. Execution time and F1-Scores w.r.t to t_{max} ($t_{min} = 0$)

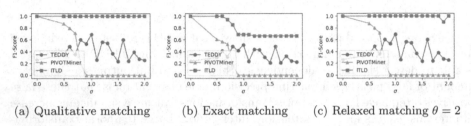

(a) Qualitative matching   (b) Exact matching   (c) Relaxed matching $\theta = 2$

Fig. 12. F1 scores over σ

three algorithms were executed with different $\Delta = [0, t_{max}]$, with $t_{max} \in [0, 60]$ (Fig. 11). ITLD runs faster than TEDDY (up to 7 times for $t_{max} = 60$) thanks to its linear complexity w.r.t Δ. As expected, running time of PIVOTMiner is marginally affected by Δ as this constraint is simply used to reduce the number of vectors for the clustering step. As this data set is sparse and noiseless, both PIVOTMiner and ITLD obtain maximum F1-Scores for the reasons discussed higher. TEDDY obtains higher recalls and a growing precision with t_{max} for this data set. That is explained by the number of true positives that grows with t_{max} due to the simulation scenario. TEDDY is capable of finding true positives but fails somehow to detect true negatives. TEDDY is designed to extract correlations based on intersection length and do not integrate directly the succession information. ITLD benefits from the intersection model assessment and, in the same time, takes into account directly temporal successions as it uses confidence elementary variations that can be seen as the number of impacted intervals.

Robustness to Temporal Variability. In real world context, a same phenomenon can occur with slight temporal variation. For example, two pedestrians can perform the same trajectory with slightly different velocities. In our motion simulation tool, for a typical speed v each simulation instance have a velocity $V \sim \mathcal{N}(v, \sigma^2)$ with σ^2 representing the variance of the distribution (inspired by [14]). Figure 12 reports on F1-score obtained on data sets with $v = 10$ and $\sigma \in [0, 2]$ ($T_{obs} = 10000$, avg. density $= 25\%$) and $\Delta = [0, 15]$, $\epsilon = 1$, $min\text{-}supp = 2/3$. The resulting F1-Scores show that ITLD is the most robust to temporal variability. Notice that even if the F1-Score with exact matching decreases with σ, the discovered dependencies remains close to expected results: the F1-Score is almost maximal using the relaxed matching ($\theta = 2$). We noticed

Table 2. Streams obtained from the experiment described in Fig. 2. #Int: number of intervals, Len: active length (s)

Occ	#Int	Den	Occ	#Int	Den	Occ	#Int	Den	Occ	#Int	Den
100	95	1.7	2000	1275	30.3	5000	1616	58.6	8000	1386	74.6
500	452	8.7	3000	1509	41.5	6000	1570	64.7	9000	1263	78.5
1000	798	16.4	4000	1606	50.5	7000	1493	69.9			

that greater values of ϵ provides better results for PIVOTMiner: as the speeds follows a normal distribution centered in v, increasing the maximal distance between interval points captures more results. The same thing can be noticed with lowering the minimum support: there are less occurrences with the exact value of v. In real world scenarios, state streams often describe several temporal phenomena with different temporal variability and density characteristics: choosing a optimal parameter for one phenomenon may not fit for the others.

5.2   Results on Real-World Data

We experimented our algorithm using real-world sensor data gathered from the motion sensor system described in Fig. 2. This data set (Table 2) describes motion activity in the office area during 18 working days between 6 am and 8 pm. First observations showed that the data set contains a significant amount of noise (e.g detection of shadows, sudden luminosity changes) and several omissions were observed (e.g when a car passes through an analysis zone with a great speed). To assess the quality of the results, a qualitative "approximate" ground truth was built based on our ground observations and the manual analysis of the resulting streams. We basically eliminated impossible dependencies w.r.t to the time lag restriction and added observed "unusual" behaviors (e.g activation of the state *4-1* then *4-2* and *3-2* is forbidden but is actually often observed).

We ran ITLD, TEDDY and PIVOTMiner (*min-support* = 0.1 and ϵ = 1) with $\Delta = [0, 40]$. Table 3 describes the obtained results based on a qualitative matching. ITLD completed the discovery process faster in comparison with the other algorithms (2.5 time faster than TEDDY and 10 times faster than PIVOTMiner). The values of precision, recall and F1-scores permits to validate our observation on synthetic data. Indeed, TEDDY obtain a good recall but suffers from the amount of false positives impacting its precision. In the other hand, PIVOTMiner is precise but have a weak recall with these parameters. ITLD is characterized by results with a good recall/precision performance and without using user-given parameters (excepting Δ). The analysis of the errors shows that their majority are false positives due to low statistical thresholds and to the large observation duration in comparison with streams length (average density of 2%). Figure 13 describes several results given by ITLD and their corresponding expansion confidence variation. For example, $101 \rightarrow 301^{(35,13)}$ is to be interpreted as follows: *The half (confidence of 0.51) of motion in 301 begin typically 35 seconds after the beginning of motion in 101 and ends typically 13 seconds after the end*

Table 3. Results on the FOX data set for $\Delta = [0, 40]$ (Qualitative matching)

	#Res	Precision	Recall	F1-Score	Accuracy	Run. time (s)
PIVOTMiner	15	**1**	0.25	0.43	0.59	1704
TEDDY	94	0.54	**0.98**	0.69	0.51	433
ITLD	64	**0.81**	**0.92**	**0.86**	**0.83**	**170**

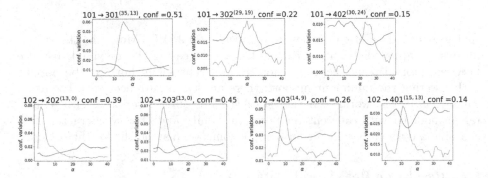

Fig. 13. Expansion confidence variation (in orange) and statistical threshold (blue) for results given by ITLD (Color figure online)

of motion in 101. One can observe that the 'characteristic form' of significant dependencies is well detected by the dynamic statistical threshold. Each dependencies row in Fig. 13 describe a way of leaving the office area that can be used to describe a typical trajectory using temporal information. One can notice that the temporal information given by temporal transformations remain consistent with the environment spatial configuration (cf. Fig. 2).

6   Conclusion and Future Work

Data streams gathered from sensor systems are challenging to temporal knowledge mining approaches. The current work investigates the problem of mining time delayed relations between sensor interval-based state streams. While most of the existing approaches focus on the problem of finding frequent patterns, our work is devised to identify quantitative delayed dependencies between interval streams on the basis of the automatic statistical test of their intersection length as introduced by [11]. We extended this dependency model by proposing a new significance definition based on the analysis of confidences elementary variations. In addition to reducing the exploration of a quadratic search space to a linear operation, this permits to better capture the time delayed transitions between streams while benefiting from the intersection assessment. Second, we designed the ITLD algorithms that uses the later property. Experiments with synthetic and real world data showed that ITLD is more robust to density of streams and temporal variability while not requiring a user-given significance parameter.

Moreover, our real-life experiment shows that it is possible to include data given by video analysis approaches in a temporal knowledge discovery process.

Our work can be extended in various directions. First, our approach may permit to define a non-ambiguous delayed relations. Indeed, the confidence variation characteristic form can be used to obtain typical intervals duration in addition to boundaries time lags. Second, pairwise dependencies can be used to build larger "patterns" involving multiple state streams.

References

1. Allen, J.F., Hayes, P.J.: A common-sense theory of time. In: IJCAI, pp. 528–531. Morgan Kaufmann (1985)
2. Casas-Garriga, G.: Discovering unbounded episodes in sequential data. In: Lavrač, N., Gamberger, D., Todorovski, L., Blockeel, H. (eds.) PKDD 2003. LNCS (LNAI), vol. 2838, pp. 83–94. Springer, Heidelberg (2003). https://doi.org/10.1007/978-3-540-39804-2_10
3. Dauxais, Y., Guyet, T., Gross-Amblard, D., Happe, A.: Discriminant chronicles mining - application to care pathways analytics. In: ten Teije, A., Popow, C., Holmes, J.H., Sacchi, L. (eds.) AIME 2017. LNCS (LNAI), vol. 10259, pp. 234–244. Springer, Cham (2017). https://doi.org/10.1007/978-3-319-59758-4_26
4. Guyet, T., Quiniou, R.: Extracting temporal patterns from interval-based sequences. In: IJCAI, pp. 1306–1311. IJCAI/AAAI (2011)
5. Hassani, M., Lu, Y., Wischnewsky, J., Seidl, T.: A geometric approach for mining sequential patterns in interval-based data streams. In: FUZZ-IEEE, pp. 2128–2135. IEEE (2016)
6. Höppner, F., Peter, S.: Temporal interval pattern languages to characterize time flow. Wiley Interdiscip. Rev. Data Min. Knowl. Discov. 4(3), 196–212 (2014)
7. Li, T., Ma, S.: Mining temporal patterns without predefined time windows. In: ICDM, pp. 451–454. IEEE Computer Society (2004)
8. Mannila, H., Toivonen, H., Verkamo, A.I.: Discovery of frequent episodes in event sequences. Data Min. Knowl. Discov. 1(3), 259–289 (1997)
9. Nakagaito, F., Ozaki, T., Ohkawa, T.: Discovery of quantitative sequential patterns from event sequences. In: ICDM Workshops, pp. 31–36. IEEE Computer Society (2009)
10. Pei, J., et al.: PrefixSpan: mining sequential patterns by prefix-projected growth. In: ICDE, pp. 215–224. IEEE Computer Society (2001)
11. Plantevit, M., Robardet, C., Scuturici, V.: Graph dependency construction based on interval-event dependencies detection in data streams. Intell. Data Anal. 20(2), 223–256 (2016)
12. Ruan, G., Zhang, H., Plale, B.: Parallel and quantitative sequential pattern mining for large-scale interval-based temporal data. In: BigData, pp. 32–39. IEEE Computer Society (2014)
13. Tang, L., Li, T., Shwartz, L.: Discovering lag intervals for temporal dependencies. In: KDD, pp. 633–641. ACM (2012)
14. Wang, W., Zeng, C., Li, T.: Discovering multiple time lags of temporal dependencies from fluctuating events. In: Cai, Y., Ishikawa, Y., Xu, J. (eds.) APWeb-WAIM 2018. LNCS, vol. 10988, pp. 121–137. Springer, Cham (2018). https://doi.org/10.1007/978-3-319-96893-3_10

Democratization of OLAP DSMS

Carlos Garcia-Alvarado$^{(\boxtimes)}$ (iD), Joy Kent, Li Liu, and Jay Hum

Autonomic LLC, Palo Alto, CA 94301, USA
{carlos,joy,li,jay}@autonomic.ai

Abstract. The expansion of IoT devices and monitoring needs, powered by the capabilities and accessibility of Cloud Computing, has led
to an explosion of streaming data and exposed the need for every organization to exploit it. This paper reviews the evolution of Data Stream
Management Systems (DSMS) and the convergence into Online Analytical Processing (OLAP) DSMS. The discussion is focused on three current solutions: Scuba, Apache Druid, and Apache Pinot in use in large
production environments that satisfy the real-time OLAP on streaming
data. Finally, a discussion is presented on a potential evolution of OLAP
DSMS and open problems.

Keywords: Streaming data · OLAP · DSMS

1 Introduction

Telecommunications, networking, and stock trading have been the prime targets
for stream processing. The improved capability of a large number of devices to
inexpensively generate and transmit data, in conjunction with the "hardware"
accessibility that Cloud Computing provides, results in a golden opportunity for
all organizations in all fields to understand and take action on their data. For
example, software companies that provide cloud solutions as part of a Software as
a Service (SaaS) strategy are required to build monitoring infrastructure that will
allow them to identify immediate problems and act on them (this also provided
the genesis of "Site Reliability Engineers"). The popularity of streaming data
processing problems and Data Stream Management Systems (DSMS) is starting
to permeate into all types of organizations at an accelerated pace.

Stream data has always been challenging to Online Analytical Processing
(OLAP) applications, given the need to perform several passes over the data
on a potentially large number of combinations for all dimensions [12]. As such,
the evolution of Data Stream Management Systems has been slow compared to
other types of data management systems that deal with "bounded data sets".
This is why it is important to bring to the OLAP community's attention the
need to provide robust, high-performance, scalable, and versatile OLAP stream
processing systems that can satisfy the needs of today's industry.

The focus of this paper is to provide a perspective on the type of problems
and triggers that lead to the development of popular and accessible data stream

ⓒ Springer Nature Switzerland AG 2019
C. Ordonez et al. (Eds.): DaWaK 2019, LNCS 11708, pp. 166–175, 2019.
https://doi.org/10.1007/978-3-030-27520-4_12

management systems that are available to everyone (democratization of DSMS), and not as an experimental evaluation on how those systems perform. Furthermore, this paper explores current systems that perform OLAP aggregations on streaming data at scale and discusses a vision of how those systems could evolve. The structure of this paper is as follows: first, the evolution of data stream management systems is explored, second the focus is on those systems that work for OLAP. Finally, we discuss the potential evolution of the architecture of current and future OLAP DSMS and discuss current open challenges.

2 Evolution of DSMS

The need to perform computations on data streams, especially driven by telecommunications and networking, gave origin to the first Data Stream Management Systems (DSMS), in which several academic prototypes [13,17] and in-house industry DSMS such as AT&T's Gigascope [11] allowed for aggregations in data streams. In the early academic prototypes such as Aurora [2], a directed acyclic graph representing streams and processors in a pull-based model was used to perform to continuous windowed aggregations. An extension of this system resulted in Borealis [3] as a distributed processing engine in a push-based model. Similarly prototypes in Telegraph-CQ [9], a Postgres-based DSMS push and pull hybrid model, and STREAM [5], a push-based model stream processing engine that handles an execution plan via a set of operators, message queues, and synopses as state stores. The result of this early research was captured and summarized in several papers, such as [6,14,18].

In 2007, two major events transformed the capability to produce and compute data in real time. In order of importance, first Apple released the first iPhone and then Amazon opened EC2 to the public. These events revolutionized the need and capabilities of companies to generate data (e.g. 'tweets'), analyze data, and transfer data wirelessly. In Fig. 1, the increased activity in the open source and industry communities is shown: In 2010 the development of Apache Storm begins to process real-time on social media. In 2011, Apache Flink, Apache Druid, and Facebook's Scuba started deploying systems that dealt with large amounts of streaming data. Flink is a generic streaming processing framework [8], Druid is a market monitoring system, and Scuba is an application and ads monitoring system. In 2012, Spark extended its RDD-based solution to provide a general purpose micro-batching approach for streaming data. The following years increased the development of streaming frameworks and a burst of new solutions appeared: Apache Apex, Linkedin-sponsored Apache Samza, Amazon's Kinesis, Apache Gearpump, Apache Ignite, Oracle Stream Explorer, Google's Cloud Dataflow, Google-sponsored Apache Beam, Linkedin-sponsored Apache Pinot, Linkedin and Confluent-sponsored Apache Kafka Streams, and Microsoft's Stream Analytics among many others. All these solutions provided incremental architectural improvements in making the frameworks cloud-native by maturing fault tolerance, scalability, throughput, and usability.

The evolution and optimization of DSMS from the early academic prototypes to serious, versatile, scalable, and robust open source solutions that operate on

Fig. 1. Data stream systems timeline.

cloud-based environments led to the democratization of stream data processing. Despite all this progress, Online Analytical Processing on streaming data is still mostly performed after a batched loading operation in a data warehouse. Facebook's Scuba [4], Apache Druid [20], and more recently Apache Pinot [16] are next-generation data stream management systems that are intended to provide true 'real-time' OLAP on streaming data at scale.

3   OLAP DSMS

OLAP DSMS are stream processing frameworks specialized in satisfying OLAP queries. As such, the incoming data streams' attributes are selected as dimensions and measures for efficient pre-computations. Three frameworks were selected. These systems allow performing OLAP aggregations on large scale streaming systems by supporting the infrastructure of companies such as Facebook, Linkedin, Yahoo, Airbnb, eBay, Paypal, Slack, Walmart, Uber, Lyft, Netflix and a large number of startups as made public by such projects. It is important to clarify that Apache Clickhouse [19] is another OLAP DSMS that is open source and claims to support similar requirements as these three systems. This system is omitted due to the limited adoption outside of Yandex, however, it is important to clarify that its architecture for data storage and processing is similar to those of distributed data warehouses such as Vertica [1] or ParAccel [10].

3.1   Scuba

Scuba is a data management system developed at Facebook, which internally is used for real time analysis of code regression analysis and ads revenue monitoring [4]. In order to provide a scalable and high-performance solution, Scuba is a Thrift-based in-memory database. Scuba exposes the data as tables with columns and supports four different data types (integers, strings, and set of integers and strings) and variable length and dictionary encoding. The data is stored in a row-oriented storage layout since most of the operations in Scuba require all the columns to be computed.

Scuba's strengths are the flexibility it allows for schema evolution by creating and populating tables across nodes only when data is available (lacks a CREATE

TABLE statement) and the capability of providing a consistent schema view of all the tuples despite the fact that it allows multiple schemas to exist within a table (since the tuples come self-described via Thrift).

Scuba enforces a data age on each table that can be configured depending on the data source. During ingestion Scuba could process data based on a selected sampling rate. Hence, it accounts for such sampling when configured and provides corresponding adjusted query results. Following a similar sampling philosophy, Scuba runs queries focused more on performance and will return query results even if not all data was processed due to timeouts or straggler brokers. Given that performance is the main focus, joins are not supported. The simplicity of Scuba has allowed it to scale to millions of rows per event per second and support major monitoring systems.

3.2   Apache Druid

Apache Druid is a distributed OLAP DSMS used in production in several major companies originally developed by Metamarkets [20]. Its architecture's major contribution is the detachment of ingestion and query processing. A Druid cluster is composed of Real-time Nodes, Historical Nodes, Coordinator Nodes, Broker Nodes, and Zookeeper. Each of these can be scaled up separately to meet the particular requirements of a user.

Druid integrates natively with message buses (e.g. Kafka, AWS Kinesis). and the Real-time Nodes handle the data ingestion, summarization, and query processing on the data not transferred yet to "deep storage" (e.g. HDFS, S3). Summarized data is then packed in an immutable segment and sent to deep storage. Segments are allocated to the Historical Nodes based on the Coordinator's discretion. Historical Nodes load and serve the data from deep storage and act as a cache and processing worker for historical data. Brokers redirect query requests to Real-time or Historical Nodes and process partial results. Zookeeper keeps the nodes' states (especially the Historical Nodes). Segment metadata is stored separately (e.g. Derby, PostgreSQL, or MySQL). Zookeeper allows the Druid cluster to handle node recovery and fault tolerance seamlessly.

Segments are the principal storage unit for Druid and maintain the measurements, dimensions, metadata, and indexes of a set of rows. Segments can be queried before being published to deep storage, allowing the operation of real-time analytics during ingestion. The storage layout within the segment is column-based and provides individual column compression within the segment. This allows for policies and mappings on segments that could result in keeping a set of Historical Nodes for hot or cold data. Segments also store inverted indexes for their elements, allowing the use of search queries on Druid data.

Finally, Druid can process a query in parallel across the entire cluster and supports a join-like functionality by using lookups. Moreover, it replaces dimension values with new values from a key-value map, and it supports JSON-based queries via POST Requests made to the Broker Node. Alternatively, Druid also supports an extended SQL dialect powered by Apache Calcite [7].

3.3 Apache Pinot

Apache Pinot is a distributed OLAP DSMS currently used in production at Linkedin [16]. The main design premises of Apache Pinot prioritize fast real-time queries, scalability, low data ingestion latency, and fault tolerance on immutable append-only data. Pinot is divided into four major components: controllers, brokers, servers, and minions. Servers store the data and provide the first level scanning, filtering, and processing; controllers maintain the data mappings to a server and retention policies; brokers process incoming queries and host partial results; and minions execute tasks assigned by the controllers for maintenance operations. Data ingestion is performed detached from querying and is executed by posting an HTTP POST to the controller. Node management and fault tolerance are backboned by Apache Helix [15].

Pinot also re-uses the notion of a segment as the main storage unit and similar to Druid, it keeps the measurements, dimensions, metadata (e.g. max and mins), and indexes of a set of rows. As pointed out in [16], Apache Pinot supports ten of thousands of queries per second over terabytes of data in production. Similar to Scuba, Pinot does not support joins or nested queries but it satisfies requests via extended SQL and integrates with the Apache Calcite query optimizer.

3.4 Comparison

In Table 1, the comparison of industry-relevant features between the three systems described in the previous sections is presented. As can be observed, Apache Druid and Apache Pinot are the systems that more closely meet the needs of a true OLAP DSMS. The main advantage of using a system such as Scuba is the simplicity that allows it to scale to process a large number of events and the capability to handle schema evolution. In the rest of the features, Scuba falls short of providing a solution that will satisfy interactive querying. Apache Druid and Apache Pinot, on the other hand, are closer in their architecture and features which allow them to efficiently satisfy a myriad of OLAP queries. Druid's focus seems to have been on dealing with massive data sets, as well as building a richer feature set and ecosystem while Pinot has a higher focus on high throughput and failure tolerance. For example, Pinot exploits vectorized execution and contains additional segment metadata that allows predicate pushdown to select faster relevant segments. In addition, Pinot is a cloud-native system while Druid originated as a Hadoop-native system. Finally, Druid has been able to mature more as a production system given that it has been adopted and used in a large set of production systems and companies, while Pinot, at the time of this writing, has mostly been used within Linkedin. It is important to highlight that the Druid committer community is larger than Pinot's community.

Table 1. OLAP DSMS comparison.

Feature	Scuba	Apache druid	Apache pinot
Storage layout	Row-based	Column-based	Column-based
Column encoding	Few	One	Several
Throughput	Extremely high	High	Very high
Failure tolerant	Supported	Supported	Supported
Joins	Not supported	Joins via lookup	Not supported
Nested queries	Not supported	Not supported	Not supported
Concurrency control	None	MVCC	None
Indexes	None	Bitmap, Inverted	Bitmap, Inverted
Execution model	None	Tuple at a time	Tuple at a time, Vectorized
Spatial support	None	Supported	Not supported
Working set management	None	User-defined data tiers	None
Predicate pushdown	None	None to brokers	Supported to brokers
Partial result	Supported	Not supported	Supported
Optimizer	None	Apache calcite	Apache calcite
Schemaless support	Per row	Per segment	Per segment
Datatype support level	Low	Medium	High

4   Future of OLAP DSMS

Section 3 showed that Apache Druid and Apache Pinot are the closest systems to true OLAP Data Stream Management Systems. Despite this, as is shown in Table 1, there are several gaps that need to be covered to provide full support. Figure 2 represents an abstraction of the overall architecture of both DSMS, from which we will focus on three components: Real-Time Nodes, Working Set Nodes, and Execution Nodes. The scalability of these components is essential for OLAP DSMS, especially the requirement of these components to scale separately in order to guarantee availability and high-performance. In other words, it is important that the nodes that focus on the stream ingestion can scale differently than the nodes that are executing a query.

Real-time nodes that are required to scale to changing ratios in volume and frequency of stream data, similarly need to be able to scale on the number of requests that they have to satisfy on "real-time" data, potentially splitting "real-time" nodes even further into high throughput data ingestion nodes that dump data into a shared memory cache (e.g. MemCached) that can be queried directly by the Execution Nodes. During the ingestion, vertical partitioning of segments must also be considered since not all dimensions need to be carried to the historical nodes. Furthermore, Real-time query nodes also need to be able to scale independently in order to satisfy a variable number of requests driven by a myriad of applications and customer use cases. As such, it is also important that the query Coordinator is also able to scale and parallelize independently to satisfy the number of query requests.

Working Set nodes need to be able to scale independently based on the queried working-set. Unfortunately, this scalability also exposes the challenges related to cache management (which include cache population and invalidation) and join co-location that could impact performance. Furthermore, Working Set nodes should be able to execute aggressive filtering (common as part of predicate pushdown) as close as possible to the data to avoid sending rows to the Execution nodes that are not needed for the evaluation of the final result. Working Set should be performing vectorized query execution and index operations to reduce retrieving unnecessary data. Also, if columnar vertical partitioning is applied to segments, working set nodes need to be able to load missing segment partitions if the query results need them. Working Set nodes interact with the deep storage system that provides the support for disaster recovery and backups.

Execution nodes are required to be fully stateless, and ephemeral nodes dedicated exclusively to the compute portion of a query. Just like Real-time and Working Set nodes, Execution nodes should be able to scale independently based on the complexity of the customer queries, the data age (hot vs cold data), concurrency, as well as "real-time" vs "historical" types of queries. The stateless property of Execution nodes allows for fast creation of computing instances and the capability to grow and shrink as the number of queries and query complexity demands. Furthermore, the execution nodes should be able to scale dynamically based on workload and data distribution in order to handle straggler nodes and data skew.

Finally, there is a real possibility that offline data warehouse systems could evolve into a "micro-service"-like architecture that resembles the one discussed in this section, potentially leading to a system that can perform efficient computations on both: bounded and unbounded data. The rationale behind this is that the continuous ingestion of data and the relational algebra semantics on unbounded data represent a superset of the problems for performing analytics on bounded data sets.

5   Open Problems and Final Remarks

While Apache Druid and Apache Pinot are driving the architectural conversation in a scalable direction, there are major open problems that need to be solved in order to handle Petabytes of stream data that require further discussion and research, such as:

- Query language: Extended SQL dialects have been adopted by Apache Druid and Apache Pinot, while Scuba excludes them. In order to build a rich ecosystem, applications are in urgent need of a standard way to consume such data (despite being called for it in early research [18] and [14]). This is especially evident and results in bespoke implementations by visualization and analytic tools.
- Full Join operation: Most DSMS operating on a real production system avoid join operations due to the major semantics and inefficiencies that accompany such operations. Co-localization of data for joins when the partitioning

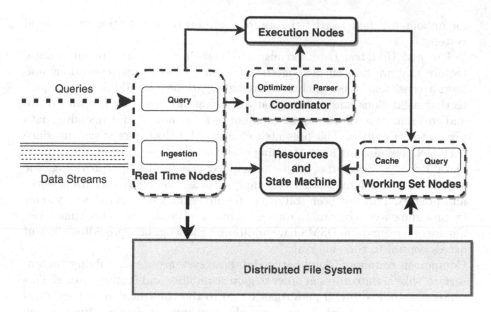

Fig. 2. DSMS for OLAP architecture.

schema allows for tables that could have an incompatible number of partitions is also a major challenge. For example, how to co-locate two stream tables if one has 3 segments and the other table has 2 segments? Despite the calls for action in [6], this issue remains open and has become more challenging at scale. As a result, several DSMS have opted to deprioritize it.

- Nested queries: Similar to the Join operation, nested queries' semantics are still incomplete, as well as the performance limitation of having to execute multiple passes over the data. Solving this issue at scale remains wide open despite being highlighted in [6].
- Nested data: A large amount of IoT and monitoring data is nested. However, current systems have difficulty dealing efficiently with nested data during storage (Druid and Pinot flatten it) and query processing.
- Full Schemaless OLAP DSMS: Scuba was the only system to attempt to tackle schema evolution as a first-class citizen, however, how to perform sub-second analytical queries on ever-changing schemas remains a topic for further exploration.
- Self-adaptive systems: while this problem is a heavy area of research across the database community, OLAP DSMS are particularly focused on auto-scaling given that the ratio and volume in which stream data arrive to real-time systems is constantly changing. Also, the system adaptability to data working set (hot vs cold data) is still an area of high relevance that remains open.
- Query optimization: There are plenty of operator semantics and query plan transformations that remain to be explored. For example, the properties

for unbounded data restrict the use of common transformation in bounded systems.

- Delete and Updates: Late arrivals, which are common in streaming data, require that we find efficient mechanisms to perform these operations and keep a consistent state (including pre-aggregations.) In addition, data protection regulations and policies that are being developed by several private and government agencies require capabilities for removing and amending data per customer request. This presents a challenge for the explored systems since they lack mechanisms for performing such tasks.
- Data Privacy: Enterprise solutions require capabilities for providing fine access control to data in order to comply with several regulatory frameworks and policies. This has been solved in "bounded data sets" database systems by providing access control to rows, columns, and cells. Due to the time series nature of stream data, DSMS have additional challenges for providing efficient access control to time intervals.
- Component coupling: Cloud-native database systems are resembling "microservice"-like architectures in order to gain scalability and fault tolerance. This presents a trade-off with performance due to the transmission and serialization costs that could penalize particularly short queries. As such, finding such balance remains an active area of research.

In conclusion, in this paper, we presented a historical overview of the evolution of the streaming processing system and how only a few of them are able to satisfy the requirements for an OLAP Data Stream Management System. While there are several open problems that remain to be solved, it is our responsibility as the OLAP community to start pushing the state of the art forward. The resolution of these problems may converge into a system that could satisfy real-time as well as interactive queries mostly focused on historical data.

Acknowledgements. We thank Jorge Orbay whose comments and feedback helped us improve this paper.

References

1. The Vertica Analytic Database: C-store 7 Years Later, vol. 5. VLDB Endowment Aug 2012
2. Abadi, D., et al.: A data stream management system. In: Proceedings ACM SIGMOD. pp. 666–666. ACM (2003)
3. Abadi, D.J., et al.: The design of the borealis stream processing engine. In: Proceedings of CIDR. pp. 277–289 (2005)
4. Abraham, L., et al.: Diving into data at facebook. vol. 6, pp. 1057–1067. VLDB Aug 2013
5. Arasu, A., et al.: Stream: the stanford stream data manager (demonstration description). In: Proceedings ACM SIGMOD. pp. 665–665. ACM (2003)
6. Babcock, B., Babu, S., Datar, M., Motwani, R., Widom, J.: Models and issues in data stream systems. In: Proceedings ACM PODS. pp. 1–16. ACM (2002)

7. Begoli, E., Camacho-Rodríguez, J., Hyde, J., Mior, M.J., Lemire, D.: Apache cal-cite: a foundational framework for optimized query processing over heterogeneous data sources. In: Proceedings ACM SIGMOD. pp. 221–230. ACM (2018)
8. Carbone, P., Katsifodimos, A., Ewen, S., Markl, V., Haridi, S., Tzoumas, K.: Apache Flink™: stream and batch processing in a single engine. vol. 38, pp. 28–38 (2015)
9. Chandrasekaran, S., et al.: TelegraphCQ: continuous dataflow processing. In: Proceedings ACM SIGMOD. pp. 668–668. ACM (2003)
10. Coffing, T., Bernier, E.: Actian Matrix (formerly ParAccel) architecture and SQL. Coffing Publ. (2015)
11. Cranor, C., Gao, Y., Johnson, T., Shkapenyuk, V., Spatscheck, O.: Gigascope: high performance network monitoring with an sql interface. In: Proceedings ACM SIGMOD. pp. 623–623. ACM (2002)
12. Garcia-Alvarado, C., Ordonez, C.: Clustering binary cube dimensions to compute relaxed group by aggregations. Inf. Syst. **53**(C), 41–59 (2015)
13. Golab, L., Johnson, T.: Data stream warehousing. In: Proceedings ACM SIGMOD. pp. 949–952. ACM (2013)
14. Golab, L., Özsu, M.T.: Issues in data stream management. SIGMOD Rec. **32**(2), 5–14 (2003)
15. Gopalakrishna, K., et al.: Untangling cluster management with helix. In: Proceedings SoCC. pp. 19:1–19:13. ACM (2012)
16. Im, J.F., et al.: Pinot: realtime OLAP for 530 million users. In: Proceedings ACM SIGMOD. pp. 583–594. ACM (2018)
17. Liu, X., Iftikhar, N., Xie, X.: Survey of real-time processing systems for big data. In: Proceedings ACM IDEAS. pp. 356–361. ACM (2014)
18. Stonebraker, M., Çetintemel, U., Zdonik, S.: The 8 requirements of real-time stream processing. vol. 34, pp. 42–47. ACM (2005)
19. Yandex: Clickhouse architecture, May 2019. https://clickhouse.yandex
20. Yang, F., Tschetter, E., Léauté, X., Ray, N., Merlino, G., Ganguli, D.: Druid: a real-time analytical data store. In: Proceedings ACM SIGMOD. pp. 157–168. ACM (2014)

Big Data Systems

Leveraging the Data Lake: Current State and Challenges

Corinna Giebler[1]([✉]) [iD], Christoph Gröger[2] [iD], Eva Hoos[2],
Holger Schwarz[1], and Bernhard Mitschang[1]

[1] University of Stuttgart, Universitätsstraße 38, 70569 Stuttgart, Germany
{corinna.giebler,holger.schwarz,
bernhard.mitschang}@ipvs.uni-stuttgart.de
[2] Robert Bosch GmbH, Borsigstraße 4, 70469 Stuttgart, Germany
{Christoph.Groeger,eva.hoos}@de.bosch.com

Abstract. The digital transformation leads to massive amounts of heterogeneous data challenging traditional data warehouse solutions in enterprises. In order to exploit these complex data for competitive advantages, the data lake recently emerged as a concept for more flexible and powerful data analytics. However, existing literature on data lakes is rather vague and incomplete, and the various realization approaches that have been proposed neither cover all aspects of data lakes nor do they provide a comprehensive design and realization strategy. Hence, enterprises face multiple challenges when building data lakes. To address these shortcomings, we investigate existing data lake literature and discuss various design and realization aspects for data lakes, such as governance or data models. Based on these insights, we identify challenges and research gaps concerning (1) data lake architecture, (2) data lake governance, and (3) a comprehensive strategy to realize data lakes. These challenges still need to be addressed to successfully leverage the data lake in practice.

Keywords: Data lakes · State of the art · Challenges · Industry case

1 Introduction

The digital transformation towards capturing and analyzing big data provides novel opportunities for enterprises to improve business and optimize processes [1]. Sensors from the Internet of Things (IoT), for example, enable the continuous gathering of production data, allowing the proactive assessment and the predictive regulation of production processes [1]. Many other novel data sources can be integrated and analyzed to generate new insights for the enterprise, using advanced analytics such as data mining, text analytics or artificial intelligence [2]. In the following, we summarize advanced analytics and traditional business intelligence as *data analytics*. The knowledge gained from data analytics represents a significant competitive advantage for enterprises [3].

Data captured for these data analytics tend to be heterogeneous, voluminous, and complex, and thus pose a challenge on traditional enterprise data analytics solutions based on data warehouses. In order to enable comprehensive and flexible data analytics

© Springer Nature Switzerland AG 2019
C. Ordonez et al. (Eds.): DaWaK 2019, LNCS 11708, pp. 179–188, 2019.
https://doi.org/10.1007/978-3-030-27520-4_13

on these complex data, the concept of the *data lake* emerged in recent years. In a data lake, any kind of data are available for flexible analytics without predefined use cases [4]. To this end, data are stored in a raw or almost raw format.

However, multiple challenges arise when building and using data lakes. Existing literature on data lakes is vague and inconsistent. Numerous approaches exist to realize selected aspects of a data lake, e.g., governance or data models, but it is unclear whether these approaches are sufficient and where additional concepts are needed.

In this paper, we address this gap. We investigate the current state of the art for data lakes and identify remaining research challenges towards a successful data lake. To this end, we make the following contributions:

- We investigate the current state of the general data lake concept.
- We discuss existing design and realization aspects.
- We identify challenges and research gaps for data lakes.

The remainder of this paper is organized as follows: Sect. 2 investigates existing data lake literature, while Sect. 3 discusses different design and realization aspects for data lakes and whether they are sufficiently covered by literature. Based on the gained insights, Sect. 4 identifies remaining challenges and research gaps. Finally, Sect. 5 concludes the paper.

2   Current State: Data Lakes in Literature

In order to put data lakes into practice, a uniform understanding of the general concept is needed. We conducted a comprehensive literature review to identify the central characteristics of a data lake. However, it showed that there is no commonly accepted concept. Instead, various definitions and views exist on data lakes, some of which are contradictory. In this section, we summarize our findings and investigate different points of view on the concept of data lakes.

The first person to use the term "data lake" was James Dixon in 2010 [5]. He defined a data lake to store data in a "natural" [5], i.e., *raw*, state compared to a traditional data mart. Large amounts of *heterogeneous data* are added from a single source [6] and users can access them for a *variety of analytical use cases*. The idea of single-source data lakes did not find much acceptance in literature. Nowadays, data lakes have been redefined to contain data from an arbitrary number of sources [7–9]. Dixon's central point of storing raw data is reflected by all investigated definitions. In some definitions, however, this characteristic is extended so that preprocessed data and results from previously performed analyses may additionally be stored in the data lake (e.g., in [10]).

To make storing large amounts of heterogeneous data financially feasible, data lakes must provide inexpensive storage [4, 8]. In many cases, data lakes are directly linked to *Hadoop*[1] and the HDFS [5, 7, 11]. However, the data lake represents a concept, while Hadoop is only one of many possible storage technologies [7, 11]. Dixon and other

---

[1] http://hadoop.apache.org/.

literature instead proposes a diverse tool landscape for data lakes, i.e., the most appropriate tool to manage and process certain data should be used [6, 12]. For example, MongoDB, Neo4J, or other data storage systems could also be used in data lakes [11]. These storage systems could be managed on premise or in the cloud [13].

Beyond data lakes being raw data repositories and the infrastructure they are built on, further characteristics of the concept depend on the exact definition, of which there are many. Often, definitions even contradict each other, especially when it comes to the *role of a data lake*, the *involved user groups*, and *governance and metadata*. In the literature, the *role of a data lake* ranges from pure central data storage [11] to the provider of services related to data management and analysis [14, 15]. For the *involved user groups*, some data lake definitions involve a wide variety of user groups (e.g., [8, 11]), while others name data scientists as the only users of a data lake (e.g., [7]). Regarding *governance and metadata*, some definitions claim a data lake comprises neither governance nor metadata (e.g., [8, 16]), while others state governance as a central aspect of data lakes (e.g., [14, 17]). Metadata and governance ensure that data are reliable and can be accessed and understood. A data lake without governance is said to risk transforming into a *data swamp* [18], where data cannot be used for value creation. Often, data lake concepts that include governance are also referred to as *data reservoir* [18, 19] to differentiate it from the ungoverned data lake. Additionally, data reservoirs are said to enforce more structure on data than a data lake. To achieve this structure, raw data are modeled using appropriate modeling techniques [19]. However, governance is mostly seen as part of a data lake and literature does not differentiate data lake governance from data reservoir governance. Furthermore, data modeling also plays a role in data lakes for data integration and facilitated use of data [15]. Therefore, we will not consider the differentiation between data lake and data reservoir here. Instead, we will use definitions that see governance as part of the data lake.

Some authors also use the term enterprise data lake in order to emphasize the data lake's application in an enterprise environment (e.g., [15, 20, 21]). However, this term is used synonymously to the general data lake term. None of the investigated sources differentiates the characteristics of an enterprise data lake from those of a general data lake. Thus, we will likewise not consider this differentiation in this work.

Overall, literature is split over the concrete characteristics of data lakes. There exists no uniform data lake concept and thus no comprehensive realization strategy. However, there is further literature focusing on particular aspects of data lakes. We examine these design and realization aspects in the following section.

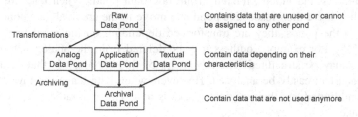

Fig. 1. In the pond architecture, data flow through the different ponds and are always available in only one of them [25].

3 Design and Realization Aspects for Data Lakes

Although the literature defining and describing data lakes contains little to no information on their practical realization, various approaches to realize selected aspects of the data lake do exist. Due to space restrictions, we focus on the realization aspects *data lake architecture, data lake modeling, metadata management,* and *data lake governance.* In the following subsections, we present literature related to these aspects and discuss whether the aspects are sufficiently covered for a practical data lake realization.

3.1 Data Lake Architecture

The data lake architecture describes how data are conceptually organized within a data lake. It facilitates the use of a data lake [15] by defining where data can be found in the condition (e.g., raw or pre-processed) needed for a particular use. There exist two variants for the overall architecture of data lakes, namely *zone* and *pond architectures.*

Various alternatives exist for *zone architectures* (e.g., [15, 17, 22–24]). They differ in multiple aspects, such as the number and characteristics of zones. Although there is no commonly accepted zone architecture, the idea always remains the same: Data are assigned to a zone according to the degree of processing that has been applied to them. When data are ingested into the data lake, they are stored in raw format in the raw zone, which is common between all zone architectures. Other zones then condition these data further. Some zones standardize data to fit a common format [15, 23], others cleanse the data [17]. Some zone architectures even include a data-mart-like zone, where data are prepared to fit certain use cases and tools [15]. The advantage of zone architectures is that even if data are available in a transformed and pre-processed format, they can still be accessed as raw data in the raw zone.

The *pond architecture* [25] is another variant of the overall architecture for data lakes (see Fig. 1). Data in the data lake are distributed across five different ponds. However, in contrast to zone architectures, data are only available in one pond at any given point in time. Upon ingestion, data are stored in the raw data pond. Only unused data and data that do not fit into any of the other ponds remain in the raw data pond, all other data flow into the analog, application, or textual data pond. Which pond they flow to depends on the data's characteristics. The analog data pond contains measurement data, such as log files or IoT data. In the application data pond, all data that are generated by applications are stored. The textual data pond contains text data. Other data, like images and videos, remain in the raw data pond. When data are not used anymore, they leave their respective pond and move to the archival data pond. As data flow through the ponds, they are transformed depending on the pond they currently belong to [25]. For example, outliers may be deleted from the analog data pond and textual data may be structured. The advantage of this approach is that data are pre-processed and can easily be analyzed. However, when data leave the raw data pond, they are conditioned and their original format is lost. This contradicts the general idea of a data lake.

In addition to those two general data lake architectures, literature suggests the *lambda architecture* [26] to organize batch and streaming data [11, 27]. The conceptual

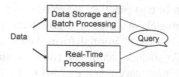

Fig. 2. The lambda architecture [26] enables separate batch and real-time processing.

idea of this architecture is depicted in Fig. 2. Incoming data are copied to two different branches. On one branch, data are stored permanently and periodically processed in batches. On the other branch, incoming data are processed in real-time to deliver quick results. However, in practice the lambda architectures often is adapted (e.g., in [12, 19]). Such adaptions are, e.g., the BRAID architecture [28] or Bolster [29].

While there exist various alternatives for data lake architectures, there is no generally accepted approach. Although zone architectures are more frequently mentioned in literature than the other variants, the definitions of the various zones differ greatly in some cases. To the best of our knowledge, there exist no assessments or comparisons of the different data lake architectures. Additionally, the data lake architectures proposed in literature only cover parts of the data lake. It is not defined how the data lake architectures interact with other aspects of the data lake. For example, it remains unclear how data lake modeling can be done, or what storage technologies can be used. Thus, defining and realizing an adequate data lake architecture is still a challenging task.

3.2 Data Lake Modeling

In the context of data lakes, literature speaks of schema-on-read [11, 15], i.e., data are only transformed when they are retrieved from the data lake for certain use cases. Transforming data requires knowledge of their schema, which in turn necessitates data modeling. However, deferring all modeling to data usage is infeasible [15] as data modeling is needed to ensure certain levels of data quality, data comprehensibility, and data integration [15, 30]. Thus, we need data models that allow data modeling with little effort and, at the same time, maintain the flexibility of the data lake.

There exist a few approaches to data lake modeling. For example, the *data droplets model* [31] models each data object, such as a single document, in the data lake as an RDF graph. These smaller graphs are then combined into an overarching data lake graph according to the relationships between the data objects.

Another approach is to model data in the data lake using *data vault* (see, e.g., in [19]). Data vault originates from the data warehouse context. It provides a flexible and simple way to model data. However, it was designed for structured data. While there exist approaches to integrate semi-structured data into data vault (e.g., [32]), integrating unstructured data is not yet covered.

Overall, literature explicitly mentions these two approaches and provides some ideas on how to model data in data lakes, but it does not offer further guidance. Multiple other data models exist in other contexts, such as 3rd normal form or head-version tables [33]. These data models might also be suitable candidates for data lake modeling but no assessments or best practices do exist so far. There also is no guidance on how to use the different models in a data lake architecture (Sect. 3.1). Additionally, many existing approaches such as 3rd normal form, head-version tables are available for structured data only. To include semi-structured or unstructured data in these models, further concepts are necessary (such as [34] or [35]). Thus, a comprehensive discussion and assessment of existing data models is still necessary for data lakes.

3.3   Metadata Management

Whenever data from different sources, contexts, and with various schemata are brought together, metadata is necessary to keep track of these data. This also applies to data lakes, where metadata management is a crucial part [36]. Metadata capture information on the actual data, e.g., schema information, semantics, or lineage [10, 14]. They ensure that data can be found, trusted, and used. There exists a large number of different approaches for metadata management. Due to space constraints, this section provides only a general overview over metadata management approaches explicitly for data lakes.

According to data lake literature, *data catalogs* [18] are used to store metadata. Whenever data is added to the data lake, the corresponding metadata has to be added to the catalog. Users are then able to search this catalog and receive additional information on the data, such as schema, relationships, or provenance [37]. However, not all metadata relevant for data lakes are covered by these catalogs [12].

Automatic extraction of metadata is an important topic in data lake environments, as vast amounts of data are ingested and stored. Tools like *GEMMS* [36] can be used that do not only extract, but also annotate the metadata with semantic information and allow querying these metadata. Some data lake concepts even provide an extensive metadata management system to store and query metadata [38]. To extract schematic metadata even from schema-free data sources, schema profiling can be used [39].

There exist various *metadata models* for data lakes (e.g., [36, 40, 41]). Some of them provide only little description and realization details [40], while others are designed for one specific application [41]. A model that is both generic and described in appropriate detail is proposed by Quix et al. [36]. It contains information on both the structure and semantical context of the data the metadata describe.

In addition to structure and semantics of data, metadata on the origin of data is just as important [15]. Lineage metadata describes where data came from, how they were produced, and how they had been processed. However, lineage metadata is not or insufficiently considered in all investigated metadata models for data lakes. Only one metadata model mentions provenance [41], but there is no further explanation on the storage or usage of provenance information.

While metadata management is crucial for data lakes, no comprehensive metadata management strategy covering all data lake metadata is available. However, metadata management is also represented in other contexts, e.g., data warehousing. Therefore, further investigation of approaches beyond the data lake literature is necessary.

3.4 Data Lake Governance

Metadata management is only a part of an overarching data lake governance. Governance comprises all kinds of policies and rules to ensure data quality and rule compliance [18]. Even though some early literature excludes governance from the data lake concept [8, 16], more recent literature views it as a very important aspect [7, 9, 10, 14]. In a data lake, governance has to compromise between control and flexibility [4]. Since many different kinds of data are managed in a data lake, data governance has to consider their differences. For example, master data managed in a data lake has a high need for governance, while IoT data typically needs less control and governance.

In literature, there exist only few approaches to data lake governance, especially for diverse data. There exists a general governance framework [18], giving some guidance on what needs to be considered in data lake governance. For example, the framework defines various roles involved in data lake governance, such as data stewards. In the various zone models (see Sect. 3.1), some governance principles are applied. For example, one zone model provides a sensitive zone, where sensitive data are encrypted [22]. Other zone models allow to encrypt or tokenize data in the raw zone [17].

However, to the best of our knowledge, none of the existing governance concepts considers the different kinds of data managed in a data lake and their governance requirements. Data lake governance is rudimentarily covered in literature, especially concerning sensitive data. Although there are governance approaches in other contexts, governance for data lakes must meet new requirements, such as compromising between control and flexibility. Thus, a governance concept specifically for data lakes is needed.

4 Challenges and Research Gaps

Even though multiple approaches exist that cover different aspects of data lakes (see Sect. 3), a comprehensive strategy to realize data lakes is missing. Additionally, it became clear that some aspects are only insufficiently covered by literature. We identified research gaps in three areas of data lakes:

Data Lake Architecture. For *data lake architecture*, the heterogeneity of concepts poses a major problem. There exist no assessments or discussions for the different alternatives. No generally accepted architecture is available, and some of the proposed architectures do not align with the data lake concept (e.g., ponds). Therefore, it is necessary to closely investigate and compare the existing alternatives to identify similarities and shortcomings. Additionally, data lake architectures cover only the conceptual organization of data. No data lake architecture exists that includes other data lake aspects, like data lake modeling or data lake infrastructure. To realize data lakes, a generalized and comprehensive data lake architecture is needed.

Data Lake Governance. The other aspect for which research gaps still remain is *data lake governance*. The data lake poses novel requirements on flexibility and open access that conflict with traditional governance approaches, for instance from data warehousing. Therefore, a comprehensive governance concept developed specifically for data lakes is required. This concept also has to consider the different kinds of data managed in the data lake, and correspond to their variant data management requirements.

Comprehensive strategy. In addition to these research gaps in the discussed data lake aspects, data lake literature is lacking a *comprehensive design and realization strategy*. Such a strategy considers interdependencies between different data lake aspects, such as data lake architecture and data lake modeling, and combines all aspects into one comprehensive and systematic data lake concept. Thus, it can provide guidance and decision support for the realization of data lakes.

The research gaps in these three areas need to be addressed to allow the definition of a holistic data lake concept and thus to leverage data lakes in practice.

5 Conclusion and Future Work

This paper summarizes our findings on the current state of data lakes. We conducted a comprehensive literature review on data lakes and existing approaches for their design and realization. It turned out that literature on data lakes is often split over the characteristics of a data lake. There exists no universal data lake concept. When it comes to realizing data lakes, research gaps concerning *data lake architecture* and *data lake governance* need to be resolved. Additionally, the lack of a *comprehensive design and realization strategy* that considers interdependencies between data lake aspects constitutes a major challenge to leverage data lakes in practice.

Our future work focuses on overcoming these challenges. Existing concepts for data lakes need to be investigated, categorized, and evaluated. Different data lake architectures have to be compared and evaluated with regard to their suitability to typical data lake use cases. Further approaches for e.g., semantic integration, data curation, or schema extraction have to be considered in a data lake context. Using all of these insights, a comprehensive design and realization strategy can be defined.

References

1. Lee, J., Kao, H.-A., Yang, S.: Service innovation and smart analytics for industry 4.0 and big data environment. In: Proceedings of the 6th CIRP Conference on Industrial Product-Service Systems (2014)
2. Russom, P.: Big data analytics. TDWI best Practices report, fourth Quarter (2011)
3. Margulies, J.C.: Data as Competitive Advantage. Winterberry Group (October) (2015)
4. Tyagi, P., Demirkan, H.: Data lakes: the biggest big data challenges. Analytics 9(6), 56–63 (2016)
5. Dixon, J.: Pentaho, Hadoop, and Data Lakes. https://jamesdixon.wordpress.com/2010/10/14/pentaho-hadoop-and-data-lakes/
6. Dixon, J.: Data Lakes Revisited. https://jamesdixon.wordpress.com/2014/09/25/data-lakes-revisited/
7. Madera, C., Laurent, A.: The next information architecture evolution: the data lake wave. In: Proceedings of the 8th International Conference on Management of Digital EcoSystems (MEDES) (2016)

8. Fang, H.: Managing data lakes in big data era: What's a data lake and why has it became popular in data management ecosystem. In: Proceedings of the 2015 IEEE International Conference on Cyber Technology in Automation, Control, and Intelligent Systems (CYBER) (2015)
9. O'Leary, D.E.: Embedding AI and crowdsourcing in the big data lake. IEEE Intell. Syst. 29(5), 70–73 (2014)
10. Terrizzano, I., Schwarz, P., Roth, M., Colino, J.E.: Data wrangling: the challenging journey from the wild to the lake. In: Proceedings of the 7th Biennial Conference on Innovative Data Systems Research (CIDR) (2015)
11. Mathis, C.: Data lakes. Datenbank-Spektrum. 17(3), 289–293 (2017)
12. Gröger, C., Hoos, E.: Ganzheitliches metadatenmanagement im data lake: anforderungen, IT-werkzeuge und herausforderungen in der Praxis. In: Proceedings der 18. Fachtagung Datenbanksysteme für Business, Technologie und Web (BTW) (2019)
13. Lock, M.: Maximizing your data lake with a cloud or hybrid approach. Aberdeen Group (2016)
14. IBM Analytics: The governed data lake approach. IBM (2016)
15. Madsen, M.: How to Build an enterprise data lake: important considerations before jumping in. Third Nature Inc. (2015)
16. Gartner Inc.: Gartner Says Beware of the Data Lake Fallacy (2014). https://www.gartner.com/newsroom/id/2809117
17. Patel, P., Wood, G., Diaz, A.: Data lake governance best practices. DZone Guide Big Data - Data Sci. Adv. Analytics 4, 6–7 (2017)
18. Chessell, M., Scheepers, F., Nguyen, N., van Kessel, R., van der Starre, R.: Governing and Managing Big Data for Analytics and Decision Makers. IBM, New York (2014)
19. Topchyan, A.R.: Enabling data driven projects for a modern enterprise. Proc. Inst. Syst. Program. RAS (ISP RAS 2016) 28(3), 209–230 (2016)
20. Stein, B., Morrison, A.: The enterprise data lake: better integration and deeper analytics. Technol. Forecast Rethinking Integr. 1, 1–9 (2014)
21. Farid, M., Roati, A., Ilyas, I.F., Hoffmann, H.-F., Reuters, T., Chu, X.: CLAMS: bringing quality to data lakes. In: Proceedings of the 2016 International Conference on Management of Data (SIGMOD) (2016)
22. Gorelik, A.: The Enterprise Big Data Lake. O'Reilly Media, Inc., Newton (2016)
23. Sharma, B.: Architecting Data Lakes - Data Management Architectures for Advanced Business Use Cases. O'Reilly Media, Inc., Newton (2018)
24. Zikopoulos, P., DeRoos, D., Bienko, C., Buglio, R., Andrews, M.: Big Data Beyond the Hype. McGraw-Hill Education, New York (2015)
25. Inmon, B.: Data Lake Architecture - Designing the Data Lake and avoiding the Garbage Dump. Technics Publications, New Jersey (2016)
26. Marz, N., Warren, J.: Big Data - Principles and Best Practices Of Scalable Real-Time Data Systems. Manning Publications Co., New York (2015)
27. Gröger, C.: Building an industry 4.0 analytics platform. Datenbank-Spektrum 18(1), 5–14 (2018)
28. Giebler, C., Stach, C., Schwarz, H., Mitschang, B.: BRAID - a hybrid processing architecture for big data. In: Proceedings of the 7th International Conference on Data Science, Technology and Applications (DATA) (2018)
29. Nadal, S., et al.: A software reference architecture for semantic-aware Big Data systems. Inf. Softw. Technol. 90, 75–92 (2017)
30. Stiglich, P.: Data modeling in the age of big data. Bus. Intell. J. 19(4), 17–22 (2014)
31. Houle, P.: Data Lakes, Data Ponds, and Data Droplets (2017). http://ontology2.com/the-book/data-lakes-ponds-and-droplets.html

32. Cernjeka, K., Jaksic, D., Jovanovic, V.: NoSQL document store translation to data vault based EDW. In: 2018 41st International Convention on Information and Communication Technology, Electronics and Microelectronics (MIPRO) (2018)
33. Schnider, D., Martino, A., Eschermann, M.: Comparison of Data Modeling Methods for a Core Data Warehouse. Trivadis, Basel (2014)
34. Gröger, C., Schwarz, H., Mitschang, B.: The deep data warehouse: link-based integration and enrichment of warehouse data and unstructured content. In: Proceedings of the 2014 IEEE 18th International Enterprise Distributed Object Computing Conference (EDOC) (2014)
35. Herrero, V., Abelló, A., Romero, O.: NOSQL design for analytical workloads: variability matters. In: Proceedings of the 35th International Conference on Conceptual Modeling (ER) (2016)
36. Quix, C., Hai, R., Vatov, I.: Metadata extraction and management in data lakes with GEMMS. Complex Syst. Inform. Model. Q. 9(9), 67–83 (2016)
37. Halevy, A., et al.: Managing Google's data lake: an overview of the goods system. IEEE Data Eng. Bullet. 39, 5–14 (2016)
38. Hai, R., Geisler, S., Quix, C.: Constance: an intelligent data lake system. In: Proceedings of the 2016 International Conference on Management of Data (SIGMOD) (2016)
39. Gallinucci, E., Golfarelli, M., Rizzi, S.: Schema profiling of document-oriented databases. Inf. Syst. 75, 13–25 (2018)
40. Walker, C., Alrehamy, H.: Personal data lake with data gravity pull. In: Proceedings of the 2015 IEEE Fifth International Conference on Big Data and Cloud Computing (BDCloud) IEEE (2015)
41. Nogueira, I., Romdhane, M., Darmont, J.: Modeling data lake metadata with a data vault. In: Proceedings of the 22nd International Database Engineering Applications Symposium (IDEAS) (2018)

SDWP: A New Data Placement Strategy for Distributed Big Data Warehouses in Hadoop

Yassine Ramdane[1(✉)], Nadia Kabachi[2(✉)], Omar Boussaid[1(✉)], and Fadila Bentayeb[1(✉)]

[1] University of Lyon, Lyon 2, ERIC EA 3083, 5, avenue Pierre Mendes, 69676 Bron-CEDEX, France
{Yassine.Ramdane,Omar.Boussaid,Fadila.Bentayeb}@univ-lyon2.fr
[2] University of Lyon, University Claude Bernard Lyon 1, ERIC EA 3083, 43, boulevard du 11 novembre 1918, 69100 Villeurbanne, France
Nadia.Kabachi@univ-lyon1.fr

Abstract. Horizontal partitioning techniques have been used for many purposes in big data processing, such as load balancing, skipping unnecessary data loads, and guiding the physical design of a data warehouse. In big data warehouses, the most expensive operation of an OLAP query is the star join, which requires many Spark stages. In this paper, we propose a new data placement strategy in the Apache Hadoop environment called "Smart Data Warehouse Placement (SDWP)", which allows performing star join operation in only one Spark stage. We investigate the problem of partitioning and load balancing in a cluster of homogeneous nodes. We take into account the characteristics of the cluster and the size of the data warehouse. With our approach, almost all operations of an OLAP query are executed in parallel during the first Spark stage, without a shuffle phase. Our experiments show that our proposed method enhances OLAP query performances in terms of execution time.

Keywords: Big data warehouse · Spark SQL · Load balancing · Parquet · Bucket · Sort-merge-bucket join

1 Introduction

Horizontal partitioning techniques are widely used to improve the performance of a data warehouse (DW) in a distributed system, such as load balancing and guiding the physical design of the DW [7,9,12,16]. We can distinguish two types of partitioning and load balancing techniques used in distributed big data warehouses (BDWs): static and dynamic. In static techniques, the system makes, in advance, a specific partition scheme (e.g., balancing split inputs in HDFS) of the DW for the nodes of the cluster, to speed up query processing. Static techniques are based on two models: a data-driven model [1,8], and a workload-driven model [3]. In dynamic techniques or on-line partitioning, the system elaborates

© Springer Nature Switzerland AG 2019
C. Ordonez et al. (Eds.): DaWaK 2019, LNCS 11708, pp. 189–205, 2019.
https://doi.org/10.1007/978-3-030-27520-4_14

the load balancing on the fly at the same time as the query processing [7,12,16]. In general this technique is based on a data-driven model.

Apache Hadoop uses partitioning and load balancing techniques to improve query performance. However, the random distribution of Hadoop blocks may slow down the query processing, especially with the OLAP query when joining several tables. An OLAP query is composed of several operations, such as filtering, projection, star joins, and grouping. Each operation can be performed in the map phase or in the reduce phase, and each one generates an execution cost, e.g. an I/O cost. The star join operation is the most expensive one, and often involves a high level of communication cost. In some cases, a star join operation will need up to $2(n-1)$ MapReduce cycles [14], where n is the number of tables involved in the query. To minimize MapReduce cycles and network traffic when performing a star join operation, some solutions have been proposed [2,6,14]. However, these solutions are not able to execute the star join operation in only one Spark stage without a shuffle phase in the Hadoop ecosystem.

In this paper, we propose a new data placement scheme for distributed BDWs upon a cluster of homogeneous nodes, using a static balancing technique based on a data-driven model. We take into account the following main parameters: the volume of data, the distribution values of the foreign and primary keys of the fact and dimension tables, and the physical characteristics of the cluster. Our strategy allows executing the filtering, projection, and star join of an OLAP query locally and in only one Spark stage. We have developed and evaluated our approach on the TPC-DS benchmark using Scala language on a cluster of homogeneous nodes, a Hadoop-YARN platform, a Spark engine, and Hive.

We tackle three main issues in our distributed BDW approach:

- How to minimize data skew during query processing to improve parallel processing in Hadoop;
- How to partition and load balance a BDW over Hadoop cluster, in order to perform the star join operation in only one Spark stage without a shuffle phase;
- How to develop a data placement strategy for a BDW which is based only on a data-driven model, to avoid the workload update issue.

The rest of this paper is organized as follows. Section 2 summarizes related work on the different types of join algorithms with the MapReduce paradigm. Section 3 presents our BDW data placement scheme and provides further details. We present our experiments in Sect. 4, and we conclude in Sect. 5.

2   Related Work

Almost all of the existing join algorithms in the literature rely on dynamic techniques of partitioning and load balancing of data, e.g., on-line partitioning and load balancing, such as repartition, broadcast join [5], multi-way join [2], and trie join [11]. Few are based on static techniques, such as trojan join [8], HadoopDB [1], and JOUM [4]. This kind of algorithm requires prior knowledge of the table schema and join conditions.

On the one hand, although the dynamic algorithm proposed by Afrati et al. [2] may perform well for the star join operation by minimizing the amount of replication of tuples, it may involve considerable communication cost during the shuffle phase, especially with heavy data skew. Also, the dynamic fully replicating tables used in the broadcast join algorithm [5], which is implemented in Spark SQL, can perform the star join operation in the Map side with minimal Spark stages if the dimension tables used in the query are small enough to fit into memory. Purdilă et al. [14] propose a dynamic algorithm that can execute the star join operation in two MapReduce iterations. However, their method still requires a shuffle phase. Zamanian et al. [16] use a predicate-based reference partitioning technique in parallel DBMS to ensure data locality automatically by co-partitioning tables that share the same join key in the same partition. They use two algorithms, based on the schema and a workload driven model, to minimize data redundancy and maximize data locality. However, we can not straightforwardly adapt their method for Hadoop.

On the other hand, the off-line partitioning technique used in Hadoop++ [8] is not suitable for the star join operation, as it can only co-partition two tables. Also, the on-line method of balancing used in CoHadoop [9] is hard to implement in the current versions of Hadoop and is suitable for log files. Moreover, the pre-Join of the star schema data used by JOUM [4] is not a realistic solution for BDW since it may generate a very big table which cannot be supported by the system. HadoopDB [1] attempts to integrate MapReduce and parallel DBMS. Some improvements have been made in both scalability and efficiency. However, the results are still not satisfactory for DW applications. Especially in star join queries, HadoopDB can lose its performance advantage. If we replicate each dimension table to all the database nodes, this will incur a high space cost. Also, HadoopDB stores the dimensions in a local DBMS and hence disrupts the dynamic scheduling and fault tolerance of Hadoop.

In this article, we propose a new data placement strategy for a BDW, called SDWP, using a static balancing technique as trojan join [8]. We tackle the problem of data skew and data duplication as in [2] and [16] by using the "balanced k-means" algorithm, as we will explain in Sect. 3.4. With our technique, and independently of the used workload, we can perform the star join operation in only one Spark stage. We used Spark as a query execution engine for the Hadoop-YARN platform. We assume that the DW is in a star schema. We used bucketing technique to partition the data and we exploit the new type of join operation embedded in Hive and Spark SQL called *Sort-Merge-Bucket* (SMB) join, which allows executing almost all the operations involved in an OLAP query (i.e., filtering, projection, and the star join operation) in only one Spark stage.

3   Our Approach SDWP

In SDMP, we start by building horizontal fragments of the fact and dimension tables using our hash-partitioning method (see Sect. 3.2). Then we distribute these fragments evenly among the cluster nodes, in which we can perform star

join and some operations (i.e., filtering and projection) of an OLAP query locally
and in only one Spark stage without a shuffle task. We suppose that we have
a prior knowledge of the size of the DW and its schema, as well as the characteristics of the cluster. Our approach has two phases: (1) building the set of
buckets of the DW tables, and (2) placing the buckets in the nodes of the cluster. Figure 1 shows the steps of our approach. Before detailing our approach, we
formulate our problem as follows.

3.1   Formalization

Let's have a star schema DW $E = \{F, D1, D2, .., Dk\}$, where F is the fact table
and the Dd, $d \in 1..k$ are the dimension tables. We denote by $Bkey$ the partition key used to bucket the fact and all dimension tables of E, such as $Bkey$
is a new added key, and we denote by NB the number of buckets that should
be created for all tables of E. We denote by $BF = \{BF_0, BF_1, ..., BF_{NB-1}\}$
the set of distinct buckets created by bucketing the fact table F with $Bkey$
into NB buckets, and by $BDd = \{BDd_0, BDd_1, ..., BDd_{NB-1}\}$ the set of
distinct buckets of each dimension Dd, $d \in 1..k$. We denote by $W_{BF} =
\{\|BF_0\|, \|BF_1\|, ..., \|BF_{NB-1}\|\}$ the set of sizes of the bucket of F, and $W_{BDd} =
\{\|BDd_0\|, \|BDd_1\|, ..., \|BDd_{NB-1}\|\}$, $d \in 1..k$, the set of bucket sizes of each
dimension Dd, $d \in 1..k$. We denote by $group$ the set of the buckets that have
the same value of $Bkey$: it is composed of one bucket of F and one bucket of
each Dd. We denote by $N = \{n_1, n_2, .., n_e\}$ the set of all homogeneous nodes of
the cluster.

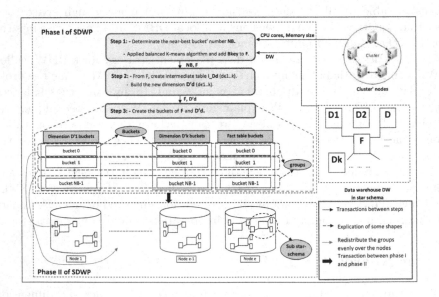

Fig. 1. The steps of building SDWP approach

Our aim is to choose $Bkey$ and NB for building the buckets of BF and all BDd, $d \in 1..k$ in such a way as to keep them roughly balanced in size, i.e., with the minimum standard deviation of W_{BF} and all W_{BDd}, $d \in 1..k$, and how to distribute them over the cluster N in order to execute the star join operation of an OLAP query locally and in only one Spark stage without a shuffle task.

3.2 Building the Buckets

This phase has three steps: (1) determining the near-best NB and $Bkey$; (2) building BF; and (3) building BDd, $d \in 1..k$.

(1) Selecting NB and Bkey. Selecting NB and $Bkey$ to guide SDWP is critical. To do this, we address some technical challenges as outlined below.

- **Selecting NB.** We should select NB as follows:

$$NB \in [min\_NB, .., max\_NB] \qquad (1)$$

Here, min_NB is the minimum value of the number of buckets and max_NB is the maximum value. To determine these values, we follow these rules:

- **Rule 1.** Our processing must be parallelized, i.e., almost all idle CPU cores in the cluster should be exploited. Hence, the minimal value of NB should be equal to N_{ct}, the total number of CPU cores assigned to execute an application task. Our aim is to assign at least one RDD partition to each CPU core (in our case, a Resilient Distributed Dataset (RDD) partition is a *group*).
- **Rule 2.** Selecting a large value for NB ($NB \gg N_{ct}$) can harm the performance of the system as a result of increasing the I/O operations, and this can incur significant overhead for processing the partition and maintaining partition-level metadata by the NameNode. Hence, and since our processing is in-memory, using Spark, we determine max_NB as follows:

$$max\_NB \leq \lfloor min\_NB \times max(1, V_E/V_M) \rfloor \text{ and } max\_NB \leq |T| \qquad (2)$$

Where V_E is the size of the DW E, V_M is the sum of all memory sizes of all slave nodes, and T is the smallest dimension in E. Our argument is based on the following: the first part of Eq. 2 means that if the total memory size of the cluster is large, i.e., $V_M \approx V_E$ and $max\_NB \approx min\_NB$, in that case we can process a large RDD partition. However, if the memory size is small, i.e., $V_M \ll V_E$, then max_NB increases and processing a small RDD partition become preferable. The second part, i.e., $max\_NB \leq |T|$, means that we must not get an empty bucket for all BF and BDd.

- **Rule 3.** We have seen in the previous rules that we should select NB from the interval $[N_{ct}, .., \lfloor N_{ct} \times max(1, V_E/V_M) \rfloor]$. However, and since the number $r\_number = \lfloor N_{ct} \times max(1, V_E/V_M) \rfloor - N_{ct}$ may reach several hundreds, it is not realistic to test all values in this interval to find the near-best solution, i.e., doing hundreds of empirical tests with different values of NB. So, if we assume that processing all tasks in each Spark wave finish roughly in the same time, i.e., the number of all job tasks is divisible by N_{ct}, therefore, to select the optimal NB value, we execute the queries with $NB = N_{ct}$, and each time we increment NB, i.e., $NB = NB + N_{ct}$, until $NB = max\_NB$ or until the execution time of the queries increases.

- **Determining** $Bkey$. The main challenge for choosing $Bkey$ is to minimize data skew and avoid obtaining stragglers[1]. Hence, a random distribution of values of $Bkey$ can give us unbalanced bucket sizes, which may slow down the query processing. In Sect. 3.4, we show how to overcome this issue.

In the following, we explain the steps for creating the buckets of the sets BF and BDd, $d \in 1..k$. After that, we show how to distribute them over the nodes of the cluster, and how to minimize data redundancy. At the end, we show how to apply some transformations to the join-predicates of an OLAP query.

(2) Building the Fact Table Buckets. The construction of BF is based on two parameters: NB and $Bkey$. To build the buckets of BF, we assign each tuple of a fact table F to the corresponding bucket BF_i, i.e., bring together into the same bucket all the tuples that have the same $Bkey$.

To choose the values of $Bkey$ we use the balanced k-means algorithm, as explained in Sect. 3.4. The main idea is to minimize the standard deviation of the sets W_{BF} and all W_{BDd}, $d \in 1..k$. Figure 2 shows an example of how to create the buckets of a fact table.

(3) Building Dimension Buckets. After getting BF, we build the BDd, $d \in 1..k$. However, in order to construct a *group*, we must also add the key $Bkey$ to all the Dd, $d \in 1..k$. To do this, we carry out the following steps. First, we create an intermediate table IDd corresponding to the dimension Dd. The IDd table is composed of two attributes, fk_d and $Bkey$, such that: (1) fk_d is the foreign key of dimension Dd in fact table F and (2) $Bkey$ is the partition key added in F. The IDd table initially has the same number of tuples as the fact table F. So, before joining IDd with Dd to obtain the new dimension $D'd$, we delete all duplicate tuples in IDd. After that, we build the final set $BD'd$ as we already did with the fact table F.

Note that the sizes of the new dimensions $D'd$, $d \in 1..k$, are large compared to the original ones. Their sizes are changed according to: (1) the value of NB and (2) our clustering method applied to choose the values of $Bkey$ and to limit data redundancy due to the duplicated tuples in the newly obtained dimensions. However, the size of the new dimensions $D'd$ remains small compared to the fact table F whatever the size of the original dimensions Dd. Indeed, in BDW, the size of the fact table is generally very high compared to the size of the dimensions, whatever the latter. In the case of small dimensions, the system can easily broadcast them, and in the case where the dimensions are too large,

---

[1] A straggler is a task that performs more poorly than similar ones due to insufficient assigned resources.

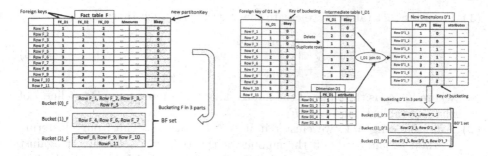

Fig. 2. Building the buckets of a fact table

Fig. 3. Building a new dimension and its buckets

they can not fit into memory. Moreover the rate of repeated values will be high with small dimensions and it may decrease with large dimensions. Furthermore, the number of distinct values of $Bkey$ is not high, since we limit their values between 0 and NB-1, using balanced k-means algorithm, and the maximum of NB is fixed according to our heuristic method detailed in Sect. 3.2. Hence, the probability of obtaining duplicated tuples in IDd, $d \in 1..k$, will be high. As a result, the size of the new dimensions $D'd$ remains small compared to F. Our experiments in Sect. 4 confirm this argument. Figure 3 shows an example of how to create a new dimension and its buckets.

3.3 Placement of the Buckets

In phase II, we evenly distribute the *group* over the cluster's nodes. Formally, we can put $group_i = BF_i \uplus_{d=1}^k BD'd_i$, $i \in 0..NB - 1$. Thus, we start to place the $group_0$ in node 1, $group_1$ in node 2,..., and the $group_{p-1}$ in the node e, such that $e = p\ mod\ NB$ and $p <= NB$. We restarted the operation with round robin technique, we put $group_p$ in node 1, $group_{p+1}$ in node 2,..., until the $group_{NB-1}$. This data placement strategy allows performing the star join operation locally and in a single Spark stage without a shuffle task.

3.4 Minimizing Data Redundancy

We have seen that the characteristics of our bucketing technique are based on the value of NB and on $Bkey$. We can select the near-best NB as recommended in Sect. 3.2. However, choosing $Bkey$ remains a challenging task. So, since the sizes of the dimensions are small compared to the fact table, we can get the BF buckets to have roughly balanced sizes by applying a simple range partitioning method. However, there is an essential factor which can affect the size of the newly constructed dimensions: the similarity of the tuples in each bucket of BF. This can increase the number of tuples in each bucket of $BD'd$, $d \in 1..k$, and use up a great deal of disk space. To overcome this issue and obtain an approximately optimal solution, we propose the following method.

(1) From the fact table F, we create the matrix MV such that:

$$MV = \begin{pmatrix} V_{FD11} & V_{FD21} & \cdots & V_{FDk1} \\ \cdots & \cdots & & \cdots \\ \cdots & \cdots & V_{FDdj} & \cdots \\ \cdots & \cdots & & \cdots \\ V_{FD1n} & V_{FD2n} & \cdots & V_{FDkn} \end{pmatrix} \qquad (3)$$

where V_{FDdj} is the value of the foreign key fk_d coming from Dd at line j of the fact table F, and $n = |F|$.

(2) After building MV, we cluster it into NB clusters. Thus, our clustering method should trade off the number of the tuples in each bucket against the similarity of the tuples in each bucket. So, we finish by using the balanced k-means algorithm [13], where the cluster size balance is a mandatory requirement that must be met, and minimize MSE^2 is a secondary criterion. The first reason to choose this kind of algorithm is to minimise the standard deviation of W_{BF} and the second one is to minimize the size of the newly built dimensions. The output of the algorithm is (n modulo NB) clusters of size $\lceil \frac{n}{NB} \rceil$, and NB-(n modulo NB) clusters of size $\lfloor \frac{n}{NB} \rfloor$.

(3) Finally, we affect the cluster values obtained from the $Bkey$ column and we apply our bucketing method, as explained before.

This clustering technique only ensures minimizing the standard deviation of W_{BF} and not all $W_{BD'd}$. However, since the of $D'd$ are small compared to F, the size of the bucket of $W_{BD'd}$ remains also small compared to the bucket of W_{BF}, and hence the sizes of the $group_i$, $i \in 0..NB-1$, remain roughly equal (see the notation in Sect. 3.3). Moreover, in our clustering algorithm we have not included another factor which can increase the size of the new dimensions: the number of the attributes of a dimension. Some dimensions can have a few attributes while others have many attributes (e.g. hundreds or thousands). However, with the "Parquet" and "ORC" storage formats, only the attributes solicited by the queries are loaded into memory and not the whole bucket of the new dimension. Furthermore, with the new compression and coding techniques in HDFS (e.g., Gzip and Snappy), the column $Bkey$ occupies negligible disk space since the number of its distinct values is not large, i.e., $values(Bkey) \in 0..NB\text{-}1$.

3.5   Query Transformation

In our approach, we do some changes in the join condition yet such that it still produces the correct join result. Namely, we add the join condition $F.Bkey = D'd.Bkey$ where $D'd$ is the newly built dimension and we activate the Sort-Merge-Bucket (SMB) join of Spark-SQL. That is to say, in our case, the join condition becomes $F.Bkey = D'd.Bkey$ and $F.fk_d = D'd.pk_d$. The following example demonstrates the execution plan of an OLAP query with our approach. We consider an OLAP query Q extracted from the TPC-DS benchmark, e.g.

---

[2] $MSE = \sum_{j=1}^{k} \sum_{X_i \in C_j} \frac{\|X_i - C_j\|^2}{n}$, Where X_i denotes the data point locations, i.e., tuples or vectors of the matrix MV, C_j denotes the centroid locations, and $n = |MV|$.

```
select c_first_name, c_last_name, d_year, sum(ss_sales_price)
from customer, date_dim, store_sales where
customer.c_customer_sk = store_sales.ss_customer_sk
and store_sales.ss_sold_date_sk = date_dim.d_date_sk and date_dim.d_year=2000
Group by c_first_name, c_last_name, d_year;
```

After transformation, we obtain the query Q':

```
select c_first_name, c_last_name, d_year, Sum(ss_sales_price)
from customer', date_dim', store_sales' where
(store_sales'.Bkey = customer'.Bkey and store_sales'.ss_customer_sk=customer'.c_customer_sk)
AND (store_sales'.Bkey= date_dim'.Bkey and store_sales'.ss_sold_date_sk = date_dim'.d_date_sk)
AND date_dim.d_year=2000 Group by c_first_name, c_last_name, d_year;
```

The execution plan of the query Q' in Spark SQL with our approach is as follow: First the system scans each bucket of the table date_dim', executes the filter d_year=2000, and retrieves only the attributes involved in the query Q', i.e., retrieves the fragments (date_dim.Bkey, date_dim.d_date_sk, date_dim.d_year). As a result, we obtain the fragment f_date_dim' (this fragment is distributed over NB RDD partitions, where NB is the number of buckets); the same process is applied to the other tables store_sales' and customer'; we obtain the fragments f_store_sales(store_sales'.Bkey, store_sales'.ss_sold_date_sk, store_sales'.ss_customer_sk, store_sales'. ss_sales_price) and f_customer' (customer'.Bkey, customer'.c_customer_sk, customer'.c_first_name, customer'. c_last_name) which are also distributed over the same number of RDD partitions (the system gathers all the buckets that have the same number into one RDD partition). Then the system executes the join condition between f_store_sales' and f_date_dim'. However, in our approach, we enforce the system to execute the two join condition (store_sales'.Bkey=date_dim'.Bkey and store_sales'.ss_sold_date_sk=date_dim'. d_date_sk), locally in each RDD partition, and avoid broadcasting join and shuffle join. As a result, we obtain the new fragment f'(store_sales'.Bkey, store_sales'. ss_customer_sk, date_dim.d_year), and finally we join f' with f_customer' to obtain f_final (c_first_name, c_last_name, d_year, ss_sales_price). At the end, the system performs Group by and aggregate function Sum(ss_sales_price).

3.6 Scalability of SDWP

The update of the data warehouses is not as the databases, the changing of dimensions and the fact table will be done slowly (e.g. update every six months). There are several methods to update the DW. In our case, to deal with updating the DW, we use our bucketing technique as a "secondary" partitioning scheme, namely, before bucketing the tables with $Bkey$, we partition some tables which can change over time, by the *date* attribute (e.g. code in Hive "..*Partitioned By (date)...Bucketed By (Bkey) into NB buckets...*"). For example, when a new partition date='2019-06-08' is added to the *customer* table we run: *ALTER TABLE customer ADD PARTITION (date='2019-06-08')....* So, with this technique, we can invoke our bucketing technique and apply balanced K-means algorithm on this newly inserted partition without affecting the existing data. The following scenario explains the procedure: Suppose that we have loaded and distributed our DW, in first time at the date='2018-12-09' with our SDWP approach.

After six months we want to update the DW tables, so, only the tuples updated and the new tuples will be inserted into the new partition (date='2019-06-08') using our data placement SDWP, and so on. Note that although this technique may increase data redundancy, however, it is necessary for a DW analysis purpose.

4   Experiments

4.1   Experimental Setup

First of all, we generate the DW using the TPC-DS benchmark, where we store the data directly in HDFS using the Parquet format. We used a cluster of 15 slave data nodes and one master node characterized by CPU Pentium I7 with 8 cores, 16 GB of memory and 2 TB of hard drive. We installed in all nodes Hadoop-YARN V-2.9.2, Hive V-2.3.3, Apache Spark V-2.3.3, TPC-DS benchmark, Scala language and Java. To the master node we added a MySQL server to store the Hive meta-data, "Maven tools" and Scala Build Tool "SBT" to build jar packages. For the HadoopDB tests (see Sect. 4.2), we added PostgreSQL to all the data nodes. In Spark, there are more than 150 configurable parameters [15] that can affect the job execution time. In our experiments, we focused of configuring the candidate parameters as recommended in [10,15]. So, we configured some Spark parameters as follows: spark.executor.instances = 30, spark.executor.memory = 6 GB, spark.executor.cores = 3 CPU cores. We kept the default block size 128 MB and kept 3 as the number of replications. For the memory size and CPU cores, we should not exploit all idle resources. Thus, for all slave nodes, we kept 4 GB and 2 CPU cores for "operating system," "executors," and also for "Application Master." With this configuration, we can run $3 \times 30 = 90$ tasks in parallel.

Data Generation. We adapted the *spark-sql-perf* application[3], using Scala language and Spark. In our experiments and since we focused of optimizing the star join operation, we used one fact table among seven and nine dimensions among seventeen of the TPC-DS benchmark (see Table 1). In view of the limitations and physical characteristics of our cluster and to avoid memory overflow during the data generation, we generated the fact table *store_sales* by partition. We chose the foreign key *ss_store_sk* of dimension *store* as "partition key" because the *ss_store_sk* key has the minimum number of distinct values compared to the other keys of other dimensions (see Table 1).

Implementation of SDWP. Before bucketing *store_sales* fact table, we add *Bkey*. The values of this key are calculated using our clustering method detailed in Sect. 3.4. We have implemented balanced k-means algorithm in SparkR. To deploy the first phase of our solution, we use three essential components: *Dataframe*, *Dataset* of Spark, and *ArrayBuffer*.

---

[3] Available from the site https://github.com/databricks/spark-sql-perf.

Table 1. Characteristics of DW tables

	Table name	Data Warehouse 1 (DW1)		Data Warehouse 2 (DW2)	
		Number of records	Parquet format	Number of records	Parquet format
1	*store_sales*	2 879 995 413	142.6 GB	28 799 954 135	1 420 GB
2	*customer*	12 000 000	607.8 MB	65 000 000	3 210 MB
3	*customer_address*	6 000 000	111.4 MB	32 500 000	603.63 MB
4	*customer_demographics*	1 920 800	7.4 MB	1 920 800	7.4 MB
5	*item*	300 000	27.3 MB	402 000	36.52 MB
6	*time_dim*	86 400	1 126 KB	86 400	1 126 KB
7	*date_dim*	73 049	1 740 KB	73 049	1 740 KB
8	*household_demographics*	7 200	30.0 KB	7 200	30.0 KB
9	*promotion*	1 500	76.0 KB	2 000	98.77 KB
10	*store*	1 002	88.0 KB	1 500	128.96 KB

To implement phase II of SDWP, we do not modify the policy placement of HDFS as in [9], since the API of Hadoop V-2.x or V-3.x would need severe modifications. Our strategy of placement is currently implemented as an external balancer tool, namely, we let Hadoop distribute the HDFS blocks of the buckets by the default placement policy (random targeting of the nodes), then with an off-line process, we rebalance the files or the blocks. This has the advantage of keeping safe the code of the HDFS default block placement policy and avoid invoking auto co-locating of the blocks when one or more nodes crash, which can harm the distributed system. Moreover, although may we cannot completely ensure placing all the buckets that have the same value of $Bkey$ in the same node, however, by exploiting SMB join and disabling the HB and SH join, the star join operation remains executing in only one Spark stage without a shuffle phase.

4.2 Experiments and Results

We carried out some experiments with two DW configurations. The first one, denoted by DW1, has about 500 GB in CSV format, and the second, DW2, has about 5 TB (equivalent to 1424 GB in Parquet format). We detail the characteristics of DW1 and DW2 in Table 1. Table 2 shows different approaches used in our experiments. We can only use the queries that solicited our DWs tables (about 32/99 queries). So, in our experiments, and for reasons of simplicity (without loss of generality), we used six queries from among the thirty-two (see Table 3) with different levels of complexity. Our aim is to show how Catalyst optimizer of Spark SQL can run the star join task in only one stage with our data placement strategy, whatever the used OLAP query. The different characteristics of the six queries are detailed in Table 4. We executed these queries with five values of NB, which are selected according to our method detailed in Sect. 3.2. We consider SHDB, SSH, and SHB as the baseline approaches, which we compare with our solutions (i.e., SDWP, SSMBR, and SSMBD). In the SSH and SHB approaches, we set the parameter $spark.sql.shuffle.partitions$ to 180 with DW1 and 630 with DW2. To deactivate the default HB join of Spark, we add the instruction $Session.conf.set("spark.sql. autoBroadcastJoinThreshold",-1)$. In this case,

Spark executes SH join. In SDWP, SSMBR, and SSMBD, since we bucket the tables with the same attribute $Bkey$, we can run SMB join in the right way. With our configuration, we can run 90 tasks in parallel. Thus, by following the rules of Sect. 3.2, we get: with DW1, $min\_NB = 90$ and $max\_NB = 180$, and with DW2, $min\_NB = 90$ and $max\_NB = 1440$.

Table 2. Notations for the compared approaches

Notation	Description
SHDB	Default partitioning and distributing scheme of HadoopDB with PostgreSQL database and Spark as execution engine
SSH	Default partitioning and distributing scheme of Hadoop and Spark, using default Spark Shuffle Hash join (SH join). (like repartition join [5] in MapReduce.)
SHB	Default partitioning and distributing scheme of Hadoop and Spark, using Hash Broadcast Join (HB join)
SSMBD$_{NB}$	Our partitioning scheme with default Hadoop distribution policy, by exploiting Spark SMB join and disabling HB and SH join, with NB buckets
SDWP$_{NB}$	Our partitioning scheme and load balancing strategy, by exploiting Spark SMB join and disabling HB and SH join, with NB buckets
SSMBR$_{NB}$	Similar to SDWP$_{NB}$ but instead of using our balanced k-means algorithm, we just create roughly equal buckets of the fact table randomly by using range partitioning

Table 3. Selected queries

Name	Code query
Q1	select dt.d_year, item.i_brand, item.i_brand_id, sum(ss_sales_price) from date_dim dt, item, store_sales where dt.d_date_sk = store_sales.ss_sold_date_sk and store_sales.ss_item_sk = item.i_item_sk and item.i_manufact_id = 128 and dt.d_moy=11 group by dt.d_year, item.i_brand, item.i_brand_id limit 100;
Q2	select dt.d_year, dt.d_month_seq, item.i_brand, item.i_brand_id, item.i_class, sum(ss_sales_price) from date_dim dt, item, store_sales where dt.d_date_sk = store_sales.ss_sold_date_sk and store_sales.ss_item_sk = item.i_item_sk and item.i_manufact_id = 128 and dt.d_moy=11 group by dt.d_year, dt.d_month_seq, item.i_brand, item.i_brand_id, item.i_class limit 100;

Table 3. (*continued*)

Name	Code query
Q3	select c_customer_id, c_first_name, c_last_name, c_preferred_cust_flag, c_birth_country, c_login, c_email_address, d_year, d_month_seq, sum(ss_sales_price) from customer, date_dim, store_sales where c_customer_sk=ss_customer_sk and ss_sold_date_sk=d_date_sk group by c_customer_id, c_first_name, c_last_name, c_preferred_cust_flag, c_birth_country, c_login, c_email_address, d_year, d_month_seq limit 100;
Q4	select a.ca_city, d.d_month_seq, i.i_brand, sum (ss_list_price) from customer_address a, date_dim d, item i, store_sales s where a.ca_address_sk = s.ss_addr_sk and s.ss_sold_date_sk = d.d_date_sk and s.ss_item_sk = i.i_item_sk and i.i_manufact_id = 128 and dt.d_moy=11 group by a.ca_city, d.d_month_seq, i.i_brand limit 100;
Q5	select a.ca_city, d.d_month_seq, i.i_brand, sum(ss_list_price) from customer_address a, date_dim d, item i, store_sales s where a.ca_address_sk = s.ss_addr_sk and s.ss_sold_date_sk = d.d_date_sk and s.ss_item_sk = i.i_item_sk group by a.ca_city, d.d_month_seq, i.i_brand limit 100;
Q6	select c.c_customer_id, c.c_first_name, c.c_last_name, c.c_preferred_cust_flag, c.c_birth_country, c.c_login, a.ca_city, a.ca_state, a.ca_country, d.d_month_seq, d.d_date, i.i_brand, i.i_class, i.i_product_name, sum(ss_list_price) from customer_address a, customer c, date_dim d, item i, store_sales s where a.ca_address_sk = s.ss_addr_sk and c.c_customer_sk = s.ss_customer_sk and s.ss_sold_date_sk = d.d_date_sk and s.ss_item_sk = i.i_item_sk group by c.c_customer_id, c.c_first_name, c.c_last_name, c.c_preferred_cust_flag, c.c_birth_country, c.c_login, a.ca_city, a.ca_state, a.ca_country, d.d_month_seq, d.d_date, i.i_brand, i.i_class, i.i_product_name limit 100;

Table 4. Characteristics of the six selected queries

Query	Characteristics
Q1	We join 2 small dimensions with the fact table, we select a few attributes and we use 2 filters
Q2	Is similar to Q1, but we select more attributes
Q3	We join 2 dimensions whose one is large with the fact table store_sales, and we selected more attributes than Q1 and Q2 without using filters
Q4	Is similar to Q1, but we add the dimension customer_address
Q5	Is similar to Q4, but without using filters
Q6	We perform star join operation with 4 dimensions of which 2 are large, and we select more attributes than Q5

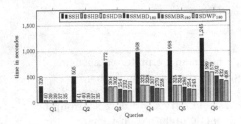

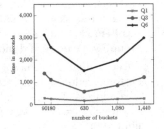

Fig. 4. Runtime of the queries with DW1   **Fig. 5.** Runtime of the queries with DW2

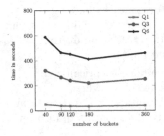

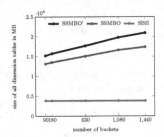

Fig. 6. Impact of NB on the query execution time with DW1

Fig. 7. Impact of NB on the query execution time with DW2

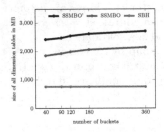

Fig. 8. Impact of NB on the size of dimensions with DW1

Fig. 9. Impact of NB on the size of dimensions with DW2

4.3   Discussion

As shown in the bar chart of Figs. 4 and 5, our strategy SDWP has improved the query execution time compared to the other approaches. In all queries, we can see that the worst results were obtained with the SSH approach. This is due to the high rate of data shuffling. In Q1 and Q2, since we have selected few attributes of only two small dimensions *item* and *date_dim*, the broadcasting of the RDD partitions become fast and we can see that the execution times of these queries in the SHB approach are roughly the same as with our approaches SDWP, SSMBR, and SSMBD. However, in Q3, Q4, Q5, and Q6, the performance of SHB suffers, especially with DW2. The reason is that in the SHB approach when the table is large, the system cannot broadcast it, and must combine with broadcast join and shuffle join (i.e., the SHB approach) to perform these queries.

We can see that the baseline approach SHDB (when using HadoopDB) is much better than the other baselines (i.e., SSH and SHB). Obviously, since in SHDB we duplicated all dimensions over the cluster nodes, the star join operation could be performed locally. However, we can see that our approach performs better than SHDB, especially in Q3 and Q6 (when using two large dimensions, i.e., *customer* and *customer_address*). The reason is that although in SHDB all the PK and FK are located in the same node, the system takes a long time to retrieve the relevant tuples from the dimensions stored in the DBMS. On the other hand, in our solutions, SDWP, SSMBR, and SSMBD, because we have bucketed all the new dimensions and the fact table with the same key $Bkey$, the optimizer of Spark SQL can easily gather all the buckets that have the same value of the join key $Bkey$ into one RDD partition. In that case, almost all the work of an OLAP query (except Group by operation and aggregate function) is performed in parallel during the first Spark stage.

Moreover, we note also that the runtime of the queries in SDWP is much better than in SSMBR, and this demonstrates the efficiency of our algorithm balanced k-means. Obviously, the random clustering applied in SSMBR can increase the size of some new dimensions and degrade the system performance. Furthermore, we can see from Figs. 4 and 5 that SDWP is up to 1.25 times faster than SSMBD. Of course, in SSMBD the buckets of the fact table *store_sales* and the nine dimensions may not be located in the same nodes, and the amount of network traffic may increase. If the nodes are located in different RACKs, the results in SDWP will become 1.4 times better or more than SSMBD.

Figures 6 and 7 show that NB has an impact on the query performance. In all queries, the best results are obtained when $NB \in [120, .., 180]$ with DW1 and $NB \in [630, .., 720]$ with DW2. This confirms the efficacy of our method for selecting the near-best value for the number of buckets. Also, as shown in Figs. 8 and 9, we can note that NB has no significant impact on the sizes of the new dimensions. The sizes of the new dimensions built in DW1 are only increased by a factor of 2.8 compared to the original ones, and by about 4.5 times with DW2. This percentage has no impact on the runtime of the queries. Moreover, with our clustering method in SSMBO, we have gained up to 30% disk space compared to SSMBR, in which we did not use the balanced k-means algorithm.

Finally we should note that if we had used the JOUM [4] approach (denormalized model), we would have generated a huge amount of data. More specifically, if we had applied this method to our DW2, which contains 179 attributes (sum of all column tables), we would have obtained a big table, with a size of 15051 GB (in Parquet format). This size is about 10 times greater than the size of the original fact table store_sales.

5   Conclusion and Future Research

In this paper, we have presented a strategy for partitioning and distributing a big data warehouse over a cluster of homogeneous nodes. Our experiments have shown that our strategy allows performing almost all the operations of an OLAP query locally and in only one Spark stage. Our approach is scalable to large clusters and for huge amounts of data. We have found that although we have roughly balanced the buckets' sizes, we have, due to the data skew, gotten unbalanced partition sizes after performing filtering and the star join operation (mapper outputs). Moreover, we have not shown how to balance the intermediate results, something which could improve the grouping operation in the reduce phase. Furthermore, we could skip loading unnecessary data blocks if we partition some tables by the frequent attributes used in the queries' filters, using the hash-partitioning technique.

In the future: (I) we plan to combine data and workload-driven models to create a new physical design of a distributed big data warehouse, and (II) we aim to propose a strategy to balance the intermediate results to speed up grouping and aggregation operations.

References

1. Abouzeid, A., Bajda-Pawlikowski, K., Abadi, D., Silberschatz, A., Rasin, A.: HadoopDB: an architectural hybrid of MapReduce and DBMS technologies for analytical workloads. Proc. VLDB Endow. **2**(1), 922–933 (2009)
2. Afrati, F.N., Ullman, J.D.: Optimizing multiway joins in a map-reduce environment. IEEE Trans. Knowl. Data Eng. **23**(9), 1282–1298 (2011)
3. Arres, B., Kabachi, N. and Boussaid, O.: Optimizing OLAP cubes construction by improving data placement on multi-nodes clusters. In: 23rd Euromicro International Conference on Parallel, Distributed and Network-Based Processing, pp. 520–524. IEEE (2015)
4. Azez, H.S.A., Khafagy, M.H., Omara, F.A.: JOUM: an indexing methodology for improving join in HIVE star schema. Int. J. Sci. Eng. Res. **6**, 111–119 (2015)
5. Blanas, S., Patel, J.M., Ercegovac, V., Rao, J., Shekita, E.J. Tian, Y.: A comparison of join algorithms for log processing in MapReduce. In: Proceedings of the 2010 ACM SIGMOD International Conference on Management of Data, pp. 975–986. ACM (2010)
6. Brito, J.J., Mosqueiro, T., Ciferri, R.R., de Aguiar Ciferri, C.D.: Faster cloud Star Joins with reduced disk spill and network communication. Procedia Comput. Sci. **80**, 74–85 (2016)
7. Chen, K., Zhou, Y., Cao, Y.: Online data partitioning in distributed database systems. In: EDBT, pp. 1–12 (2015)
8. Dittrich, J., Quiané-Ruiz, J.A., Jindal, A., Kargin, Y., Setty, V., Schad, J.: Hadoop++: making a yellow elephant run like a cheetah (without it even noticing). Proc. VLDB Endow. **3**(1–2), 515–529 (2010)
9. Eltabakh, M.Y., Tian, Y., Özcan, F., Gemulla, R., Krettek, A., McPherson, J.: CoHadoop: flexible data placement and its exploitation in Hadoop. Proc. VLDB Endow. **4**(9), 575–585 (2011)
10. Golfarelli, M., Baldacci, L.: A cost model for SPARK SQL. IEEE Trans. Knowl. Data Eng. **31**, 819–832 (2018)
11. Kalinsky, O., Etsion, Y., Kimelfeld, B.: Flexible caching in trie joins. arXiv preprint arXiv:1602.08721 (2016)

12. Lu, Y., Shanbhag, A., Jindal, A., Madden, S.: AdaptDB: adaptive partitioning for distributed joins. Proc. VLDB Endow. **10**(5), 589–600 (2017)
13. Malinen, M.I., Fränti, P.: Balanced K-means for clustering. In: Fränti, P., Brown, G., Loog, M., Escolano, F., Pelillo, M. (eds.) S+SSPR 2014. LNCS, vol. 8621, pp. 32–41. Springer, Heidelberg (2014). https://doi.org/10.1007/978-3-662-44415-3_4
14. Purdilă, V., Pentiuc, Ş.G.: Single-scan: a fast star-join query processing algorithm. Softw. Pract. Exp. **46**(3), 319–339 (2016)
15. Petridis, P., Gounaris, A., Torres, J.: Spark parameter tuning via trial-and-error. In: Angelov, P., Manolopoulos, Y., Iliadis, L., Roy, A., Vellasco, M. (eds.) INNS 2016. AISC, vol. 529, pp. 226–237. Springer, Cham (2017). https://doi.org/10.1007/978-3-319-47898-2_24
16. Zamanian, E., Binnig, C., Salama, A.: Locality-aware partitioning in parallel database systems. In: Proceedings of the 2015 ACM SIGMOD International Conference on Management of Data, pp. 17–30. ACM (2015)

Improved Programming-Language Independent MapReduce on Shared-Memory Systems

Erik G. Selin(✉) and Herna L. Viktor

University of Ottawa, Ottawa, Canada
{eseli061,hviktor}@uottawa.ca

Abstract. The MapReduce programming paradigm is a prominent model for expressing parallel computations, especially in the context of data processing of vast data sets. However, modern data processing runtimes, implementing the MapReduce programming paradigm, do not generally support the use of arbitrary programming languages. Access to programming-language independent data processing can offer great value to organizations as it enables leveraging existing programming language expertise. This paper introduces XRT, an open-source programming-language independent MapReduce runtime for shared-memory systems. We introduce a number of current optimizations and extensions to the state-of-the-art. Furthermore, this paper presents the first comparative benchmark between XRT and many popular MapReduce runtimes in order to highlight the performance capabilities of programming-language independent data processing.

Keywords: Big data · Data processing · MapReduce ·
Shared-memory systems · Parallel processing ·
Programming-language independent

1 Introduction

The MapReduce programming paradigm, popularized by Google [10] and made widely available through the Hadoop MapReduce [3] runtime, is foundational to most distributed data processing systems capable of handling vast data sets. The success of the MapReduce programming paradigm is not only due to its suitability for distributed environments but also because it provides a natural way to express parallel programs. Still, to our knowledge, the Hadoop Streaming data processing runtime [5] is the only MapReduce runtime that supports the implementation of data processing jobs in arbitrary programming languages. Unfortunately, the performance of Hadoop Streaming is sub-par [11], which has effectively left the industry without a viable programming-language independent MapReduce runtime. Yet, there is interest within the MapReduce community for supporting additional programming languages as can be witnessed by the

© Springer Nature Switzerland AG 2019
C. Ordonez et al. (Eds.): DaWaK 2019, LNCS 11708, pp. 206–220, 2019.
https://doi.org/10.1007/978-3-030-27520-4_15

numerous projects attempting to introduce additional programming-languages to the Hadoop MapReduce runtime [4, 7, 13].

Furthermore, with the advent of high core-count CPUs, increasing data volumes can be processed without the need for distributed data processing systems. Indeed, recent offerings from popular cloud providers have made shared-memory systems with over a hundred cores and terabytes of memory a reality for most organizations [1, 6]. Unfortunately, modern shared-memory MapReduce runtimes share the same shortcoming as distributed MapReduce runtimes as they do not enable a programming-language independent environment [8, 9, 14, 18, 21, 23]. Moreover, shared-memory MapReduce runtimes tend to focus on memory-only computations, which severely limits the processing of vast datasets.

In this paper, we present numerous improvements to the programming-language independent and open-source XRT runtime [20]. The XRT runtime is built from the ground up to be capable of effectively operating on high core-count and high-memory systems through a vast number of specific optimizations and architectural choices. Also, the runtime is memory aware, favouring memory-only data processing while simultaneously retaining the capacity to automatically spill to disk for processing of datasets exceeding available memory. In addition, the particular MapReduce model utilized by XRT is especially apt at handling skewed data sets, further situating XRT as a strong candidate for handling modern data processing challenges.

This paper starts by introducing the XRT paradigm, as detailed in Sects. 2 and 3. This is followed, in Sect. 4, by detailed insights into the novel XRT optimizations and extensions. Section 5 presents a range of XRT runtime benchmarks together with a comparative benchmark of XRT against many popular data processing runtimes. Finally, Sect. 6 provides concluding remarks and details future work.

2   Execution Model

This section provides a breakdown of the XRT execution model by stepping through the various stages of an XRT data processing job, illustrated in Fig. 1. On a high level, most XRT jobs transform a set of input files containing input records into a set of output files containing output records.

During an XRT job, the input files are chunked logically by the runtime into approximately 16 MB chunks which start and end at record boundaries. The runtime distributes the chunks based on demand across the running mappers resulting in near ideal resource utilization as all mappers stay active throughout the whole map stage.

Each mapper is a sub-process of the XRT process executing a user-supplied mapper program. The mapper program is expected to read input records from the stdin stream and write reducer-id prefixed records to the stdout stream. The reducer-id prefix communicates to the XRT runtime which reducer process should receive the associated record. The motivation behind using reducer-id prefixes is that it enables the implementation of custom partitioning logic within the user provided mapper program.

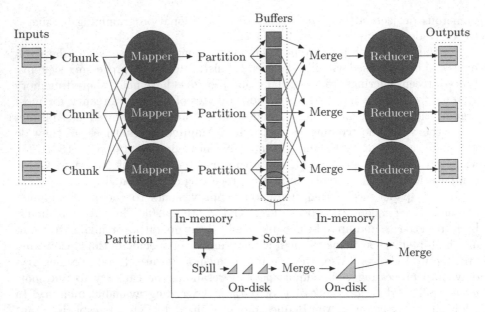

Fig. 1. Illustrative example of the XRT execution model.

The intermediate records, emitted by the mapper programs, are partitioned by the reducer-id prefix and written into a set of buffers. Each mapper program is associated, by the XRT runtime, with a set of buffers, one for each reducer. This technique, inspired by the Phoenix runtime [18], enables lock-free access to the intermediate records stored in the associated buffers. A specialized memory structure is utilized by the buffers to minimize overhead and simultaneously structure the buffer for optimal sort performance.

Once a mapper terminates, the XRT runtime immediately sorts the associated buffers. XRT uses a quicksort algorithm with all the algorithmic optimizations expected from a modern industry-ready quicksort implementation [19]. Also, a specialized prefix-based memory structure is used to reduce CPU cache-misses and memory reads during sorting.

Multiple buffer spills might occur before the start of the sort stage if the input data volume is significant enough to fill up a buffer. During a spill, the buffer is first sorted and then written into an optimized disk-based data structure. During the regular sort stage, all on-disk files produced by a mapper are merged by the runtime into a single continuous ordered sequence of records on disk. Merging spill files is essential to reduce the total number of open spill files during the final merge.

For each reducer program, all associated buffers together with all associated merged spill files are merged by the runtime to produce a single ordered stream of records. This approach, heavily inspired by classical external sort algorithms [12], enables the processing of data sets vastly larger than available memory. The in-order records are consumed over the stdin stream by the associated

```ruby
require 'digest/md5'

n = ENV['REDUCERS'].to_i
STDIN.each do |line|
  line.split.each do |w|
    d = Digest::MD5.hexdigest(w)
    r = d.to_i(16) % n
    STDOUT.write "#{r}\t#{w}\n"
  end
end
```

```python
from sys import stdin, stdout

w = stdin.readline()
c = 1
for next in stdin:
  if w != next:
    stdout.write(f'{w}\t{c}')
    w = next
    c = 0
  c += 1
stdout.write(f'{w}\t{c}')
```

Example 1. word-count mapper implemented in Ruby.

Example 2. word-count reducer implemented in Python.

reducer program which applies some transformation, generally an aggregation. The reducer program then writes the result of the transformation as output records on the stdout stream. Similarly to the buffers, one output file is used per reducer program to ensure lock-free operation and maximal disk utilization.

3   Programming Model

This section includes a brief overview of the developer experience when developing XRT data processing jobs. Data processing jobs developed for most MapReduce runtimes generally requires the developer to implement a mapper and a reducer routine. In the case of XRT, the mapper and reducer routines are stand-alone programs, written in the programming language of choice, which the runtime will execute multiple times in parallel to achieve the desired degree of parallelism. XRT mapper and reducer programs consume newline-delimited records on the stdin stream and produce newline-delimited records on the stdout stream. In addition, the XRT runtime expects input files to consists of newline-delimited records and will produce output files consisting of newline-delimited records. The decision to only operate on newline-delimited data sets is critical to enable straightforward support for arbitrary programming languages.

To show the programming-language independent capability of the XRT runtime consider a data processing job for solving the word-count problem. In the word-count problem, a data processing runtime must consume a data set containing lines of text as input and produce a data set containing the occurrence count of each word. The following example provides a solution to this problem using a mapper implemented in the Ruby programming language and a reducer implemented in the Python programming language.

Example 1 provides a simple mapper program to solve the mapper portion of the word-count problem. The mapper program consumes lines of text from the stdin stream and splits each line into words (naively assuming that there are no special characters). It then assigns a reducer-id prefix to each word through a hash of the word modulo the number of reducers, ensuring that all identical words are assigned the same reducer-id. In this example, an environment variable set by the XRT runtime is utilized by the mapper program to establish the total number of reducers. The mapper program then writes a reducer-id to word pair as a newline-delimited record to the stdout stream.

Example 2 provides a simple reducer program to solve the reducer portion of the word-count problem. It consumes an in-order stream of words from the stdin stream and counts the number of occurrences of a particular word. Since the stream is in-order, the reducer program can emit the count of the current word to the stdout stream and start over when encountering a new word. The emitted word-to-count records become the output of the job and the solution for the word-count problem.

4   Optimisations and Extensions of XRT

XRT has benefited from a significant number of novel extensions and optimizations introduced in this section. The original version of XRT focused on establishing the viability of a programming-language independent MapReduce runtime for shared-memory systems without a focus on optimal performance. Since the original version, a performance-focused approach has been pursued to further the development of XRT. In particular, the adoption of frequent CPU profiling and regular benchmarks have resulted in numerous optimization and improvements, enhancing almost every subroutine of the XRT runtime.

Open File Handle Reuse. During the mapping stage of an XRT data processing job, the XRT runtime will chunk input files into logical 16 MB chunks and distribute the chunks to the active mapper programs based on demand. Each mapper worker, receiving a logical 16 MB chunk, is responsible for accessing the underlying file and seeking to the start of the chunk before consuming the contained records. The original version of XRT employs a naive approach where each mapper worker will process a logical input chunk by first opening the underlying file containing the chunk and seeking to the start of the chunk. The mapper worker then consumes the records associated with the chunk before closing the file handle of the underlying file. Lastly, the mapper worker waits for the next chunk and the process repeats.

In the extended version of XRT, each mapper worker will instead process chunks by first verifying if the underlying file of the chunk is the same file as the underlying file of the previous chunk. If the underlying file of the current chunk is the same file as the underlying file of the previous chunk, then the improved XRT runtime will re-use an open file handle. If the underlying file of the current chunk is not the same file as the underlying file of the previous chunk, then the improved XRT runtime will close the open file handle from the previous chunk and open the new underlying file. In both of the above cases, the mapper worker will then seek to the start of the chunk and processes it before waiting for the next chunk to process. Since chunks from input files often end up being distributed sequentially across the mapper workers by the XRT runtime, this approach can result in a significant reduction in file open calls and shorter seeks.

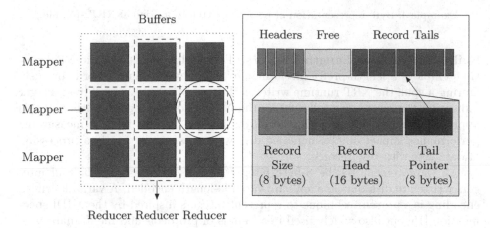

Fig. 2. Detailed illustration of the in-memory data layout.

Prefix-Pointer Buffers and Sort. A performant sort is of such importance that the original version of XRT already included a memory layout optimization for efficient sort. In the original version of XRT, the writing of a record to an in-memory buffer consist of prepending an 8-byte pointer to the buffer which points to an appended 8-byte length integer which is followed by the N-byte record. During a sort, only the prepended 8-byte pointers need to be moved, resulting in zero memory fragmentation and significantly reduced memory writes compared to moving the records. This optimization, however, has relatively bad CPU-cache awareness as comparing two records require accessing the relatively distant appended record bytes.

A better solution, inspired by AlphaSort [16,17], as illustrated in Fig. 2 and implemented in the extended version of the XRT runtime, is to append small record headers instead of pointers. In the extended version of the XRT runtime, writing a record to an in-memory buffer consists of prepending a record header containing an 8-byte integer identifying the record length, the first 16 bytes of the record and an 8-byte pointer to the start of the remaining N-16 bytes of the record. Similarly to the pointer based approach, each record's N-16 tail bytes are appended to the in-memory buffer. This technique has the same benefits as the method in the original version of XRT since, while sorting, the XRT runtime will only move the 32-byte record headers. However, this solution has significantly better CPU-cache awareness since the 16-byte record prefix, stored in the header, enable record comparisons without accessing record data stored outside of the header. Specifically, during a compare, the appended N-16 record bytes are only accessed for two records where the first 16 bytes are identical. Thus, the number of CPU-cache misses are reduced significantly during sorting resulting in an overall performance improvement.

```
000000000000000d 33372c323031392d30312d3230 | 37,2019-01-20
000000000000000d 33372c323031392d30312d3234 | 37,2019-01-24
000000000000000d 33372c323031392d30322d3032 | 37,2019-02-02
```

Example 4. Byte representation of records written to a original XRT spill file.

Spill with Variable Length Integers. The original version of the XRT runtime employs a length-prefix file format to store intermediate records on disk. During a spill, the XRT runtime writes each record to a file as an 8-byte integer indicating the record size followed by the N-bytes of the record. Although this approach works relatively well, the 8-byte integer prefixes incur a measurable performance impact due to increased disk writes, especially for data processing of data sets with tiny records.

An improvement to the 8-byte integer prefixes is to ensure that all integers written to disk during a spill are variable-length encoded. A variable-length encoding is a proven technique to represent integers inspired by the MIDI specification [15] but also widely used in all kinds of protocols and file formats. The benefit of the variable-length encoding is that smaller integers can be encoded using fewer bytes. This encoding strategy is of particular value to the XRT runtime since integers written to XRT spill files, representing record lengths, will generally be relatively small.

As a result, the number of bytes written per record during a spill to disk is decreased, especially for data processing jobs with smaller records, where the integer overhead becomes significant. The tradeoff is a slight increase in overall CPU utilization during writing and reading. Nonetheless, the decrease in disk I/O significantly outweigh this overhead.

Spill with Front Coding. As was discussed in the previous section on variable-length encoding, the original version of XRT writes spilled records to disk using a length-prefixed format. Example 4 illustrates this format by showing the content of a spill file from the original XRT runtime. An important observation from Example 4 is that the records are in-order, resulting in a significant amount of repeated prefixes. Indeed, since a sort of the in-memory buffer precurses any spill, records in spill files will generally have repeated prefixes. In the XRT model, this is especially likely as the runtime sorts records in byte-order.

Compression techniques taking advantage of in-order streams of records has been widely studied and is precisely what the improved XRT runtime leverages [22]. When the extended version of the XRT runtime writes a record to disk two variable-length encoded integers are written followed by a tail component of the record. The first integer indicates the size of the shared record prefix with the preceding record while the second integer indicates the size of the non-shared tail component of the record. Example 5 illustrates the significant reduction in bytes written to disk, for the same spill file as was presented in Example 4, when using both variable-length encoding and front coding.

```
00  0d  33372c323031392d30312d3230  | 37,2019-01-20
0c  01  34                          | 37,2019-01-24
09  04  322d3032                    | 37,2019-02-02
```

Example 5. Byte representation of records written to an improved XRT spill file.

Improved Merge. In the original version of XRT, the merge routine combines in-order streams of records into a single in-order stream of records through a heap-based k-way merge. In particular, the merge keeps a min-heap of in-order streams where the next record of a stream establishes the minimum of the stream. The merge routine produces a single in-order stream from the k in-order streams by removing the stream with the smallest next record from the heap, reading the record, advancing the stream, returning it to the heap and restoring the heap property before repeating the process. This approach is efficient and similar to strategies used in many parallel and external sorting algorithms [12,19].

The extended version of XRT introduces a slight improvement by not returning the stream with the smallest next record to the heap after reading one record, but instead keep reading records as long as the records are identical to the first record. Once the merge routine reads a different record, the XRT runtime returns the stream to the heap and restores the heap property before repeating the process. This minor modification can significantly reduce the amount of CPU time spent on heap rebalancing, especially for skewed data sets or data sets with multiple identical records.

Assembly Integer Encoding and Decoding. For the XRT runtime to store integers in the in-memory buffers, it must serialize and deserialize Go typed integers to and from bytes. The non-assembly solution for serializing and deserializing a Go integer to and from bytes is already extremely performant and responsible for only a small amount of total CPU usage. Actually, from a CPU usage perspective, it does not make much sense to implement these operations in hand-optimized assembly.

The reason for hand-optimizing integer encoding and decoding is that the most performance critical section of assembly code, presented last in this section, requires reading record-associated integers mid-execution. Thus, the implementation of integer encoding and decoding in assembly is mostly to ensure that the same subroutine handles integer encoding and decoding in Go and assembly routines. The hand-optimized assembly routines for integer encoding and decoding are straightforward implementations that work by bypassing the type system and directly accessing the bytes underlying the Go typed integer.

As can be observed in Fig. 3, the reduction in CPU utilization following this optimization is minor. In fact, due to the assembly optimization CPU time spent on integer encoding and decoding is decreased from 2.02% to 1.64%.

Assembly Swap. The native Go version of the swap routine consists of first allocating a temporary 32-byte buffer and copying one of the two record headers into it. The swap routine then copies the second record header into the position of the first record header. Finally, the routine copies the first record header, stored in the temporary buffer, into the original position of the second record header. A slight overhead exists in this approach as the use of a temporary buffer not only introduces a memory allocation but also results in three record header memory reads, and three record header memory writes.

In the hand-optimized assembly implementation of the swap routine better CPU register utilization is possible. Specifically, the 32-byte record header swap is significantly improved by entirely executing the swap using CPU registers. The hand-optimized swap routine starts by first moving the first record header into the XMM1 and XMM2 registers followed by moving the second record header into the XMM3 and XMM4 registers. The routine then moves the first record header from the XMM1 and XMM2 registers into the original memory location of the second record header. Finally, it moves the second record header from the XMM3 and XMM4 registers into the original memory location of the first record header. As a result, the hand-optimized assembly implementation only requires two record header memory reads, and two record header memory writes.

An XRT data processing job running without hand-optimized assembly optimizations requires approximately 3.02% of total CPU time for the swap routine. In the improved version of XRT, with assembly optimizations, this has been decreased so that, as is illustrated in Fig. 3, only 0.69% of total CPU utilization is spent on the swap routine.

Assembly Compare. The native Go implementation of the compare operation starts by reading the record lengths from the record headers of the two records for which the runtime wants to establish the ordering. The compare operation then compares the record prefixes stored in the record header and returns the result if it is capable of determining the ordering. If the compare operation cannot establish the order, meaning the prefix components are equal, then the compare operation will read the tail pointers from the record headers, access the tail components of the two records and finally compare the tail components to establish the order. Throughout the execution of the compare operation, a few performance issues are present. In particular, multiple function calls are required which the Go compiler fails to inline. Moreover, the Go compiler fails to fully utilize CPU registers, leading to some unnecessary memory allocations. Besides, byte-slice operations are required to access parts of the in-memory buffer which results in the introduction of various out-of-bounds checks by the Go compiler, adding to the overall overhead.

The hand-optimized assembly version of the compare operation solves all the performance shortcomings. First, the entire compare is written in assembly, removing any overhead associated with function calls. Second, there is no memory allocation required, and the entire compare operation can be completed through CPU registers. Finally, there are no out-of-bounds checks or non-optimal assembly as the compare operation can leverage various assumptions around the shape of the XRT in-memory buffer.

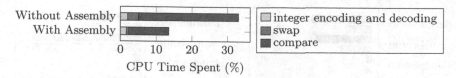

Fig. 3. CPU time spent on sub-routines for the same data processing job before and after the introduction of hand-optimized assembly to the XRT runtime.

At a high level, the hand-optimized assembly version of the compare operation progresses in the same manner as the native Go version. It starts by loading the record lengths of the two records into registers RBX and RDX followed by a move of the record prefixes from the record headers into registers XMM0 and XMM1. The compare operation then attempts to establish the order using the prefixes stored in registers XMM0 and XMM1. If the compare operation establishes the order, then the result is returned. Otherwise, the hand-optimized assembly version of the compare operation enters a partially unrolled loop that uses wide registers to effectively compare as large segments of the record tails as possible, resulting in a minimal number of instructions. The partially unrolled loop also reduces the number of jump instructions, further reducing the total number of instructions required to complete the comparison.

As is illustrated in Fig. 3, XRT will spend approximately 28.11% of total CPU time in the compare sub-routine for the native Go implementation, even with the record header optimization. However, with the hand-optimized assembly version of the compare sub-routine, the XRT runtime is spending significantly less time in the compare sub-routine at approximately 11.21%.

5 Results and Discussion

All tests in this section were conducted on the Graham cluster [2] using a single bigmem3000 64 core server with 3 TB of memory and high-performance networked storage. The input for all benchmarks came from one of two generated 100 GB data sets containing 1 billion 100-byte records. The first data set contained a perfectly uniform distribution of records while the second data set contained a significant skew where half the records were identical. All XRT benchmarking jobs were implemented using Python 3.7 and executed on CentOS Linux 7.4 using an XRT runtime compiled with Go 1.10.

5.1 Improvements

The algorithmic and implementation extensions, together with the introduction of hand-optimized assembly, has resulted in a significant overall increase in XRT performance. To illustrate, Fig. 4 demonstrates the speedup of the improved version of XRT in various benchmarking scenarios relative to the original version of XRT. Three typical MapReduce tasks were implemented and benchmarked

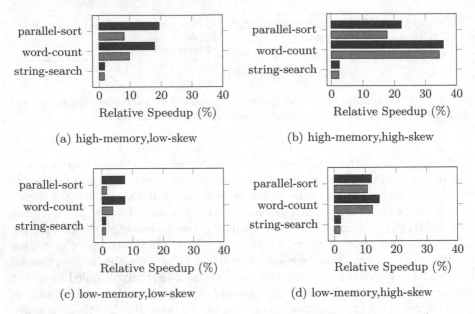

(a) high-memory,low-skew

(b) high-memory,high-skew

(c) low-memory,low-skew

(d) low-memory,high-skew

Fig. 4. Relative speedup of the improved version ■ and improved version with assembly ■ of XRT compared to the original version of XRT for various data processing tasks on various environments.

against uniform and skewed data sets in two different memory environments. The first memory environment offered 500 GB of available memory while the second, low-memory environment, was limited to 1 GB of available memory.

The string-search task is a map-only data processing job which filters through input records and produces an output data set containing records matching the search criteria. Since this data processing job does not require any sorting or in-memory buffering the various improvements made to the XRT runtime are not as applicable. The only optimization responsible for a slight relative speedup, observable in all scenarios in Fig. 4, is the implementation of input-chunk file-handle re-use. The word-count task is a MapReduce data processing job which consumes a data set constituting of lines of text and returns a word to occurrence count mapping. As is observable in Fig. 4, this type of simple aggregation dramatically benefits from the various novel optimizations and improvements. The parallel-sort task is another MapReduce data processing job which consumes an unordered data set of records and returns the same data set in order. As is apparent in Fig. 4, this data processing task also dramatically benefits from the various novel optimizations and improvements.

Figure 4 (a) details the results from benchmarks against a data set with a uniform distribution in a high-memory environment. Processing a uniform data set in a high-memory environment presents an ideal scenario as computations are guaranteed to be spread evenly across all active cores while a high-memory environment enables all processing to occur without disk spills. In this environment, word-count and parallel-sort benefit mainly from the record-header algorithmic improvement together with the hand-optimized assembly.

Figure 4 (b) depicts the results from benchmarks against a data set with a skewed distribution in a high-memory environment. This situation shows the best overall improvements due to many of the novel developments in the improved version of XRT. A high-skew will result in multiple disk spills since a subset of the in-memory buffers will receive a considerable number of records. As a result, this scenario significantly benefits from the variable-length integers, spill with front coding and the improved merge optimizations. Also, since this environment is a high-memory environment, the spills contain a large number of records. A large number of records per spill in a high-skew scenario is especially beneficial since front coding will compress a large number of identical records, significantly reducing disk writes. Hand-optimized improvements are only slightly beneficial in this scenario as disk writes quickly becomes the primary bottleneck.

Figure 4 (c) shows results from benchmarks against a data set with a uniform distribution in a low-memory environment. This environment shows the least benefits from the various improvements with only a tiny speedup without assembly optimizations. The tiny speedup comes mainly from the variable-length integers, improved merge and spill with front coding, which assists in reducing the amount of disk writes. Nevertheless, since the data set is uniform and the spills are small, the compression ratio is not as high as for a skewed data set, resulting in the lower overall performance increase. With assembly optimization, the benchmark is slightly better due to the faster sort.

Figure 4 (d) details result from benchmarks against a data set with a skew distribution in a low-memory environment. Similarly to the skewed distribution benchmark in a high-memory environment, this benchmark benefits greatly from the variable-length integers, spill with front coding and merge improvements. Above all, a high number of similar records are tightly compressed in the spills, resulting in a significant decrease in disk writes. The improvement is not as significant as the high-memory environment since each spill contains fewer records resulting in fewer opportunities for compression and more spill files to merge. More spill files to merge results in more disk activity which quickly becomes the bottleneck for this benchmark.

5.2   Comparisons

This section presents a comparative benchmark of the word-count problem to situate the performance of the XRT runtime amongst widely used MapReduce runtimes. The Phoenix [18,23], Metis [14] and Phoenix++ [21] runtimes represent academic shared-memory MapReduce runtimes while the Apache Spark [24] and Hadoop Streaming [5] runtimes represent distributed runtimes which are widely used in industry.

Although the XRT runtime borrows ideas from Phoenix, Metis and Phoenix++ these three shared-memory runtimes are significantly different. MapReduce tasks for these three runtimes must be written in C or C++ through complex programming interfaces and compiled into a single binary. Notably, the

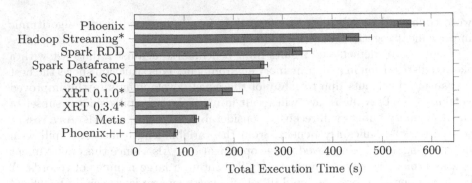

Fig. 5. Total execution time in seconds for various data processing runtimes solving the word-count data processing problem in the same environment. Programming-language independent runtimes are identified by a star*.

word-count data processing job benchmarked in this section requires 341 lines of C for the Phoenix runtime, 196 lines of C for the Metis runtime and 236 lines of C++ for the Phoenix++ runtime. In contrast, the XRT word-count data processing job spans 42 lines of approachable Python, similar to the code presented earlier in Example 1 and Example 2. Writing MapReduce tasks using any programming language while simultaneously requiring less code is a key advantage of the XRT runtime when compared to Phoenix, Metis and Phoenix++.

Apache Spark has quickly become one of the most widely used data processing runtimes utilized in industry. More importantly, Hadoop Streaming, although less utilized, is the only industry-ready runtime that provides a programming-language independent alternative to the XRT runtime. Both of these runtimes target a distributed environment of networked commodity hardware. However, for this benchmark, the runtimes were configured for a shared-memory environment to achieve a closer comparison.

Figure 5 illustrates the results of the benchmarks and serves as a clear indication of the performance of the XRT runtime. As expected, Hadoop Streaming performs poorly. The older Python Spark RDD API performs significantly worse than the novel Spark Dataframe and SQL API's, which perform nearly identically. The old Phoenix runtime performs exceptionally poorly. However, the original Metis paper reports a four to five times speedup over Phoenix, which is corroborated by these results. The Phoenix++ and Metis runtimes perform exceptionally well as expected due to their compiled nature and use of C or C++ without inter-process communication. Finally, the XRT runtime is capable of significantly outperforming Phoenix, Hadoop Streaming and Apache Spark. Indeed, being nearly twice as fast as the popular Spark Dataframe and SQL API's. More importantly, XRT can outperform Hadoop Streaming, the only genuinely programming-language independent runtime, by a factor of 3. Metis and Phoenix++ achieve slightly better performance than XRT since their lack of support for programming-language independent MapReduce allows them to avoid certain performance overheads. In particular, Metis and Phoenix++ do not use multiple processes or data transfers over standard streams, fundamental techniques for the XRT programming-language independent model.

6 Conclusions

To our knowledge, the open-source XRT runtime is the only programming-language independent MapReduce solution for shared-memory systems. Furthermore, following the new extensions, presented in this paper, XRT has quickly become one of the fastest MapReduce runtimes available. Through the various benchmarks, it is clear to us that XRT generally provides a more efficient data processing environment when compared to distributed data processing runtimes, such as Hadoop Streaming or Apache Spark. Besides, when comparing the improved version of XRT to the original version of XRT, it is also clear that the new improvements and optimizations have been a success.

Going forward, we intend to continue evolving the XRT runtime and pursue the ideas and philosophies motivating the XRT project. In particular, hand-optimized assembly routines for additional architectures, especially ARM, are currently in progress. The ARM architecture is increasingly popular and widely used in resource-constrained systems, where data processing using the XRT runtime works very well. Also, on the opposite end, we intend to continue evaluating XRT on high-core count deployments and tune various subroutines further. In particular, we are increasingly interested in replacing the sorting algorithm with an implementation designed especially for the kind of sorts that the XRT runtime undertakes.

Overall, we believe the improved version of XRT further illustrates the great performance capabilities of large shared-memory systems and that programming-language independent MapReduce can be extremely performant while remaining flexible and straightforward.

References

1. The aws x1 instance type. https://aws.amazon.com/ec2/instance-types/x1
2. Graham. https://docs.computecanada.ca/wiki/Graham
3. Hadoop. http://hadoop.apache.org
4. Hadoop pipes. https://hadoop.apache.org/docs/r1.2.1/api/org/apache/hadoop/mapred/pipes/package-summary.html
5. Hadoop streaming. http://hadoop.apache.org/docs/current/hadoop-streaming/HadoopStreaming.html
6. List of google cloud instance types. https://cloud.google.com/compute/docs/machine-types
7. Abuín, J.M., Pichel, J.C., Pena, T.F., Gamallo, P., García, M.: Perldoop: efficient execution of perl scripts on hadoop clusters. In: 2014 IEEE International Conference on Big Data (Big Data) pp. 766–771 (2014)
8. Arif, M., Vandierendonck, H., Nikolopoulos, D.S., de Supinski, B.R.: A scalable and composable map-reduce system. In: IEEE International Conference on Big Data (Big Data) pp. 2233–2242 (2016)

9. Chen, R., Chen, H.: Tiled-mapreduce: efficient and flexible mapreduce processing on multicore with tiling. ACM Trans. Archit. Code Optim. **10**(1), 3:1–3:30 (2013). https://doi.org/10.1145/2445572.2445575

10. Dean, J., Ghemawat, S.: Mapreduce: simplified data processing on large clusters. In: OSDI Proceedings of the 6th Symposium on Operating Systems Design and Implementation (2004)

11. Ding, M., Zheng, L., Lu, Y., Li, L., Guo, S., Guo, M.: More convenient more overhead: the performance evaluation of hadoop streaming. In: Proceedings of the 2011 ACM Symposium on Research in Applied Computation pp. 307–313 (2011)

12. Knuth, D.E.: The art of computer programming. Sorting and Searching, vol. 3, 2nd edn. Addison Wesley Longman Publishing Co., Inc., Redwood (1998)

13. Leo, S., Zanetti, G.: Pydoop: a python mapreduce and hdfs api for hadoop. In: Proceedings of the 19th ACM International Symposium on High Performance Distributed Computing pp. 819–825 (2010)

14. Mao, Y., Morris, R., Kaashoek, M.F.: Optimizing mapreduce for multicore architectures. Tech. Rep, Computer Science and Artificial Intelligence Laboratory, Massachusetts Institute of Technology (2010)

15. MIDI Association: The Complete MIDI 1.0 Detailed Specification (2014)

16. Nyberg, C., Barclay, T., Cvetanovic, Z., Gray, J., Lomet, D.: Alphasort: a risc machine sort. In: Proceedings of the 1994 ACM SIGMOD International Conference on Management of Data. pp. 233–242. SIGMOD 1994, ACM, New York, NY, USA (1994). https://doi.org/10.1145/191839.191884

17. Nyberg, C., Barclay, T., Cvetanovic, Z., Gray, J., Lomet, D.: Alphasort: a cache-sensitive parallel external sort. VLDB J **4**(4), 603–628 (1995). http://dl.acm.org/citation.cfm?id=615232.615237

18. Ranger, C., Raghuraman, R., Penmetsa, A., Bradski, G., Kozyrakis, C.: Evaluating mapreduce for multi-core and multiprocessor systems. In: IEEE 13th International Symposium on High Performance Computer Architecture pp. 13–24 (2007)

19. Sedgewick, R.: Algorithms in C, 3rd edn. Addison-Wesley Longman Publishing Co. Inc, Boston (2002)

20. Selin, E., Viktor, H.: Xrt: programming-language independent mapreduce on shared-memory systems. In: 2018 IEEE International Congress on Big Data (BigData Congress). pp. 182–189 (July 2018). https://doi.org/10.1109/BigDataCongress.2018.00031

21. Talbot, J., Yoo, R.M., Kozyrakis, C.: Phoenix++: modular mapreduce for shared-memory systems. In: Proceedings of the Second International Workshop on MapReduce and its Applications pp. 9–16 (2011)

22. Witten, I.H., Moffat, A., Bell, T.C.: Compressing and indexing documents and images. Managing Gigabytes, 2nd edn. Morgan Kaufmann Publishers Inc., San Francisco (1999)

23. Yoo, R.M., Romano, A., Kozyrakis, C.: Phoenix rebirth: scalable mapreduce on a large-scale shared-memory system. In: IEEE International Symposium on Workload Characterization pp. 198–207 (2009)

24. Zaharia, M., Chowdhury, M., Franklin, M.J., Shenker, S., Stoica, I.: Spark: cluster computing with working sets. In: Proceedings of the 2nd USENIXConference on Hot Topics in Cloud Computing. pp. 10–10. HotCloud 2010, USENIX Association, Berkeley, CA, USA (2010).http://dl.acm.org/citation.cfm?id=1863103.1863113

Evaluating Redundancy and Partitioning of Geospatial Data in Document-Oriented Data Warehouses

Marcio Ferro[1,3](✉) [ID], Rinaldo Lima[2] [ID], and Robson Fidalgo[1] [ID]

[1] Federal University of Pernambuco, Recife, Brazil
{mrcf,rdnf}@cin.ufpe.br
[2] Federal Rural University of Pernambuco, Recife, Brazil
rinaldo.jose@ufrpe.br
[3] Federal Institute of Alagoas, Viçosa, Brazil

Abstract. A Geospatial Data Warehouse (GDW) is a repository of historical and geospatial data used in the decision-making process. These systems manage large volumes of data, and their dimensions are usually denormalized to increase query performance. Many studies have analyzed the impact of geospatial data redundancy on a relational GDW. However, to the best of our knowledge, no previous study performed a similar analysis considering the NoSQL scenario. In this context, to design a scalable document-oriented GDW (DGDW) with low storage cost and low query response time, it is important to identify which geospatial fields should be normalized (referenced) or denormalized (embedded), as well as how the documents should be partitioned among collections. In this study, we exhaustively evaluated 36 DGDWs in the MongoDB document-oriented database with different levels of geospatial redundancy and different approaches to partitioning documents among collections. Our experimental results indicate that both the normalization of low-selectivity geospatial fields and the partitioning of documents into homogenous collections provide better query performance and lower storage space. The performance evaluation presented in this paper provides strong evidence that can help guide the creation of a DGDW.

Keywords: Geospatial Data Warehouse · NoSQL · Experimental evaluation

1 Introduction

A Data Warehouse (DW) is a repository of historical data used in the decision-making process [8]. Recent studies have proposed the use of NoSQL databases to implement DW [1–4,6,10,11,16,20–22]. The main reason for this is to achieve

This work was supported by Fundação de Amparo à Pesquisa do Estado de Alagoas (FAPEAL).

C. Ordonez et al. (Eds.): DaWaK 2019, LNCS 11708, pp. 221–235, 2019.
https://doi.org/10.1007/978-3-030-27520-4_16

high scalability with low hardware costs, where the data is distributed and processed by machines in a cluster. However, given the growing volume of data in the Big Data scenario, it is important to identify how DW schemas should be modeled to achieve good performance and low storage cost.

A DW can store geospatial objects, which consist of a descriptive component (e.g., name or address) and a geospatial component (e.g., vector geometry or raster surface) [7]. In this paper, the geospatial component is restricted to geometric data. Therefore, this study uses "geospatial data" and "geospatial documents" as synonyms for geometric data. Given the unconventional nature of these data, the design of a Geospatial DW (GDW) provides several challenges for storing and processing large volumes of data [9]. Studies in GDW using a relational database [12,17] indicates how fact tables, conventional dimensions, and geospatial dimensions can be modeled to ensure good performance. However, conducting this type of study of a GDW with a NoSQL database to support large volumes of data is still an unexplored topic. In this context, we present the following research problem: *how should document-oriented GDW (DGDW) schemas be modeled to achieve both low storage cost and high query performance?*

In this study, we analyze the query performance and the storage cost of DGDW schemas that have: (i) several levels of geospatial data redundancy; and (ii) partitioning of documents among different collections. For that, we use MongoDB, because it is the most popular NoSQL database used in business applications with geospatial data support [5]. We also define a UML-based notation to represent the DGDW schemas. Using this notation, we exhaustively generated 36 DGDW schemas that have different models for (i) fact tables and their dimensions, (ii) conventional dimensions and geospatial dimensions, and (iii) geospatial dimensions and themselves. Finally, we evaluate these 36 schemas based on query execution time and data volume. In this context, our key contributions are: (1) a performance evaluation that identifies the level of redundancy and partitioning of documents into collections that provides good performance and low-cost storage; and (2) a UML-based notation that can model DGDW schemas in the MongoDB database.

The organization of this paper is as follows: Sect. 2 presents related work, Sect. 3 introduces the UML-based notation and the normalization/partitioning approaches used to model the 36 GDW schemas of our experiment, Sect. 4 details our experiment, Sect. 5 discusses the results, and finally, Sect. 6 presents the paper's conclusions.

2   Related Work

A study on the impact of geospatial data redundancy on relational GDW (RGDW) is presented in [18], which compares two schemas containing normalized and denormalized geospatial data. The data redundancy cost (in query performance and storage consumption) in the schema with denormalized geospatial data was found to be higher than the cost of having a greater number of joins to perform in the schema with normalized geospatial data. A similar study [12] proposed that geospatial dimensions with only one relationship should be denormalized in the conventional dimension. This study also investigated whether the complexity of the geospatial data (i.e., point vs. polygon) influences the performance of the queries, and identified that the denormalization of geospatial data of low complexity (i.e., points) improves query performance. Although both studies [12,18] examined the impact of data redundancy on RGDW, no analysis was performed to determine whether selectivity of a geospatial field must influence its normalization/denormalization. It is worth noting that the geometry of a geospatial object is not a good factor to define whether it has low or high selectivity, and therefore, whether the data should be normalized or not. For example, considering a GDW containing costs or revenues for a particular country's embassies, its geospatial fields for country, state, city, and address have high selectivity, regardless of the type of geometries, because there is only one embassy per country. That is, each piece of geospatial data is related to only one embassy.

Reference [13] discusses the use of a structure in cloud computing to provide horizontal scalability in an RGDW. In that study, the use of multiple machines connected in a cluster increases both the storage and processing capacity, factors that contribute to improved RGDW performance. However, the authors did not address the use of a NoSQL database.

In short, to the best of our knowledge, no previous study has analyzed the use of NoSQL databases in GDW. However, some studies address the use of NoSQL for conventional DW. Among those studies, we highlight: [21] which presents a methodology based on a key-value database to model a DW to support Big Data applications; references [1,4,6,11,16,20,22] investigate the use of column-oriented databases; references [2–4,22] investigate the use of document-oriented databases; and [10] proposes a graph-oriented approach. Among these studies, a performance analysis between relational DW and NoSQL DW [1] shows that NoSQL databases have better performance in a cluster. Reference [1] also shows that the use of NoSQL databases is more acceptable when it is intended to achieve horizontal scalability.

3   DGDW Design

Given the popularity and expressiveness of the Unified Modeling Language (UML) class diagram, we adopted its package, class, composition, and association constructors to represent, respectively, the following constructors: collection (i.e. a collection of documents), document type (i.e., documents having the same field structure), embedded documents (i.e., denormalized documents), and referenced documents (i.e., normalized documents). We highlight that a document type can correspond to a fact table (i.e., Fact Document Type, graphically represented with a thick line) or a dimension table (Dimension Document Type, graphically represented with a thin line). Figure 1 illustrates our notation for specifying the different relationship forms (i.e., normalized/referenced vs. denormalized/embedded) and different partitioning forms (i.e., homogeneous collection vs. heterogeneous collection) in DGDW. That is, our notation allows us to represent relationships between: (i) fact table and dimensions (Fig. 2); (ii) conventional and geospatial dimensions (Fig. 3); and (iii) geospatial dimensions and themselves (Fig. 4).

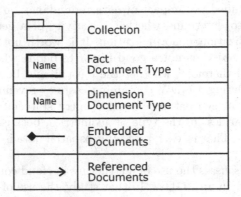

Fig. 1. Notation used to model DGDW schemas.

Figure 2 introduces four models for facts and dimensions. In model M1, there is a single collection of documents. Each fact is stored in a single document, which has a set of embedded documents, one for each dimension (D1 and D2 represent dimensions). Model M2 also has a single collection of documents. However, in this collection, facts and dimensions are normalized into distinct documents and referenced by an identifier. Model M3 also normalizes facts and dimensions into separate and referenced documents, but this model partitions these documents into two collections: one to store the facts, and another to store the dimensions. Model M4 differs from the M3 only in the number of collections, because there is one collection for each dimension.

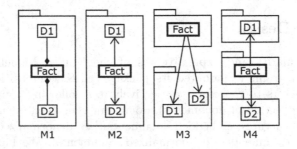

Fig. 2. Relationships between facts and dimensions.

Figure 3 shows distinct forms of modeling relationships between the conventional and geospatial dimensions. We define each conventional or geospatial dimension as, respectively, a conventional document type (CD) or a geospatial document type (GD). Thus, model A represents a GD embedded/denormalized into the CD, i.e., if a geospatial document needs to be used by more than one CD, it must be replicated for each CD. Model B represents a reference from a CD to a GD. In this model, if a geospatial document needs to be used by more than one CD, there will be no redundancy. Model C is a hybrid representation of the two previous models, in which a geospatial field with low-selectivity (e.g., *"nation":aNation*) is mapped to a GD and is normalized/referenced by a CD, whereas a geospatial field with high-selectivity (e.g., *"address":anAddress*) is mapped to a GD and is denormalized/embedded into a CD.

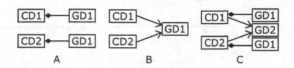

Fig. 3. Relationships between CD and GD.

Figure 4 presents three forms of representing the relationships between GD and themselves. In model G1, there is no relationship between GDs, because these are related only to the CD. In model G2, the GD is embedded/denormalized into other GD, so that the GD with higher selectivity embeds/contains the GD with lower selectivity (e.g., *"city": aCity* ⊃ *"nation": aNation* ⊃ *"region": aRegion*). Model G3 normalizes the GD into separate and referenced documents.

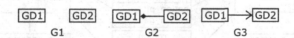

Fig. 4. Relationships between GD.

4   Experimental Evaluation

To the best of our knowledge, there is no benchmark for DGDW. Therefore, based on the models defined in Sect. 3, we used the RGDW schema presented in Fig. 5 to generate the DGDW schemas that were evaluated in our experiment. The RGDW is based on the SSB Benchmark [14] and adapted to support geospatial data [19]. We generated the RGDW with a scale factor of 1, because this factor was sufficient to identify which DGDW schemas are prohibitive or have both higher performance and lower storage cost. The RGDW was stored in PostgreSQL 10.3 with PostGIS 2.4.3. As can be seen in Fig. 5, the RGDW has a fact table called *lineorder*; four conventional dimensions called *part*, *supplier*, *date*, and *customer*, as well as five geospatial dimensions called *c_address*,

Table 1. RGDW used in numbers.

Table	Records	Columns	Size in MB
Lineorder	6,001,171	17	981.0
Customer	30,000	12	27.0
Date	2,556	17	0.3
Part	200,000	9	27.0
Supplier	2,000	11	1.4
c_address	30,000	2	2.8
s_address	2,000	2	0.2
City	250	2	7.7
Nation	25	2	5.7
Region	5	2	2.3

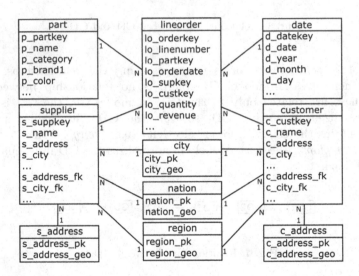

Fig. 5. RGDW schema used to generate the DGDW schemas.

s_address, city, nation, and region. Table 1 displays the number of records, number of columns, and the table size, which corresponds to 1.03 GB, in total.

We used Pentaho Data Integration 8.0 to exhaustively transform the RGDW into 36 DGDWs (i.e., {M1, M2, M3, M4} × {A, B, C} × {G1, G2, G3}) (Fig. 6). The name of each DGDW schema is a concatenation of the model names presented in Sect. 3 (i.e., M1 + A + G1 = M1AG1, M1 + A + G2 = M1AG2, and so on). Although *part* and *date* are not shown in Fig. 6, these dimensions was included in the experiment and are related to *lineorder* as just like *supplier* and *customer*. We decided to highlight only the CD (i.e., *supplier* (sup) and

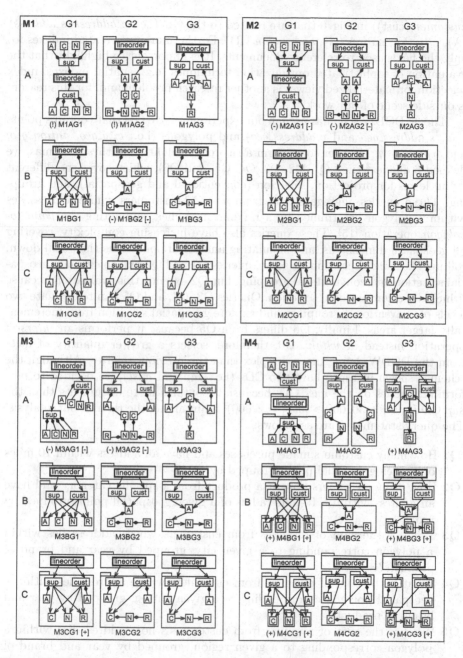

Fig. 6. DGDW schemas. Symbols used to highlight the schemas discussed in the Experimental Results Section: (+) Best volume size results; (−) Worst volume size results; [+] Best query performance results; [−] Worst query performance results; (!) Schemas discarded from the experimental evaluation.

customer (cust)) in Fig. 6 that are related to the GD (i.e., *c_address/s_address*[1] (A), *city* (C), *nation* (N), and *region* (R)). Furthermore, we created indexes for only the fields used to reference documents. It is important to highlight that the evaluation of sophisticated optimization techniques (e.g., spatial indexes, index selection, or materialized views) is another research problem and, for this reason, is outside scope of our work.

The 36 DGDW schemas used in our experiment have GD that contains points (i.e., *s_address_geo* and *c_address_geo*) and polygons (i.e., *city_geo*, *nation_geo*, and *region_geo*). In order to perform a rigorous evaluation of these schemas, we defined 6 queries[2] (i.e., Q1, Q2, Q3, Q4, Q5, and Q6), which cover the different detail levels for projection, selection (conventional and geospatial), and grouping. These queries aggregate data from a set of customers whose address lies within a polygon determined by a random user click, simulating the *point-to-polygon* operation [15]. These queries also have increasing complexity, covering all geospatial dimensions, and simulating analytical operations (i.e., drill down, roll up, slice, and dice). In this sense, queries Q1 and Q2 aim to evaluate geospatial selection and conventional grouping, but Q2 extends this evaluation because it includes a conventional selection. Queries Q3, Q4, and Q5 aim to evaluate two levels of conventional groupings and explore geospatial selection from incrementally larger areas. Finally, Q6 differs from Q5 because it performs an *intersect* operation instead of *within*. Q6, therefore, selects a greater quantity of facts from the DGDW, being more complex than the previous queries. Although the schemas shown in Fig. 6 have two CDs (i.e., *sup* and *cust*), we decided to perform all queries on the *cust* CD, because it contains more documents than the *sup* CD ($|cust| = 30,000 > |sup| = 2,000$), maintaining the rigor of our analysis. The query statements are as follows:

Q1 How many customers made purchases and have an address within 10 miles of a given geospatial point, grouped by year?

Q2 How many customers bought a product from a specific category and have an address within a 10-mile radius of a given geospatial point, grouped by year?

Q3 What is the sum of revenues from customers whose address lies within a polygon corresponding to a given city, grouped by year and brand of products?

Q4 What is the sum of revenues from customers whose address lies within a geospatial polygon corresponding to a given nation, grouped by year and brand of products?

Q5 What is the sum of revenues from customers whose address lies within a polygon corresponding to a given region, grouped by year and brand of products?

---

[1] The *c_address* and *s_address* dimensions will be mentioned as *address*, since they have the same document type (i.e., the same field structure).

[2] The queries are available in https://github.com/mrcferro/gdw.

Q6 What is the sum of revenues from customers whose address intersects a polygon corresponding to a given region, grouped by year and brand of products?

We perform the queries five times for each DGDW schema shown in Fig. 6. The test environment was composed of four computers. The central node was an Intel Xeon with 1 TB HD, 8 GB RAM, and a 1 Gbit/s network, and three other nodes were equipped with an Intel Core i3 processor having 500 GB HD, 4 GB RAM, and a 1 Gbit/s network. The computers used a CentOS 7.1 operating system and MongoDB 3.6.4 in sharding mode, which distributes the data across the machines. This setup favors horizontal scalability with low cost, by adding more hardware to the cluster.

The experimental evaluation investigated each schema shown in Fig. 6, analyzing the data volume and the arithmetic mean of five execution times for each query. Furthermore, we took special care to control the experiments. Each computer node was used exclusively for the proposed assessment and no other background processes were allowed to run while executing the queries. We rebooted all machines before executing the queries for each schema, cleaning any cached data used by the previous query.

5 Experimental Results

During the transformation from the RGDW to the 36 DGDWs, we noted that the M1AG1 and M1AG2 schemas would each need approximately 16 TB of storage. These schemas were therefore discarded from the experimental evaluation, because their data volume was much larger than that supported by the test environment (≈ 2.5 TB). This high amount of storage was a consequence of considerable data redundancy, which follows the successive nesting of documents (i.e., $Fact \supset CD \supset GD$).

Table 2 presents the results of the experimental evaluation, showing the following columns: *Schema*, the name of the schema; *Size*, the disk space required to store each schema (in GB); *Q1* to *Q6*, the average of the five execution times for each query in seconds, followed by its respective standard deviation; $Avg(Q_{1-6})$, the average of the execution times for the six queries, also in seconds. The table is sorted in ascending order on column $Avg(Q_{1-6})$.

The schemas were evaluated based on data volume (*Size* column), and the average runtime of the 6 queries ($Avg(Q_{1-6})$ column). Figure 7 depicts the volume vs. runtime relationship, showing each schema by its identifier (M1, M2, M3, and M4). Among the 34 schemas in Fig. 7, 18 have very similar results, with volume and runtime below 3 GB and 400 s, respectively. Figure 8 shows these 18 results at a larger scale.

Due to the great number of schemas, we will discuss the features of the top 5 (best) schemas and the bottom 5 in order to have some insight regarding the two ends of the results spectrum. Section 5.1 discusses the results based on data volume while Sect. 5.2 takes the average query execution time into account.

Table 2. Results of the experimental evaluation. Size in GB and query execution time in seconds. Symbols used to highlight the DGDW schemas: (+) Best results in volume size; (−) Worst results in volume size; [+] Best results in query performance; [−] Worst results in query performance; (!) Schemas discarded from experimental evaluation.

Schema	Size	Q1	Q2	Q3	Q4	Q5	Q6	$Avg(Q_{1-6})$
M4CG1	0.95(+)	0.59(0.00)	0.59(0.00)	1.33(0.00)	12.87(0.02)	61.12(0.07)	61.00(0.08)	22.92[+]
M4CG3	0.94(+)	0.57(0.00)	0.59(0.00)	1.33(0.00)	12.94(0.03)	61.76(0.08)	61.84(0.12)	23.17[+]
M4BG1	0.95(+)	0.58(0.01)	0.60(0.01)	1.36(0.02)	13.06(0.05)	62.26(0.16)	62.19(0.15)	23.34[+]
M4BG3	0.96(+)	0.59(0.01)	0.60(0.01)	1.36(0.02)	13.18(0.03)	62.49(0.16)	62.64(0.19)	23.48[+]
M3CG1	1.28	0.68(0.00)	0.69(0.00)	1.48(0.00)	13.67(0.03)	64.85(0.20)	65.01(0.14)	24.40[+]
M3CG3	1.15	0.67(0.01)	0.69(0.00)	1.47(0.00)	13.68(0.05)	65.19(0.14)	65.19(0.10)	24.48
M3BG3	1.20	0.69(0.00)	0.71(0.01)	1.51(0.03)	13.87(0.04)	65.28(0.12)	65.50(0.07)	24.59
M3BG1	1.14	0.70(0.00)	0.73(0.03)	1.53(0.02)	14.03(0.01)	65.57(0.21)	65.81(0.13)	24.73
M4AG3	0.94(+)	0.70(0.01)	0.65(0.03)	1.59(0.04)	15.64(0.03)	75.26(0.16)	75.51(0.08)	28.22
M3AG3	1.19	0.77(0.01)	0.72(0.03)	1.66(0.01)	15.71(0.07)	75.34(0.23)	75.77(0.43)	28.33
M2CG1	1.49	5.32(0.11)	5.33(0.21)	6.20(0.10)	18.92(0.09)	71.80(0.29)	71.53(0.13)	29.85
M2CG3	1.74	5.48(0.13)	5.59(0.12)	6.43(0.06)	19.22(0.33)	72.62(0.37)	70.41(0.37)	29.96
M4CG2	1.10	0.59(0.00)	0.61(0.02)	1.38(0.02)	13.72(0.01)	73.76(0.12)	90.78(0.14)	30.14
M3CG2	1.18	0.98(0.01)	0.95(0.04)	1.70(0.07)	13.65(0.04)	73.45(0.20)	90.56(0.11)	30.22
M2BG3	1.32	5.01(0.25)	5.24(0.20)	6.02(0.22)	19.04(0.40)	73.70(0.87)	73.18(1.10)	30.37
M2BG1	1.46	5.30(0.11)	5.44(0.05)	6.20(0.11)	19.14(0.14)	74.10(0.27)	73.95(0.19)	30.69
M2AG3	1.23	5.33(0.08)	5.51(0.22)	6.29(0.19)	20.91(0.35)	82.91(0.22)	83.24(0.25)	34.03
M2CG2	1.38	5.38(0.11)	5.72(0.13)	6.51(0.17)	18.73(0.06)	80.87(0.93)	98.46(0.26)	35.94
M1BG1	4.15	74.63(0.83)	56.99(0.95)	95.72(3.24)	391.71(6.75)	311.12(1.94)	102.51(1.64)	172.11
M1CG1	4.23	60.55(1.03)	59.72(0.93)	59.04(1.63)	417.80(10.98)	356.54(2.66)	119.12(1.52)	178.80
M1AG3	4.00	54.71(0.58)	53.89(0.47)	53.99(1.36)	410.78(11.61)	1978.58(33.86)	2016.48(34.39)	761.40
M1CG3	3.96	54.55(1.37)	56.11(2.42)	56.22(1.42)	413.32(6.86)	2039.98(19.72)	2050.72(18.22)	778.48
M1CG2	3.97	57.24(1.21)	56.03(1.84)	57.25(1.55)	454.27(5.40)	2159.45(17.65)	2188.14(20.80)	828.73
M4AG1	23.95	277.31(1.62)	282.22(2.92)	291.47(3.70)	439.22(8.45)	1654.81(10.47)	3631.29(5.74)	1096.05
M4BG2	23.95	296.27(0.50)	297.90(0.44)	312.31(1.14)	485.87(3.53)	1702.12(8.47)	3652.17(2.71)	1124.44
M1BG3	3.67	66.50(1.99)	52.48(2.34)	92.89(2.83)	565.32(11.80)	3013.13(49.15)	3096.65(56.61)	1147.83
M4AG2	24.34	353.25(0.88)	354.44(0.87)	364.86(3.16)	527.89(2.59)	1725.18(7.17)	3704.93(9.00)	1171.76
M3BG2	24.12	297.31(7.38)	299.04(6.18)	314.91(6.62)	674.98(11.82)	1907.40(58.06)	3814.66(12.06)	1218.05
M2BG2	24.37	360.39(10.07)	373.56(8.05)	397.82(6.87)	589.98(24.62)	1823.89(43.51)	3772.91(24.40)	1219.76
M3AG1	24.47(−)	299.21(7.13)	293.12(2.03)	329.84(7.88)	691.96(31.36)	1847.49(57.95)	4682.02(192.67)	1357.27[−]
M2AG1	25.66(−)	508.84(6.84)	508.41(9.92)	533.87(7.33)	901.93(7.50)	3023.39(40.32)	5061.43(36.19)	1756.31[−]
M3AG2	24.88(−)	309.38(3.61)	306.38(2.68)	347.55(4.29)	832.18(8.78)	3664.03(150.74)	5604.45(43.73)	1843.99[−]
M2AG2	25.72(−)	509.16(1.35)	504.56(4.62)	556.20(3.47)	1177.88(58.13)	4196.18(107.57)	6247.27(127.26)	2198.54[−]
M1BG2	26.48(−)	455.58(1.49)	441.85(3.47)	502.59(3.79)	1325.18(25.19)	5635.84(66.01)	7725.80(105.64)	2681.14[−]
M1AG1	≈16000(!)	-	-	-	-	-	-	-
M1AG2	≈16000(!)	-	-	-	-	-	-	-

5.1 Data Volume

The top 5 and bottom 5 schemas are indicated in Fig. 6 by the symbols (+) and (-) at the left side of its name, respectively. In Table 2, these symbols are also found in the *Size* column.

Top Schemas. The five schemas with the lowest storage cost are M4CG3 (0.94 GB), M4AG3 (0.94 GB), M4BG1 (0.95 GB), M4CG1 (0.95 GB), and M4BG3 (0.96 GB). In Fig. 8, these schemas used a data volume of less than 1 GB.

The schemas with low storage costs tend to normalize their GDs. For example, all GDs of the M4BG1 and M4BG3 schemas were normalized. Except for the *address* GD, all other GDs of M4CG3, M4AG3, and M4CG1 schemas are also normalized. It is easy to see that the normalization of GDs with low selectivity

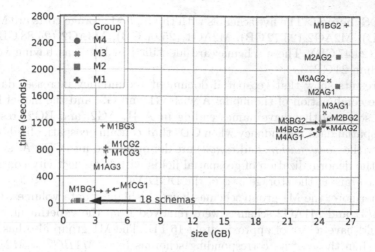

Fig. 7. Volume vs. execution time for the 34 evaluated schemas.

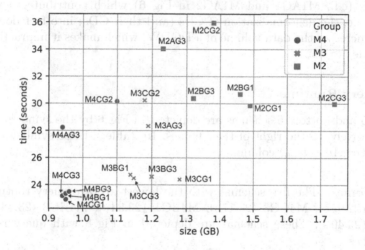

Fig. 8. Volume vs. execution time for the 18 schemas having similar results.

geospatial fields (i.e., *city, nation,* and *region*) strongly contributes to reducing the DGDW storage cost.

Analyzing the five schemas with regard to document partitioning shows that partitioning documents into specific and homogeneous collections also contributes to reduced data volume. This happens because the storage volume of the index structure is smaller in schemas holding the same document type in the collections. That is, all schemas in the M4 group (which have a collection for each dimension, i.e., homogeneous collections) have a lower volume than their corresponding schemas (e.g., $M4CG1 < M3CG1 < M2CG1$) (c.f., Table 2).

Bottom Schemas. The five schemas with the highest storage costs are M1BG2 (26.48 GB), M2AG2 (25.72 GB), M2AG1 (25.66 GB), M3AG2 (24.88 GB), and M3AG1 (24.47 GB). These schemas are identified in Fig. 7 as having volumes greater than 24 GB.

The results show that geospatial document redundancy increases data volume. The combination of the model A with G1 and G2, and the model B with G2 (i.e., schemas that have names ending in AG1, AG2, and BG2) results in high geospatial data redundancy when GDs that contain geospatial fields of low-selectivity (i.e., *city*, *nation*, and *region*) are denormalized into CDs. As shown in Table 2, the denormalization of geospatial fields with low-selectivity contributes strongly to raising the storage cost of the DGDWs.

Furthermore, the M1 group schemas have the highest data volumes. Indeed, the M1AG1 and M1AG2 schemas were removed from the experiment because they would have a size of approximately 16 TB. The M1 group also has higher volumes than those of its corresponding schemas (e.g., $M1BG2 > M2BG2 > M3BG2 > M4BG2$). Although the M1 group has schemas with homogeneous collections (c.f., M1AG1 and M1AG2 in Fig. 6) which contribute to reduced data volume, the denormalization of CDs (and their GDs) into fact documents strongly increases the data volume of a DGDW, which makes it impractical and not feasible.

5.2 Query Runtime

The top 5 and bottom 5 schemas are depicted in Fig. 6 by the symbols [+] and [-], respectively, to the right of their names. In Table 2, these symbols are also found in the $Avg(Q_{1-6})$ column.

Top Schemas. The five schemas with the lowest average query runtimes are M4CG1 (22.92 s), M4CG3 (23.17 s), M4BG1 (23.34 s), M4BG3 (23.48 s), and M3CG1 (24.40 s). These schemas are depicted in Fig. 8 with query runtimes close to 24 s.

The top 5 schemas have their low-selectivity geospatial fields normalized into GDs (i.e., *city*, *nation*, and *region*). That is, the computational cost of joins to perform the queries is less than the cost of high data redundancy in the document-oriented database. However, the denormalization of the GD that contains a high-selectivity geospatial field (i.e., *address*) into CDs (i.e., M4CG1, M4CG3, and M3CG1 schemas) positively influences the query runtimes, by reducing the number of query joins (i.e., they do not need to perform joins to obtain *address*). Both the normalization of low-selectivity geospatial fields and the denormalization of high-selectivity geospatial fields contribute positively to the performance of the DGDWs.

Regarding schema partitioning, it can see in Fig. 6 that, except the M3CG1 schema, the best schemas all have one collection for each dimension (i.e., these schemas partition the fact documents, CDs, and GDs into specific and homogeneous collections). The M4 schemas also have better performance than their corresponding M3 and M2 schemas (e.g., $M4CG3 < M3CG3$ and $M4CG3 < M2CG3$). The partitioning of documents into homogeneous collections therefore contributes to improved DGDW performance.

Bottom Schemas. The five schemas with the highest average query runtimes are also the schemas with highest storage cost: M1BG2 (2681.14 s), M2AG2 (2198.54 s), M3AG2 (1843.99 s), M2AG1 (1756.31 s), and M3AG1 (1357.27 s). These schemas are depicted in Fig. 7, having query runtimes greater than 1300 s.

These schemas have high data redundancy, because their GDs that contain low-selectivity geospatial fields (i.e., *city, nation,* and *region*) are denormalized into CDs. When performing queries, selection operations and joins among dimensions require more processing power to manipulate the large volumes of data, greatly increasing the query runtimes.

Regarding the partitioning of these schemas, we noticed that they have heterogeneous collections (i.e., more than one dimension stored in the same collection). Therefore, there is a loss of query performance in collections that contain different document types, and the more heterogeneous the collection, the worse the performance. This finding can be shown by comparing, in Table 2, the results of corresponding schemas that have different partitioning models (e.g., M2BG2 vs. M3BG2).

6   Conclusion

In this paper, we investigated the use of a document-oriented NoSQL database for a GDW project. We sought to identify which level of geospatial data redundancy and document partitioning among collections provides low storage cost and good query performance. We proposed a UML-based notation for specifying collections, documents, and relationships between documents in DGDW. Using our notation, we exhaustively transformed one RGDW schema in 36 DGDW schemas. The 36 DGDWs contain different levels of redundancy in their conventional and geospatial data, as well as different forms of partitioning documents among collections. Next, we performed an experimental evaluation based on a computational infrastructure that can be scaled horizontally with the addition of new nodes in a cluster.

Considering the evidence obtained by experimental analysis, we identified that schemas with low-selectivity geospatial fields normalized into GDs have low storage cost and good query execution performance. Also, the denormalization of GDs that contain low-selectivity geospatial fields into CDs strongly contributes to an increase in the size of the schemas, as well as reduced query performance. This observation resembles that identified in [18]. However, in addition to the normalization of its high-selectivity geospatial fields, partitioning documents into collections also influences the volume and performance of the DGDW. We also identified that the denormalization of high-selectivity geospatial fields into CDs contributes to improved query performance because it reduces the number of joins in the queries and does not substantially affect the data volume.

Partitioning documents into homogeneous collections is another factor that positively influences the DGDW. The indexes structure of the collections has less volume and higher performance when the collections have the same document type (i.e., documents with the same fields structure). Thus, modeling the DGDW dimensions into distinct collections reduces data volume and increases system performance.

Therefore, the normalization of low-selectivity geospatial fields into GDs, the denormalization of high-selectivity geospatial fields into CDs, and the modeling of dimensions into distinct collections contribute to achieving a high level of performance at a low cost of storage for the DGDW schema.

For future studies, we will evaluate the performance of the best schemas based on volume size or query performance (i.e., M4CG1, M4CG3, M4BG1, M4BG3, M3CG1, and M4AG3) with higher data volumes (e.g., at scale factors of 10 and 100). Following this, we will compare the performance of these schemas with their corresponding schemas in a relational database.

References

1. Almeida, R., Bernardino, J., Furtado, P.: Testing SQL and NoSQL approaches for big data warehouse systems. Int. J. Bus. Process. Integr. Manag. **7**(4), 322–334 (2015). https://doi.org/10.1504/IJBPIM.2015.073656. https://www.inderscienceonline.com/doi/abs/10.1504/IJBPIM.2015.073656
2. Chavalier, M., Malki, M.E., Kopliku, A., Teste, O., Tournier, R.: Document-oriented data warehouses: models and extended cuboids, extended cuboids in oriented document. In: 2016 IEEE Tenth International Conference on Research Challenges in Information Science (RCIS), pp. 1–11, June 2016. https://doi.org/10.1109/RCIS.2016.7549351
3. Chevalier, M., El Malki, M., Kopliku, A., Teste, O., Tournier, R.: Document-oriented models for data warehouses. In: Proceedings of the 18th International Conference on Enterprise Information Systems, ICEIS 2016, vol. 1, pp. 142–149, December 2016
4. Chevalier, M., El Malki, M., Kopliku, A., Teste, O., Tournier, R.: Implementing multidimensional data warehouses into NoSQL. In: 17th International Conference on Enterprise Information Systems, Proceedings 1, ICEIS 2015, pp. 172–183 (2015)
5. DB-Engines: DB-engines ranking, September 2018. https://db-engines.com/en/ranking. Accessed 20 Sept 2018
6. Dehdouh, K., Bentayeb, F., Boussaid, O., Kabachi, N.: Using the column oriented NoSQL model for implementing big data warehouses. In: 21st International Conference on Parallel and Distributed Processing Techniques and Applications (PDPTA), pp. 469–475 (2015)
7. Fidalgo, R.N., Times, V.C., Silva, J., Souza, F.F.: GeoDWFrame: a framework for guiding the design of geographical dimensional schemas. In: Kambayashi, Y., Mohania, M., Wöß, W. (eds.) DaWaK 2004. LNCS, vol. 3181, pp. 26–37. Springer, Heidelberg (2004). https://doi.org/10.1007/978-3-540-30076-2_3
8. Kimball, R., Ross, M.: The Data Warehouse Toolkit: The Complete Guide to Dimensional Modeling. Wiley, Hoboken (2011)
9. Lee, J.G., Kang, M.: Geospatial big data: challenges and opportunities. BigData Res. **2**(2), 74 – 81 (2015). https://doi.org/10.1016/j.bdr.2015.01.003. http://www.sciencedirect.com/science/article/pii/S2214579615000040, visions on Big Data

10. Liu, Y., Vitolo, T.M.: Graph data warehouse: steps to integrating graph databases into the traditional conceptual structure of a data warehouse. In: 2013 IEEE International Congress on Big Data, pp. 433–434, June 2013. https://doi.org/10.1109/BigData.Congress.2013.72

11. Luo, W., Liu, B., Watfa, A.K.: An open schema for XML data in hive. In: 2014 IEEE International Conference on Big Data (Big Data), pp. 25–31, October 2014. https://doi.org/10.1109/BigData.2014.7004409

12. Mateus, R., Siqueira, T., Times, V., Ciferri, R., Ciferri, C.: How does the spatial data redundancy affect query performance in geographic data warehouses? J. Inf. Data Manag. **1**(3), 519 (2010)

13. Mateus, R.C., Siqueira, T.L.L., Times, V.C., Ciferri, R.R., de Aguiar Ciferri, C.D.: Spatial data warehouses and spatial OLAP come towards the cloud: design and performance. Distrib. Parallel Databases **34**(3), 425–461 (2016). https://doi.org/10.1007/s10619-015-7176-z

14. O'Neil, P.E., O'Neil, E.J., Chen, X.: The star schema benchmark (SSB). Pat 200(0)50 (2007)

15. Rigaux, P., Scholl, M., Voisard, A.: Spatial Databases: With Application to GIS. Morgan Kaufmann Publishers Inc., San Francisco (2001)

16. Scabora, L.C., Brito, J.J., Ciferri, R.R., Ciferri, C.D.d.A., et al.: Physical data warehouse design on NoSQL databases OLAP query processing over HBase. In: International Conference on Enterprise Information Systems, XVIII. Institute for Systems and Technologies of Information, Control and Communication-INSTICC (2016)

17. Siqueira, T.L.L., de Aguiar Ciferri, C.D., Times, V.C., de Oliveira, A.G., Ciferri, R.R.: The impact of spatial data redundancy on SOLAP query performance. J. Braz. Comput. Soc. **15**(2), 19–34 (2009)

18. Siqueira, T.L.L., Ciferri, R.R., Times, V.C., de Aguiar Ciferri, C.D.: Investigating the effects of spatial data redundancy in query performance over geographical data warehouses. In: Proceedings of the 10th Brazilian Symposium on Geoinformatics, pp. 1–12 (2008)

19. Siqueira, T.L.L., Ciferri, R.R., Times, V.C., de Aguiar Ciferri, C.D.: Benchmarking spatial data warehouses. In: Bach Pedersen, T., Mohania, M.K., Tjoa, A.M. (eds.) DaWaK 2010. LNCS, vol. 6263, pp. 40–51. Springer, Heidelberg (2010). https://doi.org/10.1007/978-3-642-15105-7_4

20. Song, J., Guo, C., Wang, Z., Zhang, Y., Yu, G., Pierson, J.M.: HaoLap: a hadoop based olap system for big data. J. Syst. Softw. **102**, 167–181 (2015)

21. Tria, F.D., Lefons, E., Tangorra, F.: Design process for big data warehouses. In: 2014 International Conference on Data Science and Advanced Analytics (DSAA), pp. 512–518, October 2014. https://doi.org/10.1109/DSAA.2014.7058120

22. Yangui, R., Nabli, A., Gargouri, F.: Automatic transformation of data warehouse schema to nosql data base: comparative study. Procedia Comput. Sci. **96**, 255–264 (2016)

Graphs and Machine Learning

Scalable Least Square Twin Support Vector Machine Learning

Bakshi Rohit Prasad and Sonali Agarwal[✉]

Indian Institute of Information Technology Allahabad, Allahabad, India
{rsl51,sonali}@iiita.ac.in

Abstract. Machine Learning (ML) on massive scale datasets, called Big Data, has become a challenge for traditional computing and storage technologies. Henceforth, massive scale ML is an emerging domain of research. Least Square Twin Support Vector Machine (LSTSVM) is a faster variant of Support Vector Machine (SVM). However, it suffers from scalability issues and shows computational and/or storage bottlenecks on massive datasets. Proposed work designs a scalable solution to LSTSVM called Distributed LSTSVM (DLSTSVM). DLSTSVM is designed using distributed parallel computing on top of cluster of multiple machines. After applying horizontal partitioning on massive datasets, DLSTSVM trains it in distributed parallel fashion and finds two non-parallel hyper-planes as decision boundaries for two different classes. MapReduce paradigm is utilized to execute parallel computation on partitioned data in a way that averts memory constraints. Proposed technique achieves computational and storage scalability without losing prediction accuracy.

Keywords: Big Data · MapReduce · Cluster computing ·
Distributed machine learning · Supervised learning · LSTSVM ·
Parallel processing

1 Introduction

1.1 Background

In any supervised ML technique, classification rules are built on training data instances with known class labels. The task is to classify new data instances using the learned classification model [1]. Support Vector Machine (SVM) is a well-known supervised ML algorithm proposed by Vapnik et al. [2]. During training, SVM constructs parallel hyper-planes separating data from two classes by formulating and solving a Quadratic Programming Problem (QPP). For SVM, overall training time is order of $O(r^3)$, where r is the number of training data samples. Another variant of SVM called Twin Support Vector Machines (TWSVM), solves two smaller QPPs instead of a single large QPP [3] which fastens the training four times. Still, solving QPP is a complex task. Hence, a new faster version is developed by Kumar et al. [4] known as LSTSVM. It solves two systems of linear equations for training which makes it significantly faster than that of TWSVM. Figure 1 represents the categorization of two classes by using LSTSVM.

© Springer Nature Switzerland AG 2019
C. Ordonez et al. (Eds.): DaWaK 2019, LNCS 11708, pp. 239–249, 2019.
https://doi.org/10.1007/978-3-030-27520-4_17

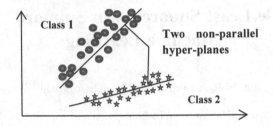

Fig. 1. Geometric representation of LSTSVM for classification

Although LSTSVM is much faster, when the dataset size grows, the computations involved in its training, lead to computational and storage bottlenecks which limit the capability and applicability of LSTSVM. In order to tackle these problems, recently developed Big Data frameworks are very suitable. Apache Spark is one of them and provides data-parallel processing for faster computation. These computations are done on clusters of commodity machines. Spark is based on MapReduce [5] model of computing whereas its data flow types are generalized based on Dryad [6] and Map-Reduce Merge [7]. Applications involving reuse of working set of data many times, such as iterative computing and interactive analytics, are supported by Apache Spark [8]. RDDs are the main abstraction of Spark and are the read-only collection of objects. Spark driver is responsible for internally distributing RDDs over several machines, receiving and aggregating responses from multiple machines in the cluster [9].

1.2 Research Objective and Contribution

The challenges associated with LSTSVM put a motivation for the design of large scale LSTSVM which would be capable enough to handle a huge number of instances, features, and classes. This research work targets scalable design of LSTSVM with the help of recent distributed framework named Apache Spark. Proposed DLSTSVM exploits the data parallel computing feature of Spark along with its optimized library for vector and matrix calculations. Thus, proposed design not only handles scalability for large input patterns, it achieves speed-ups too, due to data parallel processing. Section 1 of the research work gives a brief introduction about the need for large-scale machine learning in context of LSTSVM. In Sect. 2, state-of-the-art is presented in this regard. Section 3 describes the formulation of appropriate mapper and reducer tasks for DLSTSVM and detailed analysis of its run time and storage requirements. Detailed experimental results are given in Sect. 4. Finally, Sect. 5 gives the conclusion with probable future prospects of the work.

2   Literature Survey

In order to utilize the advantages of distributed computing to enable machine learning systems to deal with large-scale data, several efforts are made to provide distributed ML frameworks and libraries. Apache Mahout [10] in Hadoop, GraphLab [11] for graph related operations in parallel, Microsoft developed DryadLINQ [12] using LINQ on top of Dryad and IBM Parallel ML Toolbox [13] are some of them. Spark [14], a distributed computing engine, well suited for iterative nature of ML, contains a library for supporting vivid ML tasks. Advancements in multi-core processing and cloud computing has caused widely accessibility to distributed and parallel computing systems and parallel machine learning [15]. A comprehensive study regarding distributed and parallel learning is carried out and discussed in the work [16].

In ML techniques, SVM has an advantage that it provides a global solution for the data classification in binary class problem [2] and multi-class problems [17]. Jayadeva et al. proposed TWSVM, which is a binary classifier with reduced time complexity [3]. Kumar et al. proposed LSTSVM which is the least squares variant of TWSVM [4] and determines two non-parallel hyper-planes by solving two linear equations. Some researchers extended the TWSVM and LSTSVM to multi-class, multi-instance and multi-label problem scenarios [18]. However, these variants show incapability in handling large problem sizes. Some research efforts have been done to deal with it. Collobert et al. [19] divide the dataset into multiple subsets and trains SVM on each subset in parallel. Zanghirati et al. [20] used Message Passing Interface to avoid re-computations of already computed results. In Hazen et al. [21] approach, each computing node finds individual sub-solutions which are combined and communicated to other nodes which results in slow training time. The approach given by Graf et al. [22], partitions dataset and solve SVM for each. Support vectors previous two classifiers are combined to form new training set. This approach is cascaded till final set of support vectors. With distributed computation on commodity nodes using Map-Reduce paradigm, Chu et al. [23] exploited them to propose distributed linear SVM. The approach is reported slow convergence of batch gradient descent. Catanzaro et al. [24] employed high-speed costly Graphics Processing Unit utilizing multi-threading to speed up training. Limitation of this technique is its tedious setting of specialized configurations.

3   Proposed Methodology - Design of DLSTSVM

Proposed work gives a general design of DLSTSVM algorithm. However, for the sake of implementation and experimental analysis, Apache Spark framework is used. As we know, the LSTSVM finds two hyper-planes which are not necessary to be parallel. Necessary computations for determining those non-parallel hyperplanes involve solving two linear equations given by (1) and (2).

$$\min(w_1, b_1, \xi)\frac{1}{2}\|X_1w_1 + e_1b_1\|^2 + \frac{c_1}{2}\xi^T\xi \qquad s.t. -(X_2w_1 + e_2b_1) + \xi = e_2 \quad (1)$$

$$\min(w_2, b_2, \eta)\frac{1}{2}\|X_2w_2 + e_2b_2\|^2 + \frac{c_2}{2}\eta^T\eta \qquad s.t \quad (X_1w_2 + e_1b_2) + \eta = e_1 \quad (2)$$

Here, X_1 and X_2 are dataset matrix for class -1 and class $+1$ and containing l_1 and l_2 data instances respectively. Moreover, w_1, b_1 and w_2, b_2 represents hyperplane parameters (i.e. weights and bias) for either class, $e_1 \in R^{l_1}$ and $e_2 \in R^{l_2}$ are vectors with entries of 1 only, c_1 and c_2 are non-negative penalty parameters. Here, $\xi \in R^{l_2}$ and $\eta \in R^{l_1}$ work as slack variables, added corresponding to class -1 and class $+1$ respectively. LSTSVM has optimization Eqs. (1) and (2) with equality constraints only. After, solving Eqs. (1) and (2) hyper-plane parameters of LSTSVM are attained as given in Eqs. (3) and (4) where, $G = [X_1e_1]$ and $H = [X_2e_2]$. Using these parameters, corresponding hyper-planes can be written as per Eqs. (5) and (6).

$$u = \begin{bmatrix} w_1 \\ b_1 \end{bmatrix} = -\left(H^TH + \frac{1}{c_1}G^TG\right)^{-1}H^Te_2 \tag{3}$$

$$v = \begin{bmatrix} w_2 \\ b_2 \end{bmatrix} = \left(G^TG + \frac{1}{c_2}H^TH\right)^{-1}G^Te_1 \tag{4}$$

$$x^Tw_1 + b_1 = 0 \tag{5}$$

$$x^Tw_2 + b_2 = 0 \tag{6}$$

Now, design of DLSTSVM is illustrated in three major phases. First phase identifies serial/parallel segments of problem to decompose it into modules. Second phase specifies Mapper and Reducer tasks for each module. Third phase provides theoretical analysis of runtime and storage requirements in case of LSTSVM and DLSTSVM.

3.1   Problem Decomposition and Data Partitioning

Equations 3 and 4 involves series of matrix multiplications. Following sub-sections segregate the serial-parallel computations and specify data partitioning approach.

Identification of Serial-Parallel Segments of Processing. In this work, computations of Eq. 3 are divided into two sub-parts; matrix R and vector S where, $R = -((H^TH) + (1/c_1)(G^TG))^{-1}$ and $S = H^T e_2$. These parts are solved separately in sequence. However, computation in each sub-part is done in parallel fashion as depicted in Fig. 2. Vector u is obtained as multiplication of matrix $R_{c \times c}$ and vector $S_{c \times 1}$, i.e. $u = R \times S$.

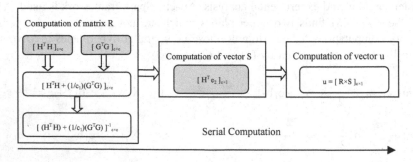

Fig. 2. Serial-parallel computation flow (shaded box signify parallel execution)

Data Representation and Partitioning. Computation of matrix R involves two major components; G^TG and H^TH which is needed to be distributed to avert memory bottleneck. Hence, matrix G is transformed as RDD of rows implicitly partitioned and stored in distributed fashion on multiple machines in Spark (see Fig. 3). Consider matrix $K_{r \times c}$ where, r and c represents instances and its dimension. Each k_i represents i^{th} row vector of K, where $1 <= i <= r$. Thus, the transpose of matrix K can be formulated as $K^T = [k_1^T \; k_2^T \; .... \; k_i^T \; .... \; k_r^T]_{c \times r}$, where each k_i^T is a column vector.

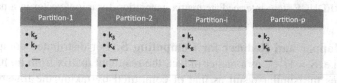

Fig. 3. Partitioning example for r instances each having c attributes

3.2   Design of MAP and REDUCE Phase for Distributing Computations

Design of Mapper and Reducer for Computing Matrix R. Map-Reduce paradigm is very effective in large-scale distributed computation frameworks like Spark. It involves two phases; MAP phase and REDUCE phase. To compute G^TG, the MAP and REDUCE steps are designed as below:

MAP Step: MAP step multiplies j^{th} column of G^T to j^{th} row of G, i.e. $g_j^T \times g_j$ as shown in Fig. 4. Each $g_j^T \times g_j$ yields an intermediate matrix t_j of dimension c × c. This map step is applied for each g_j in each partition i residing on multiple machines.

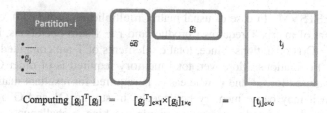

Computing $[g_j]^T[g_j]$ = $[g_j^T]_{c \times 1} \times [g_j]_{1 \times c}$ = $[t_j]_{c \times c}$

Fig. 4. MAP step: Intermediate matrix generation for a row-vector in a partition

REDUCE Step: Let a partition possesses d number of instances (row vectors), then that partition produces d intermediate matrices. In REDUCE step, all intermediate temporary matrices t_j are summed up cumulatively, in order to yield matrix $G^TG = T = \sum_{j=1}^{r} t_j$. Alternatively, we say that intermediate matrices are reduced on the *index(i, j)* via sum operation. In this way, final matrix T is computed. Considering p number of partitions, Fig. 5 shows the diagrammatic representation of REDUCE step.

Similarly, computation of H^TH is carried out in distributive parallel fashion. Further, G^TG is multiplied to $1/c_1$, added to H^TH and inversed to produce matrix R.

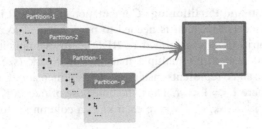

Fig. 5. REDUCE step: Intermediate matrix generation for a row-vector in a partition

Design of Mapper and Reducer for Computing S. For distributed computation of second part, i.e. $S = H^T e_2$, as a matter of fact, the resultant matrix from multiplying H^T to e_2 produces the similar result as that of computed by taking the transpose of the vector achieved after summing up the elements of H column-wise. Thus, the MAP and REDUCE steps are designed as follows:

MAP Step: The MAP step just copies the input matrix instances as output without any modification. Thus, it simply allows the matrix H to be fed as input to the reducer.

REDUCE Step: This phase cumulatively computes the sum of elements in the columns of matrix H. We can say that column-vectors are reduced on *index(j)* via sum operation where j is column index. Thus, S is computed in distributed parallel way.

3.3 Storage and Run-Time Analysis

Data instances from two different classes are segregated and stored in matrix G and H. Let the dimension of matrix G be $r \times c$, then the dimension of transpose matrix G^T will be $c \times r$. Hence, the dimension of resultant matrix $T = G^T \times G$ will be $c \times c$.

Analysis of LSTSVM. In case of usual matrix multiplication (of $G^T G$), computation of each element of matrix T requires r product and $r - 1$ sum operations, i.e. order of $O(r + r - 1) = O(r)$ operations. Since, total c^2 elements of T are computed, it takes an order of $O(rc^2)$ operations. Moreover, total memory required is of order $O(2rc + c^2)$. Here, 2rc is required for G^T and G whereas c^2 is required for resultant matrix T. If r is too high, then it may cause memory bottlenecks in storing G^T and G. Also, $O(rc^2)$ computations is also time-consuming on a single machine, a challenge for LSTSVM.

Analysis of DLSTSVM for Computing Matrix R and Vector S
Storage and Run-Time Analysis of Computing Matrix R: In distributed parallel fashion, computation of each t_j requires c^2 products and c^2 storage, and are under the limits of the single machine. Thus, for each machine order of computation is $O(mc^2)$ and of storage is $O(mc^2)$ where, $m = r/n$, is the number of data instances in the data partition residing on that machine and n is the number of machines. Hence, $O(rc^2)$ computation required in LSTSVM is now distributed among n machines in case of DLSTSVM. Also, storage bottleneck is resolved by reducing storage on machine from $O(rc + c^2)$ in LSTSVM to $O(c^2)$ in DLSTSVM. Further, adding $(1/c_1(G^T G))$ and $H^T H$ takes $O(c^2)$ additions and $O(c^2)$ and resulting in matrix of order $c \times c$. Now, taking its inverse will

take $O(c^3)$ operation and $O(c^2)$ storage only. Thus, matrix R is computed in $O(mc^2 + c^3 + c^2) = O(mc^2 + c^3)$ operations and $O(c^2 + c^2 + c^2) = O(c^2)$ storage on each machine.

Storage and Run-Time Analysis of Computing S: Consider computation of $S = H^T e_2$. As discussed under Subsect. 3.2, the REDUCE step for computing S sums corresponding elements of two vectors which requires c additions. Since, m such vectors are present on each machine, total $(m - 1) \times c$ sum operations would be required, i.e. $O(mc)$ operations. Moreover, on each machine, $O(mc)$ storage is required to store m vectors.

Thus, computation of matrix R requires $O(mc^2 + c^3)$ operations and $O(c^2)$ storage whereas S requires $O(mc)$ operations and $O(mc)$ storage that are under the limit of a single machine. Moreover, the resultant matrix in the proposed approach is same as that in normal matrix multiplication. Thus, hyper-plane parameters u and v achieved in DLSTSVM are same as we get in LSTSVM, thereby the accuracy is also same.

4 Results and Discussion

Experimental setup involves computer systems (nodes) equipped with Intel(R) Core (TM)-i3 CPU 3.30 GHz Processor having 4 cores, 6 GB of RAM and 64-bit Linux (Ubuntu 12.04). The cluster is setup containing 8 such nodes, 24 cores (3 cores from each node), 32 GB of executor memory (4 GB from each node) and Spark 1.5.0.

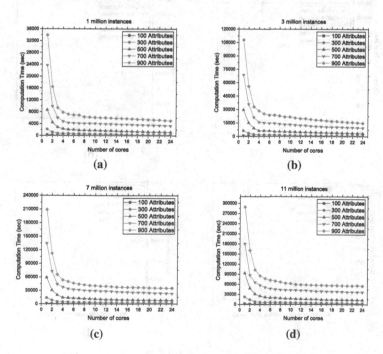

Fig. 6. DLSTSVM computation time for different cores with fixed number of instances

4.1   Scalability Analysis of DLSTSVM

A set of experiments have been performed on large scale synthetically generated data instances to test the scalability of DLSTSVM in the light of several key parameters such as number of cores, attributes and instances at large scale. Experiments are performed. Following sub-sections provide detailed discussion on these observations.

Performance Analysis Based on #cores Versus Computation Time. Figure 6(a)–(d) depicts plots drawn between number of cores and computation time and corresponds to 1, 3, 7, and 11 million data instances respectively.

In each figure, multiple curves correspond to the graph for 100, 300, 500, 700 and 900 attributes. For initial increments in number of cores downfall in computation time is fast and becomes gradual later due to addition of more machines thereby causing extra communication overheads and task scheduling delays. More attributes result in more computation time which is expected as per run-time analysis in Subsect. 3.3.

Performance Analysis Based on #data Instances Versus Computation Time. Figure 7(a)–(d) specifies plots drawn between number of data instances and computation time. Each figure depicts graph for a fixed value of number of cores (between 24 cores and 1 core) while multiple curves depict graph for 100, 300, 500, 700, 900 attributes.

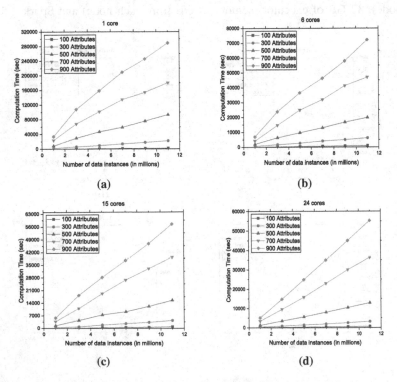

Fig. 7. DLSTSVM computation time for different instances with fixed cores

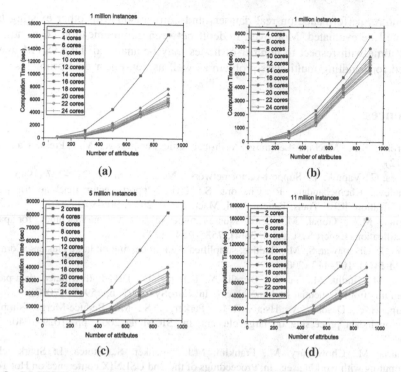

Fig. 8. DLSTSVM computation time for different number of attributes with fixed instances

Performance Analysis Based on #attributes Versus Computation Time. Figure 8 (a)–(d) shows the plot drawn between number of attributes and computation time with respect to a fixed number of cores (varied from 1 to 24). Computation time for lesser number of cores (<=3) is dominant over that of a higher number of cores as shown in the Fig. 8(a). For clearer visibility of the gap in computation time, a plot is separately drawn for 1 million instances as depicted in Fig. 8(b). Figure 8(c) and (d) corresponds to the plot for a fixed number of data instances (i.e. 5 and 11 million). Greater the number of attributes, higher is the computation time.

5 Conclusion and Future Work

A novel scalable LSTSVM, called DLSTSVM, is designed and implemented in using distributed matrix operations via MapReduce paradigm. Underlying memory constraints of LSTSVM with extremely large datasets are averted in DLSTSVM and due to distributed parallel processing on partitioned data, significant speedup is also gained. Increasing number of cores puts more communication and meta-information management task overheads, and hence, reduction in computation time gets slower.

In future, performance on real datasets and comparison with other existing techniques will be evaluated. Moreover, tradeoff between communication cost and computation time with respect to number of nodes may be analyzed. DLSTSVM may be extended for handling multi-class problem as well as non-linear data problems.

References

1. Murphy, K.P.: Machine Learning: A Probabilistic Perspective. MIT Press, Cambridge (2012)
2. Cortes, C., Vapnik, V.: Support-vector networks. Mach. Learn. **20**, 273–297 (1995)
3. Jayadeva, Khemchandani, R., Chandra, S.: Twin support vector machine for pattern classification. IEEE Trans. Pattern Anal. Mach. Intell. **29**(5), 905–910 (2007)
4. Kumar, M.A., Gopal, M.: Least squares twin support vector machines for pattern classification. Expert Syst. Appl. **36**(4), 7535–7543 (2009)
5. Dean, J., Ghemawat, S.: MapReduce: simplified data processing on large clusters. Commun. ACM **51**(1), 107–113 (2008)
6. Isard, M., Budiu, M., Yu, Y., Birrell, A., Fetterly, D.: Dryad: distributed data-parallel programs from sequential building blocks. In: EuroSys 2007, pp. 59–72 (2007)
7. Yang, H.-c., Dasdan, A., Hsiao, R.-L., Parker, D.S.: Map-Reduce-Merge: simplified relational data processing on large clusters. In: SIGMOD 2007, pp. 1029–1040. ACM (2007)
8. Zaharia, M., Chowdhury, M., Franklin, M.J., Shenker, S., Stoica, I.: Spark: cluster computing with working sets. In: Proceedings of the 2nd USENIX conference on Hot Topics in Cloud Computing (HotCloud 2010), p. 10. USENIX Association, Berkeley (2010)
9. Zaharia, M., et al.: Resilient distributed datasets: a fault-tolerant abstraction for in-memory cluster computing. In: Proceedings of the 9th USENIX conference on Networked Systems Design and Implementation (NSDI 2012), p. 2. USENIX Association (2012)
10. Owen, S., Anil, R., Dunning, T., Friedman, E.: Mahout in Action. Manning Publications, Greenwich (2011)
11. Low, Y., Gonzalez, J., Kyrola, A., Bickson, D., Guestrin, C., Hellerstein, J.M.: Distributed GraphLab: a framework for machine learning and data mining in the cloud. Proc. VLDB Endow. **5**(8), 716–727 (2012)
12. Budiu, M., Fetterly, D., Isard, M., McSherry, F., Yu, Y.: Large-scale machine learning using DryadLINQ. In: Bekkerman, R., Bilenko, M., Langford, J. (eds.) Scaling Up Machine Learning. Cambridge University Press, Cambridge (2012)
13. Pednault, E., Tov, Y.E., Ghoting, E.: IBM parallel machine learning toolbox. In: Bekkerman, R., et al. (eds.) Scaling Up Machine Learning. Cambridge University Press, Cambridge (2012)
14. Apache Spark: Apache Spark: lightning-fast cluster computing (2016)
15. Upadhyaya, S.R.: Parallel approaches to machine learning—a comprehensive survey. J Parallel Distrib. Comput. **73**(3), 284–292 (2013)
16. Peteiro-Barral, D., Guijarro-Berdiñas, B.: A survey of methods for distributed machine learning. Prog. Artif. Intell. **2**(1), 1–11 (2012)
17. Hsu, C.-W., Lin, C.-J.: A comparison of methods for multiclass support vector machines. IEEE Trans. Neural Netw. **13**(2), 415–425 (2002)
18. Tomar, D., Agarwal, S.: A comparison on multi-class classification methods based on least squares twin support vector machine. Knowl. Based Syst. **81**, 131–147 (2015)
19. Collobert, R., et al.: A parallel mixture of SVMs for very large scale problems (2002)

20. Zanghirati, G., Zanni, L.: A parallel solver for large quadratic programs in training support vector machines. Parallel Comput. **29**, 535–551 (2003)
21. Hazan, T., Man, A., Shashua, A.: A parallel decomposition solver for SVM: distributed dual ascend using Fenchel duality. In: IEEE Conference on Computer Vision and Pattern Recognition, CVPR 2008, pp. 1–8 (2008)
22. Graf, H., Cosatto, E., Bottou, L., Durdanovic, I., Vapnik, V.: Parallel support vector machines: the cascade SVM. In: Neural Information Processing Systems (2004)
23. Chu, C., et al.: Map-reduce for machine learning on multicore. In: NIPS, pp. 281–288. MIT Press (2006)
24. Catanzaro, B., Sundaram, N., Keutzer, K.: Fast support vector machine training and classification on graphics processors. In: Proceedings of the 25th International Conference on Machine Learning, Helsinki, Finland, pp. 104–111. ACM (2008)

Finding Strongly Correlated Trends in Dynamic Attributed Graphs

Philippe Fournier-Viger[1]([⊠]), Chao Cheng[2], Zhi Cheng[3], Jerry Chun-Wei Lin[4], and Nazha Selmaoui-Folcher[3]

[1] School of Natural Sciences and Humanities,
Harbin Institute of Technology, Shenzhen, China
philfv8@yahoo.com
[2] School of Computer Sciences and Technology,
Harbin Institute of Technology, Shenzhen, China
tidescheng@gmail.com
[3] University of New Caledonia, ISEA, BP R4, 98851 Noumea, New Caledonia
{zhi.cheng,nazha.selmaoui}@univ-nc.nc
[4] Department of Computing, Mathematics and Physics,
Western Norway University of Applied Sciences (HVL), Bergen, Norway
jerrylin@ieee.org

Abstract. Discovering patterns in dynamic attributed graphs allows to capture how attribute values and graph structures changes over time. This allows to understand how a graph has evolved and may change in the future, to support decision-making. But an important limitation of current studies is that they mainly select patterns based on their frequency. Thus, they may find many frequent but weakly correlated patterns. To discover strongly correlated patterns, this paper proposes a novel significance measure named *Sequence Virtual Growth Rate*. It allows evaluating if a pattern represents entities that are correlated in terms of their proximity in a graph over time. Based on this measure a novel type of graph patterns is defined called *Significant Trend Sequence*. To efficiently mine these patterns, an algorithm named TSeqMiner is proposed, which relies on a novel upper bound and pruning strategy to reduce the search space. Experiments show that the algorithm is efficient and can identify interesting patterns in social network and spatio-temporal data.

Keywords: Dynamic attributed graph · Significant sequential patterns

1 Introduction

In the last decades, analyzing graphs has received an increasing amount of attention from the data mining community. The reason is that graphs naturally capture the structure of data in many domains such as in social network and wireless sensor network analysis, and bioinformatics [1,7,12,16,18]. Discovering patterns in a dynamic attributed graph such as trends is useful to understand how the graph evolved in terms of attribute values and graph structure, and how it may

© Springer Nature Switzerland AG 2019
C. Ordonez et al. (Eds.): DaWaK 2019, LNCS 11708, pp. 250–265, 2019.
https://doi.org/10.1007/978-3-030-27520-4_18

change in the future. Several algorithms have been proposed for mining patterns in dynamic attributed graphs (graphs evolving other time where nodes have multiple numeric attributes) [3–5,14]. However, most algorithms select patterns based on their occurrence frequency. Using the frequency as main criterion to select patterns can filter some noise but it can result in discovering many weakly correlated patterns whose constituting elements apppeared together by chance. To illustrate this issue, consider Fig. 1, which will be used as running example. It depicts the evolution of a dynamic attributed graph over six timestamps. In this graph, nodes are labelled using numbers from 1 to 5 and attributes are labelled as a_1, a_2 and a_3. Figure 1(a), (b), (c), (d) and (e) respectively depict the evolution of the graph from the 1st to 2nd, 2nd to 3rd, 3rd to 4th, 4th to 5th, and 5th to 6th timestamps. Attribute values are represented using the "+", or "−" trend symbols to indicate whether each attribute value has increased or decreased since the previous timestamp, respectively. Now consider that one wants to find patterns indicating that a vertex influences the trends of its neighbors for the following timestamp. If only the frequency is considered, patterns such as $\langle \{a_1+\}, \{a_3+\} \rangle$ can be found. It indicates that an increase of attribute a_1 for a vertex is followed by an increase of a_3 for a neighbor. However, since $\{a_3+\}$ appears almost everywhere in the graph, the pattern $\langle \{a_1+\}, \{a_3+\} \rangle$ is weakly correlated, and uninteresting. Conversely, the pattern $\langle \{a_1+, a_2+\}, \{a_3-\} \rangle$ has a relatively low support but is a strongly correlated trend. In fact, the trend $\{a_3-\}$ does not frequently appear globally but it often locally appears, followed by the trend $\{a_1+, a_2+\}$. To filter the many spurious weakly correlated patterns and present only the strongly correlated ones to the user, a novel significance measure is needed. Such measure would allow to find that $\{a_3-\}$ is more likely to appear after the observation $\{a_1+, a_2+\}$ than $\{a_3+\}$ is likely to follow $\{a_1+\}$.

To address this issue, this paper proposes the novel problem of mining significant trend sequences in dynamic attributed graphs. It defines a novel significance measure called *Sequence Virtual Growth Rate* to identify correlated patterns. Furthermore, an algorithm named TSeqMiner is presented to find these correlated patterns efficiently using novel search space pruning techniques.

The rest of this paper is organized as follows. Section 2 reviews related work on pattern mining in dynamic attributed graphs. Section 3 presents the proposed problem. Section 4 describes the algorithm and its pruning techniques. Section 5 presents results of a performance evaluation and discusses patterns found in real-life social network and flight data. Finally, Sect. 6 draws the conclusion.

2 Related Work

This section gives an overview of relevant related work about mining patterns in dynamic attributed graphs, emerging patterns, and spatio-temporal patterns.

Recently, much attention has been given to mining patterns representing trends in dynamic attributed graphs, which capture how entity relationships and entity attributes change over time. Identifying such trend patterns is useful for both the commercial and scientific communities for many applications such as social network analysis [2]. To analyze a vertex-weighted dynamic network (a dynamic attributed graph with a single attribute), Jin et al. [14] mined connected subgraphs whose vertices show the same trend during a time interval. Such pattern can reveal important events occurring in a dynamic system. However, an important limitation of this work is that it considers a single attribute with only two trend types, that is an increase ($+$) or decrease ($-$), which restricts its applications. A second limitation is that it enforces a strict constraint that timestamps must be consecutive in each time interval. Desmier et al. [4,5] generalized this work by considering patterns with multiple attributes and nonconsecutive timestamps. This type of patterns is called *co-evolution patterns*. A co-evolution pattern is a graph representing attribute variations between two timestamps. A co-evaluation pattern may appear multiple times for different timestamps of a dynamic attributed graph. Besides, several constraints were introduced to obtain an efficient algorithm and find more useful patterns. Cheng et al. [3] then proposed a more general type of patterns called *recurrent patterns*, which are frequent sequences of graphs rather than a single graph. Each graph of a recurrent pattern may contain different vertices and trends. This type of patterns allows to capture the relationships between trends in a graph over time. Although finding such patterns is useful, many recurrent patterns may contain graphs that are weakly correlated. For example, a pattern containing a sequence of two graphs may be frequent only because the second graph appears very frequently over time (similarly to the example presented in the introduction with a_3+). Hence, a significance measure is required to find patterns containing correlated graph entities. Such measure will be presented in this paper.

Another related work is the discovery of *emerging patterns* in databases. In emerging pattern mining, the *growth rate* measure is used to find patterns having a frequency that is largely different from one database to another [6]. Kaytoue et al. [15] used the growth rate to measure the confidence that variations of attribute values will trigger topological changes. A triggering pattern is a sequence of attribute variations, followed by a single topological variation. However, a limitation of this work is that the growth rate is only calculated to assess the influence of the last attribute variations of a triggering pattern on the topology variation. Thus, this approach may find patterns where attribute variations are weakly correlated over time, and thus not meaningful. The novel significance measure presented in this paper addresses this limitation.

Spatio-temporal (ST) data mining is another related field. The relationship between graph mining and ST data mining is that (1) graphs are often used to model spatial information [18], and (2) graphs are often interpreted from a spatial perspective (e.g. the shortest path between two vertices can be considered as a spatial distance) [17]. In ST data mining, Huang et al. [13] proposed to mine significant sequential patterns in spatio-temporal data using a significance measure similar to the growth rate, called *sequence index*, to ensure that events in sequences are strongly correlated. Although the concept of pattern significance as defined by the sequence index is useful, it cannot be directly applied to mine patterns in a dynamic attributed graph. The main reason is that this work considers that two events cannot co-occur. In dynamic attributed graphs, time and locations are discrete and multiple trend types often appear simultaneously for different vertices and timestamps. Allowing simultaneous events is desirable but greatly increases the size of the search space.

To address this problem, reasonable constraints and efficient pruning techniques must be designed to avoid exploring the whole search space. Another challenge for applying the sequence index in dynamic graph minng is that ST mining mainly considers the distance relationship between events as spatial information. But in a dynamic graph, vertices and edges provide much richer information as the topology of a graph may dynamically change over time. This enables users to adopt various neighborhood definitions suia$_1$ for specific applications.

In summary, pattern significance is an important concept in data mining but has not been fully introduced in dynamic attributed graph mining. In some related work, the concept of significant sequential patterns was defined for spatio-temporal mining [13]. But it is not trivial to extend these concepts for dynamic attributed graph mining.

3   Notations and Problem Definition

To address the aforementioned limitations of previous work, this section defines a novel problem of mining significant patterns in dynamic attributed graphs. Definitions are presented, and then the problem.

Dynamic Attributed Graph. Let there be a set of timestamps $\mathcal{T} = \{1, 2, \ldots, t_{max}\}$, a set of vertices $\mathcal{V}$ and a set of attributes $\mathcal{A}$. A dynamic attributed graph is a sequence of attributed graphs $\mathcal{G} = (\mathcal{V}, \mathcal{A}, \langle E_1, E_2, \ldots, E_{t_{max}} \rangle, \langle \lambda_1, \lambda_2, \ldots, \lambda_{t_{max}} \rangle)$, where the attributed graph for timestamp $t \in \mathcal{T}$ is defined by a set of edges $E_t \subseteq \mathcal{V} \times \mathcal{V}$ and a function $\lambda_t : \mathcal{V} \times \mathcal{A} \to \mathbb{R}$ that associates a real value to each vertex-attribute pair. In the following, the attributed graph at timestamp t of a dynamic attributed graph $\mathcal{G}$ is denoted as $G_t = (\mathcal{V}, \mathcal{A}, E_t, \lambda_t)$. Notice that the vertices in the dynamic graph are not changing. Figure 1 illustrates a dynamic attributed graph where trends were obtained by preprocessing an original dynamic attributed graph with six timestamps.

Trend Set, Trend Sequence and Point. A *trend* is an attribute variation, such as a_1+, a_2-. A *trend set* is a set of attribute variations, such as $\{a_1+, a_2+\}$, $\{a_1+, a_2-, a_3+\}$. A k-trend-sequence tss is an ordered list of k trend sets. Let the notation $tss[i]$ denotes the i-th trend set of tss. Moreover, let $tss[i : j]$ denotes the subsequence consisting of the i-th trend set to the $(j-1)$-th trend set. A point $p = (t, v)$ is a tuple consisting of a timestamp t and a vertex v. For example, $tss = \langle \{a_1+, a_2+\}, \{a_3-\}, \{a_2+, a_3+\} \rangle$ is a 3-trend-sequence. $tss[2] = \{a_3-\}$, $tss[2 : 4] = \langle \{a_3-\}, \{a_2+, a_3+\} \rangle$. In the graph of Fig. 1, there are 5 points, i.e. $\{(t_1, 1), (t_1, 2), (t_1, 3), (t_1, 4), (t_1, 5)\}$.

Neighboring Space. The neighboring space $Ns(p)$ of a point $p = (t, v)$ in a dynamic attributed graph is defined as a set of points that are neighbors of p,

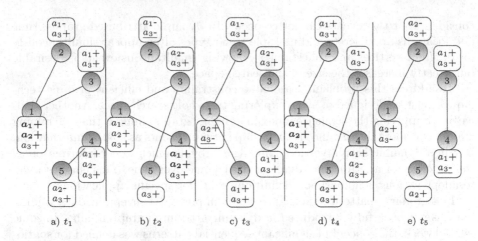

a) t_1 b) t_2 c) t_3 d) t_4 e) t_5

Fig. 1. Trend graphs obtained by preprocessing a dynamic attributed graph originally having 6 timestamps. Occurrences of the pattern $< \{a_1+, a_2+\}, \{a_3-\} >$ and $< \{a_3-\} >$ are highlighted in bold and with an underline, respectively.

i.e. $Ns(p) = \{p' \mid neighborhood(p', p) = true\}$, where $neighborhood$ is a boolean function indicating if two points p' and p are neighbors. The neighboring space of a set of points ps is defined as the union of the neighboring spaces of each point in ps, i.e. $Ns(ps) = \{ p' \mid \exists p, neighborhood(p', p) = true$ and $p \in ps\} = \bigcup_{p \in ps} Ns(p)$. The above definition depends on that of the $neighborhood$ predicate, which can be parameterized for various applications. A simple setting is to consider $neighborhood(p', p) = true$ for two points p and p' if only if p' appears at the timestamp following p, and the vertex of p' is connected to that of p at previous timestamp. This is the definition that will be used in the rest of this paper. For the above setting, in Fig. 1, $Ns((t_1, 1)) = \{(t_2, 2), (t_2, 3)\}$, $Ns((t_4, 1)) = \{(t_5, 2), (t_5, 4)\}$, $Ns\{(t_1, 1), (t_4, 1)\} = \{(t_2, 2), (t_2, 3), (t_5, 2), (t_5, 4)\}$.

Supporting Points. Let $ts(p)$ denotes the trend set of a point p. The supporting points of a trend set ts are the points that have a trend set that is a superset of ts, i.e. $SP(ts) = \{p \mid ts \subseteq ts(p)\}$. For a k-trend sequence tss, if $k = 1$, the tail supporting points of tss are the supporting points of $tss[1]$. If $k > 1$, the tail supporting points of tss are the points that support $tss[k]$ in the neighboring space formed by all tail supporting points of $tss[1 : k]$, i.e. $TSP(tss) = \{p \mid p \in Ns(TSP(tss[1 : k])) $ and $ tss[k] \subseteq ts(p)\}$ for $k \geq 2$. For example, $TSP(tss[1 : 2]) = SP(tss[1]) = \{(t_1, 1), (t_2, 4), (t_4, 1)\}$, $Ns(TSP(tss[1 : 2])) = \{(t_2, 2), (t_2, 3), (t_3, 1), (t_5, 2), (t_5, 4)\}$, $TSP(tss[1 : 3]) = \{(t_2, 2), (t_3, 1), (t_5, 2)\}$.

Significance Measure. The support of a trend set ts in a neighboring space Ns is defined as: $supp(ts, Ns) = \frac{|\{p \mid p \in Ns \ and \ ts \subseteq ts(p)\}|}{|Ns|}$. For example, $supp(\{a_3-\}, Ws) = \frac{5}{25}$. Thus, the *Virtual Growth Rate(VGR)* of ts in Ns is defined as: $VGR(ts, Ns) = \frac{supp(ts, Ns)}{supp(ts, Ws)}$, where Ws is the whole space, i.e. set of all points in $\mathcal{G}$.

For example, $VGR(\{a_3-\}, Ns(\{a_1+, a_2+\})) = \frac{supp(\{a_3-\}, Ns(\{a_1+, a_2+\}))}{supp(\{a_3-\}, Ws)} = \frac{\frac{3}{5}}{\frac{1}{25}} = 3$. The above measure can be viewed as an adaptation of the growth rate measure used in traditional emerging pattern mining. Next we adapts the VGR measure to obtain a more general significance measure that can be applied for a k-trend sequence tss.

The *Sequence Virtual Growth Rate (SVGR)* of a k-trend-sequence $(k \geq 2)$ tss is defined as: $SVGR(tss) = \min_{i \in [2,k]} VGR(tss[i], Ns(TSP(tss[1:i])))$ $k \geq 2$. The proposed significance measure is partially anti-monotone, i.e. for any k-trend-sequence tss such that $k \geq 2$, it follows that $SVGR(tss[1:i]) \geq SVGR(tss)$ for $i \in [3, k+1]$. The proof is omitted due to space limitation. The anti-monotonicity property is important for reducing the search space. To filter some noise, two frequency constraints are applied. A k-trend-sequence $tss(k \geq 2)$ is said to be frequent if and only if $|SP(tss[1])| \geq minInitSup$ and $|TSP(tss[1:i])| \geq minTailSup$ for $\forall i \in [3, k+1]$, where $minInitSup$ and $minTailSup$ are user specified thresholds. Due to the anti-monotone property of the support, the frequency constraints can be used to prune the search space. Then, a k-trend-sequence is said to be significant if and only if tss is frequent and $SVGR(tss) \geq minSig$, where $minSig$ is a parameter specified by the user.

Problem Setting. Let there be a dynamic attributed graph, a significance threshold $minSig$, and two support thresholds $minInitSup$ and $minTailSup$. The problem of discovering all significant k-trend sequences is to find each significant trend sequence $SigTSeq$ such that $SVGR(SigTSeq) \geq minSig$, $|SP(SigTSeq[1])| \geq minInitSup$ and $|TSP(SigTSeq[1:i])| \geq minTailSup$ for $\forall i \in [3, k+1]$ (where k is the length of $SigTSeq$).

4   The TSeqMiner Algorithm

The proposed algorithm, named TSeqMiner, performs a depth-first search to find the significant trend sequences, and applies two pruning strategies. In the following, the dynamic attributed graph of Fig. 1 is considered as example, with $minInitSup = 3$, $minTailSup = 1$, $minSig = 3$, and the neighborhood definition presented in Sect. 3.

4.1   The Search Space

The proposed algorithm first applies a frequent itemset mining algorithm such as *Eclat* [19] with the parameter $minInitSup$ to find all frequent trend sets. They are used to build the search space, as shown in Fig. 2. The search space can be viewed as containing two parts called *inner-levels* and *outer-levels*. The i-th outer-level contains all i-trend sequences, where some of them share a same (i-1)-trend sequence as prefix. The inner-level contains all frequent trend sets (organized in a certain way, which will be explained in Sect. 4.3). For example, the first outer-level is depicted in the top of Fig. 2, for the running example.

It consists of a single inner-level, represented by the topmost dashed box. This box contains all frequent trend sets (1-trend-sequences), represented as a tree. The second outer-level consists of all 2-trend-sequences. In the illustration, part of this level is represented by the two bottommost boxes. Each box represents an inner-level of the second outer-level. For example, the leftmost box of Fig. 2 shows trend sets that can be appended to the prefix 1-trend sequence $\langle\{a_1+\}\rangle$ to form 2-trend sequences. Similarly, the rightmost bottom box of Fig. 2 shows trend sets that can be appended to the prefix 1-trend sequence $\langle\{a_3+, a_1-\}\rangle$ to form 2-trend sequences. In the illustration, each dashed arrow represent extension(s) of a k-sequence to form $(k+1)$-trend sequence(s) having a same prefix.

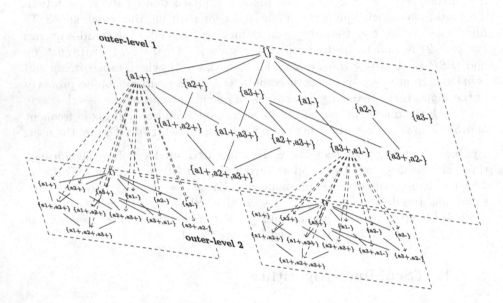

Fig. 2. The search space

4.2 Two Pruning Strategies

Outer-Level Pruning. Because $SVGR$ is anti-monotone, if a k-trend-sequence is insignificant, all the $(k+1)$-trend-sequences that extend that trend-sequence are also insignificant. Moreover, if a k-trend-sequence is infrequent, all its extensions are also infrequent.

Inner-Level Pruning. Pruning using $minTailSup$ can be done at the inner-level and is trivial. Hence, it is not discussed here and this paper focuses on inner pruning using $minSig$.

For the VGR measure, there is no pruning property. Consider a trend sequence tss and two trend sets R and T such that $T \subseteq R$. Let $VGR(tss \rightarrow T) = VGR(T, Ns(TSP(tss)))$. Then, $VGR(tss \rightarrow T) = \frac{supp(T, Ns(TSP(tss)))}{supp(T, Ws)}$. $VGR(tss \rightarrow R) = \frac{supp(R, Ns(TSP(tss)))}{supp(R, Ws)}$. Since $supp(R, Ws) \leq supp(T, Ws)$ and

$supp(R, Ns(TSP(tss))) \leq supp(T, Ns(TSP(tss)))$ always hold, $VGR(tss \to T)$ can be larger than, equal to, or smaller than $VGR(tss \to R)$.

To address the lack of search space pruning property for the VGR measure, a lower bound on the VGR named lbs is proposed for pruning. Let there be a set of trend sets I_s for which the support in Ws has already been measured. Besides note that in the following Ws is ignored to improve readability. Moreover, assume that $T \in I_s$. A lower bound on the support of T, named lbs is defined as: $lbs(T) = \min\limits_{T \subset R \in I_s} lbs(R)$ if $\exists R \in I_s$ and $T \subset R$; $lbs(T) = supp(T)$ otherwise. The lbs measure is a lower bound on the support and is monotone. In other words, for two trend sets $T_1, T_2 \in I_s$ such that $T_1 \subseteq T_2$, $lbs(T_1) \leq supp(T_1)$ and $lbs(T_1) \leq lbs(T_2)$. *Proof of monotonicity*: if (a) $T_1 = T_2$, $lbs(T_1) = lbs(T_2)$. (b) else $lbs(T_1) = \min\{lbs(T_2), \min\limits_{T_1 \subset R \in I_s, R \neq T_2} lbs(R)\} \leq lbs(T_2)$.

Proof of lower bound: (a) if $\nexists R \in I_s$ and $R \supset T$, then $lbs(T) = supp(T)$. (b) else there must $\exists R \in I_s, R \supset T$ such that $\nexists Q \in I_s, Q \supset R$. Then, $lbs(T) \leq lbs(R) = supp(R) \leq supp(T)$.

Next we introduce an Upper bound on the Virtual Growth Rate ($UVGR$), defined as $UVGR(tss, Ns) = \frac{supp(ts, Ns)}{lbs(ts, Ws)}$. According to properties of lbs and anti-monotonicity of $supp$, it is evident that $UVGR$ is an upper bound on the VGR and is anti-monotonous. Based on properties of $UVGR$, pruning techniques can be developed to accelerate the search for patterns.

4.3   A Structure to Support Searching for Patterns

During the search, quickly finding trend sets is necessary. To achieve this, a map is used. First, a total order $\prec$ of trend sets is defined. Without loss of generality, assume that the elements of any trend set ts are sorted according to the lexicographical order. Let $ts[i]$ denotes the i-th element of ts. For two trend sets $ts1$ and $ts2$, let $minL = \min\{|ts1|, |ts2|\}$. The trend set $ts1$ is said to precede $ts2$, denoted as $ts1 \prec ts2$, if and only if one of the following constraints is satisfied. 1) $\exists i \in [1, minL]$ such that $ts1[j] = ts2[j]$ for $j < i$ and $ts1[i] < ts2[i]$; 2) $\forall j \in [1, minL]$, $ts1[j] = ts2[j]$ and $|ts1| < |ts2|$. Then, $ts2$ is said to be a dominated superset of $ts1$ (and $ts1$ is said to be a dominating subset of $ts2$) if $ts1 \subset ts2$, $|ts1| + 1 = |ts2|$ and $ts1 \prec ts2$. For the running example, assume that the lexicographical order is $a_1+ < a_2+ < a_3+ < a_1- < a_2- < a_3-$. The search structure stores all dominance relationships, as shown in Fig. 3. The dominated supersets

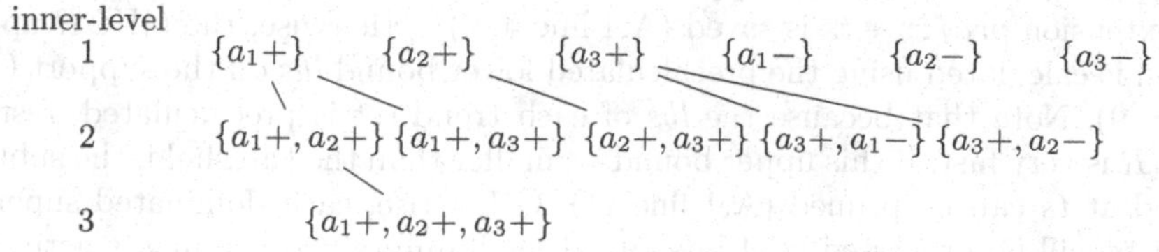

inner-level

1    $\{a_1+\}$ $\{a_2+\}$ $\{a_3+\}$ $\{a_1-\}$ $\{a_2-\}$ $\{a_3-\}$

2    $\{a_1+, a_2+\}$ $\{a_1+, a_3+\}$ $\{a_2+, a_3+\}$ $\{a_3+, a_1-\}$ $\{a_3+, a_2-\}$

3    $\{a_1+, a_2+, a_3+\}$

Fig. 3. The search structure.

of $\{a_1+\}$ are $\{a_1+, a_2+\}$, $\{a_1+, a_3+\}$. When calculating $lbs(\{a_1+\}, Ws)$, the considered range is $Is = \{\{a_1+\}, \{a_1+, a_2+\}, \{a_1+, a_3+\}, \{a_1+, a_2+, a_3+\}\}$.

4.4 The Algorithm

The algorithm performs two steps, and consists of four main procedures presented in A.1, A.2, A.3 and A.4, where "A.x" denotes "Algorithm x".

Step 1. Generate 1-Trend-Sequence and Maps. The algorithm scans the set of trend graphs once and acquires all trend sets with their supporting points. The *Eclat* algorithm is applied to find all frequent trend sets with their supporting points (*trendsetMapSP* in A.1). Since this part is not the main contribution of this paper and can be easily implemented, details are omitted. Then, three maps are created. *levelMapTrendset* maps each level to its trend sets. *trendsetMapDom* maps each trend set to all its dominated supersets. *trendsetMapMin* maps each trend set to its *lbs* value. Notice that to compute the lower bound on the support of *ts*, only trend sets in the subtree rooted at *ts* are considered. Calculating *lbs* in this way for pruning is called large-grained pruning. Another way of calculating *lbs* called medium-grained pruning is presented in Algorithm 4, to improve efficiency.

Step 2. Pattern Extension. The algorithm performs a depth-first search. For each visited k-trend-sequence *tss* and its tail supporting points, a neighboring space *neighborSP* is constructed based on the provided neighborhood definition $\mathcal{NH}$. Then, trend sets from this space are tested to generate (k + 1)-trend-sequences with their supporting points. The algorithm starts the search from 1-trend-sequences, which are processed one by one (A.1 line 2–6). For each k-trend-sequence *tss*, if its size is larger than one, it is output (A.2 line 2–3). Then, the method *acquireNeighbors* constructs the neighboring space based on the tail supporting points of *tss* (A.2 line 4–5). In this neighboring space, all trend sets with supporting points that have a large significance will be found (A.2 line 7). Then, the depth-first search is continued (A.2 line 8–9).

At the inner-level, trend sets are visited using a depth-first search on the search structure presented in Sect. 4.3 (A.3 1–3). The goal is to find trend sets that can extend a trend-sequence *prefix* to generate larger trend-sequences. For each trend set *ts*, the method *computeLocalSup* is applied to calculate the local support and supporting points of *ts* in the neighboring space (A.4 line 1). Then, the *VGR* of *ts* is derived (A.4 line 2–3). If the *VGR* of *ts* is not smaller than *minSig*, and the number of supporting points is not smaller than *minTailSup*, the extension *prefix* + *ts* is saved (A.4 line 4–6). Otherwise, the *UVGR* upper bound is calculated using the precalculated lower bound *lbs* on the support (A.4 line 8–9). Note that because the *lbs* of each trend set is precalculated, testing *UVGR* is very fast. If this upper bound is smaller than the threshold, the subtree rooted at *ts* can be pruned (A.4 line 11). Otherwise, each dominated superset ts' of *ts* will be processed (A.4 line 19). If medium-grained pruning is activated (A.4 line 15), then *lbs* of *ts* is the *lbs* of ts' (A.4 line 16). It means that the trend sets of the subtree rooted at ts', together with *ts*, are used to calculate *lbs*.

Algorithm 1. TSeqMiner$_{dfs}$

---

    input : a set of trend graphs(tGs), $minSig$, $minInitSup$, $minTailSup$;
           $\mathcal{NH}$: a neighborhood definition;
           $trendsetMapSP$: map from each trend set to its set of supporting points;
           $levelMapTrendset$: map from each level to its trend sets;
           $trendsetMapDom$: map from each trend set to its dominated supersets;
           $trendsetMapMin$: map from each trend set to its lbs lower bound on the support;
    output: all significant trend sequences

1 $addedTsList, addedSPList, prefix \leftarrow \{\}$;
2 **foreach** $(ts, SP) \in trendsetMapSP$ **do**
3     $addedTsList.add(ts)$;
4     $addedSPList.add(SP)$;
5 **end**
6 $outerDFSHelper(prefix, addedTsList, addedSPList)$;

Algorithm 2. OuterDFSHelper

---

    input : $prefix$: a sequence of trend sets that form a sequence;
           $newTrendsetList$: a list of trend sets to be appended to $prefix$;
           $tailSPList$: a list of sets of supporting points corresponding to trend sets in
    $newTrendsetList$;

1   **foreach** $ts \in newTrendsetList$ **do**
2     $tss \leftarrow prefix + ts$;
3   **if** $tss.size > 1$ **then** $outputPattern(tss)$;
4     $SP \leftarrow$ corresponding set of supporting points in $tailSPList$;
5     $neighborSP \leftarrow acquireNeighbors(SP, \mathcal{NH})$;
6     $addedTsList, addedSPList \leftarrow \{\}$;
7     $innerDFS(neighborSP, addedTsList, addedSPList)$;
8     $prefix \leftarrow tss$;
9     $outerDFSHelper(prefix, addedTsList, addedSPList)$;
10   remove the last trend set of $prefix$;
11 **end**

Algorithm 3. InnerDFS

---

    input : $neighborSP$: neighboring space consisting of points;
           $addedTsList$: a list of trend sets that are significant in the neighboring space;
           $addedSPList$: supporting points of each trend set in $addedTsList$
1 **foreach** $trend\ set\ ts \in levelMapTrendset.get(1)$ **do**
2     $innerDFSHelper(ts, neighborSP, addedTsList, addedSPList)$
3 **end**

In that case, if $UVGR$ of ts is smaller than $minSig$, the subtree rooted at ts' can be pruned (A.4 line 17–18).

4.5 Discussion

One of the main interest of discovering the type of patterns proposed in this paper is that patterns that contain strongly correlated entities can be found, while filtering other spurious uncorrelated patterns. Thus, a user can obtain a small subsets of correlated patterns that is significant.

    A question about the proposed problem is why using two minimum frequency thresholds in addition to the minimum significance threshold. There are two reasons. First, by allowing multiple attribute variations to co-occur, the search space

---

Algorithm 4. InnerDFSHelper

---

input : ts: current processed trend set;
          $neighborSP, addedTsList, addedSPList$: same as in Algorithm 3.

1   $localSup, SP \leftarrow computeLocalSup(neighborSP, ts)$;
2   $supp \leftarrow trendsetMapSP.get(ts).size$;
3   $vGR \leftarrow \frac{localSup}{supp}$;
4   **if** $vGR \geq minSig$ **and** $|SP| \geq minTailSup$ **then**
5 | $addedTsList.add(ts)$;
6 | $addedSPList.add(SP)$;
7   **else**
8 | $lbs \leftarrow trendsetMapMin.get(ts)$;
9 | $uVGR \leftarrow \frac{localSup}{lbs}$;
10 | // Large-grained pruning
11 | **if** $uVGR < minSig$ **then** return;
12   **end**
13   **foreach** $ts' \in trendsetMapDom.get(ts)$ **do**
14 | // Medium-grained pruning
15 | **if** mGP **then**
16 | | $lbs \leftarrow trendsetMapMin.get(ts')$;
17 | | $uVGR \leftarrow \frac{localSup}{lbs}$;
18 | | **if** $uVGR \geq minSig$ **then**
                   $innerDFSHelper(ts', neighborSP, addedTsList, addedSPList)$;
19 | **else** $innerDFSHelper(ts', neighborSP, addedTsList, addedSPList)$;
20   **end**

---

becomes quite large on real datasets. Thus, we cannot afford to explore all combinations of all possible variations. The minimum frequency thresholds thus act as a filter to avoid exploring the whole search space. Second, it is often undesirable to explore very low frequency patterns because some combinations are just noise and appear in very few situations. Therefore, it is suggested to set the two proposed frequency parameters to low values to filter noise to some extent, while not eliminating too many patterns. Unlike previous work who aim to mine frequent patterns using only a frequency threshold, the proposed algorithm allow to set the frequency thresholds quite low because the significance measure is also used for search space pruning using the proposed $UVGR$ upper-bound.

In [15], a dynamic attributed graph is transactionized to get a sequential database. Then, the classical PrefixSpan sequential pattern mining algorithm is applied to find sequential patterns [10] in the graph. For the proposed problem, we argue that this approach is inappropriate. The reason is that a pattern as in [15] can be described using a single transaction item. However, in this study, for the neighborhood definition of the running example, a pattern involve at least two transaction items. And this is the case for most complicated neighborhood definitions. Thus, it is not trivial to modify sequential pattern mining algorithms for the proposed problem. Besides, a research goal was to propose a general algorithm that is flexible (can be used with various neighborhood definitions).

The complexity of the TSeqMiner algorithm is analyzed as follows. The algorithm first discovers frequent trend sets using frequent itemset mining. Frequent itemset mining algorithms generally have a time complexity that is linear to the number of possible patterns (trend sets). If there are w possible trend values in the original database, in the worst case $2^w - 1$ possible trend sets are considered, although in real-life that number is generally much smaller than $2^w - 1$ because not all trends co-occur. After the set M of frequent trend sets has been identified, the three mappings *levelMapTrendset*, *trendsetMapSuperset*, *trendsetMapMin* are generated, which require $O(M)$, $O(M^2)$, $O(M^2)$ time, respectively. Suppose that the maximal depth of the tree is $\overline{D}$. For each node in the tree up to M extensions are attempted. The main cost to evaluate an extension is the intersection of its neighboring space and supporting points of a trend set, whose average cost is assumed to be $\overline{I}$. Therefore, the cost for considering all extensions is $(M + M^2 + ... + M^{\overline{D}}) \cdot M \cdot \overline{I}$. The total time cost is $O(M^{\overline{D}+1} \cdot \overline{I})$. In the worst case, $\overline{D} = T_{max} - 1$.

5   Experimental Results

To evaluate the proposed TSeqMiner algorithm, quantitative and qualitative experiments have been conducted on two real world datasets. Note that because the proposed problem is considerably different from previous pattern mining problems, no algorithm can be directly compared with the proposed algorithm. Thus, performance experiments have been done by studying how parameter settings influence the algorithm's performance, and the algorithm was compared with a basic version without optimizations to show that optimizations are useful. The **DBLP dataset** consists of references published from 1990 to 2010. It involves 2723 authors and 43 conferences/journals(more than 10 publications). The corresponding dynamic attributed graph has 9 timestamps([1990–1994][1992–1996]...[2006–2010]). Each vertex denotes an author and has an attribute vector indicating the number of publications for 43 conferences or journals. The **Domestic US Flight dataset** contains records of US airport traffic from 01/08/2005 to 25/09/2005. Three hurricanes, namely Irene(04/08–18/08), Katrina(23/08–31/08) and Ophelia(06/09–17/09), occurred during this time. Data are aggregated by weeks for a total of 8 timestamps. Each attributed graph has 280 vertices (airports) having 8 attributes (e.g. number of arrivals/departures, average arrival/departure delay). Edges denote flight connections.

The proposed algorithm was implemented in Java. Experiments were conducted on a computer with Ubuntu 16.04 (Intel(R) Xeon(R) CPU E3-1270 3.60 GHz) an 64 GB of RAM. Source code can be downloaded from the SPMF [9] software at http://philippe-fournier-viger.com/spmf/.

Quantitative Evaluation. In the following, "DFS-OP" denotes TSeqMiner, "NO-PRUNING" denotes TSeqMiner without pruning, and "DFS" denotes TSeqMiner without medium-grained pruning. The algorithm takes as input three parameters, *minInitSup*, *minTailSup* and *minSig*. *minInitSup* determines the number of frequent trend sets, and thus also the size of the inner search space. It is observed in Fig. 4(a)(b) that when the number of frequent trend sets increases, the execution time and the number of patterns grow slowly (note that values of the x-axis are not equidistant). It is also observed in Fig. 4(a)(b) that pruning and optimizations improve the execution time. Figure 4(c)(d) indicate that both

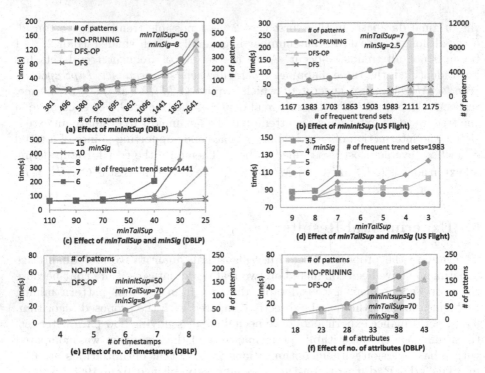

Fig. 4. Influence of parameters and scalability on performance

minSig and *minTailSup* have great impact on pruning. If any of them is set to a large value (e.g. *minTailSup* = 90 or *minSig* = 15 in Fig. 4(c)), the search stops quickly. Figure 4(e)(f) reports results about scalability. Number of found patterns and execution time grow relatively fast when the number of timestamps is increased, since it increases both the number of edges and vertices. However, the proposed algorithm have great scalability with respect to the number of attributes. Hence, TSeqMiner works well on graphs where entities have many attributes, which is one of the main challenges for mining dynamic attributed graphs.

Qualitative Evaluation. Consider the neighborhood definition of Sect. 3. For the DBLP dataset, let *minInitSup* = 35, *minTailSup* = 100, and *minSig* = 10. For these parameters, TSeqMiner generates 1610 frequent trend sets and 3 patterns (show in Table 1) in 45 s. The first pattern indicates that a co-author of someone who has published more papers in *ICDE*, *PVLDB*, and *ACMTransDBSys* in recent years, is more likely to publish more papers in *PVLDB* and less in *VLDB*. In this pattern, the first trend set is supported by 45 authors in the whole space and the second by 107 authors in the neighboring space of the first trend set. This pattern has a strong correlation as measured by the significance measure. The significance value is 11.5. This is a pattern that provides interesting insight on authorship behavior.

Table 1. Three significant trend sequences mined from the DBLP dataset, with their number of supporting points and significance.

Patterns	Supporting point count	Signif
$\{\{ICDE+, PVLDB+, ACMTransDBSys+\}, \{PVLDB+, VLDB-\}\}$	$\{45, 107\}$	$\{11.5\}$
$\{\{PVLDB+, VLDB+\}, \{PVLDB+, VLDB-\}\}$	$\{57, 120\}$	$\{11.5\}$
$\{\{JMLR =\}, \{JMLR-\}\}$	$\{283, 147\}$	$\{10.3\}$

For the US Flight dataset, $minInitSup = 8$, $minTailSup = 7$ and $minSig = 6$, TSeqMiner$_{dfs}$ returned 768 patterns in 9 s. A significant trend sequence is $\{\{NbCancel--, NbDivert--, DelayDepart-\}, \{NbDepart-, NbCancel-, NbDivert-, DelayDepart-, DelayArriv+\}\}$. In the first trend set, $NbCancel--$ and $NbDivert--$ indicate a significant decrease of the number of cancelled and diverted flights after recovering from the damage caused by a hurricane. Most supporting points of that trend set are at timestamp 3 or 4 on the east coast where the most severe damage occurred. In the second trend set, $NbCancel-$ and $NbDivert-$ indicate moderate recovery from damage. The supporting points of the second trend are all at timestamp 5 and mostly not located on the east coast. That is, these airports experienced moderate problems caused by delays and diverted flights due to the hurricane on the east coast (more specifically at timestamp 4). Therefore, these airports are moderately influenced by the hurricane and recover moderately quickly after the recovery of east coast airports. $DelayArriv+$ indicates an increase of the average delay time of arriving flights. It may indicate that rescheduled flights are not completely operating as usual. The above patterns capture an interesting aspect of spatio-temporal data. Thus, on overall, the proposed algorithm can be deemed as finding interesting patterns in real-life data. It can also be applied to other domains where data is modelled as graphs.

6   Conclusion

This paper proposed a novel type of pattern called significant trend sequence. It is a sequence of attribute variations, where correlations between elements are measured by a novel significance measure named *Sequence Virtual Growth Rate*. To efficiently mine these patterns, an algorithm named TSeqMiner is proposed, which relies on a novel upper bound and search space pruning strategy. Experiments on real-life datasets have shown that the proposed algorithm is efficient and can identify interesting patterns in social network and flight data.

For future work, we will extend the proposed model for mining sequences of significant subgraphs, and design methods to find other types of significant patterns [8, 11].

References

1. Aggarwal, C.C., Wang, H. (eds.): Managing and Mining Graph Data. Springer, Boston (2010). https://doi.org/10.1007/978-1-4419-6045-0
2. Ahmed, R., Karypis, G.: Algorithms for mining the evolution of conserved relational states in dynamic networks. Knowl. Inf. Syst. **33**(3), 603–630 (2012)
3. Cheng, Z., Flouvat, F., Selmaoui-Folcher, N.: Mining recurrent patterns in a dynamic attributed graph. In: Kim, J., Shim, K., Cao, L., Lee, J.-G., Lin, X., Moon, Y.-S. (eds.) PAKDD 2017. LNCS (LNAI), vol. 10235, pp. 631–643. Springer, Cham (2017). https://doi.org/10.1007/978-3-319-57529-2_49
4. Desmier, E., Plantevit, M., Robardet, C., Boulicaut, J.-F.: Trend mining in dynamic attributed graphs. In: Blockeel, H., Kersting, K., Nijssen, S., Železný, F. (eds.) ECML PKDD 2013. LNCS (LNAI), vol. 8188, pp. 654–669. Springer, Heidelberg (2013). https://doi.org/10.1007/978-3-642-40988-2_42
5. Desmier, E., Plantevit, M., Robardet, C., Boulicaut, J.-F.: Cohesive co-evolution patterns in dynamic attributed graphs. In: Ganascia, J.-G., Lenca, P., Petit, J.-M. (eds.) DS 2012. LNCS (LNAI), vol. 7569, pp. 110–124. Springer, Heidelberg (2012). https://doi.org/10.1007/978-3-642-33492-4_11
6. Dong, G., Li, J.: Efficient mining of emerging patterns: Discovering trends and differences. In: Proceedngs of the 5th ACM SIGKDD International Conference on Knowledge Discovery and Data Mining, pp. 43–52. ACM (1999)
7. Fassetti, F., Rombo, S.E., Serrao, C.: Discovering discriminative graph patterns from gene expression data. In: Proceedings of the 31st Annual ACM Symposium on Applied Computing, pp. 23–30. ACM (2016)
8. Fournier-Viger, P., Li, X., Yao, J., Lin, J.C.-W.: Interactive discovery of statistically significant itemsets. In: Mouhoub, M., Sadaoui, S., Ait Mohamed, O., Ali, M. (eds.) IEA/AIE 2018. LNCS (LNAI), vol. 10868, pp. 101–113. Springer, Cham (2018). https://doi.org/10.1007/978-3-319-92058-0_10
9. Fournier-Viger, P., et al.: The SPMF open-source data mining library version 2. In: Berendt, B., et al. (eds.) ECML PKDD 2016. LNCS (LNAI), vol. 9853, pp. 36–40. Springer, Cham (2016). https://doi.org/10.1007/978-3-319-46131-1_8
10. Fournier-Viger, P., Lin, J.C.W., Kiran, U.R., Koh, Y.S.: A survey of sequential pattern mining. Data Sci. Pattern Recogn. **1**(1), 54–77 (2017)
11. Fournier-Viger, P., Zhang, Y., Lin, J.C.W., Fujita, H., Koh, Y.S.: Mining local and peak high utility itemsets. Inf. Sci. **481**, 344–367 (2019)
12. Holder, L.B., Cook, D.J., et al.: Learning patterns in the dynamics of biological networks. In: Proceedings of the 15th ACM SIGKDD International Conference on Knowledge Discovery and Data Mining, pp. 977–986. ACM (2009)
13. Huang, Y., Zhang, L., Zhang, P.: A framework for mining sequential patterns from spatio-temporal event data sets. IEEE Trans. Knowl. Data Eng. **4**, 433–448 (2007)
14. Jin, R., McCallen, S., Almaas, E.: Trend motif: a graph mining approach for analysis of dynamic complex networks. In: Proceedings of the 7th IEEE International Conference on Data Mining, pp. 541–546. IEEE (2007)
15. Kaytoue, M., Pitarch, Y., Plantevit, M., Robardet, C.: Triggering patterns of topology changes in dynamic graphs. In: Proceedings of the 6th IEEE/ACM International Conference on Advances in Social Networks Analysis and Mining, pp. 158–165. IEEE/ACM (2014)
16. Lv, T., Gao, H., Li, X., Yang, S., Hanzo, L.: Space-time hierarchical-graph based cooperative localization in wireless sensor networks. IEEE Trans. Sig. Process. **64**(2), 322–334 (2016)

17. Sanhes, J., Flouvat, F., Selmaoui-Folcher, N., Pasquier, C., Boulicaut, J.F.: Weighted path as a condensed pattern in a single attributed dag. In: Proceedings of the 23rd International Joint Conference on Artificial Intelligence (2013)
18. Wen, Y.-T., Fan, Y.Y., Peng, W.-C.: Mining of location-based social networks for spatio-temporal social influence. In: Kim, J., Shim, K., Cao, L., Lee, J.-G., Lin, X., Moon, Y.-S. (eds.) PAKDD 2017. LNCS (LNAI), vol. 10234, pp. 799–810. Springer, Cham (2017). https://doi.org/10.1007/978-3-319-57454-7_62
19. Zaki, M.J.: Scalable algorithms for association mining. IEEE Trans. Knowl. Data Eng. **12**(3), 372–390 (2000)

Text-Based Event Detection: Deciphering Date Information Using Graph Embeddings

Hilal Genc[1]([✉]) and Burcu Yilmaz[2]([✉])

[1] Department of Computer Engineering, Gebze Technical University, Gebze, Turkey
hgenc@gtu.edu.tr
[2] Institute of Information Technologies, Gebze Technical University, Gebze, Turkey
byilmaz@gtu.edu.tr

Abstract. Event detection is increasingly gaining attention within the fields of natural language processing and social network analysis. Graph models have always been integral to social media analysis literature. Owing to the long processing time and time complexities of graph-based algorithms, these models were initially very difficult to improve upon. Over the past few years, researchers proposed many approaches to create representations such as word2vec and doc2vec [11]. With the emergence of graph embedding techniques in recent years using deep learning techniques such as node2vec, it is possible to extract node embeddings that can be used to embed graph information into machine learning methods. We introduce SnakeGraph, a new model which uses the sequences of words making up each body of text along with key representations such as the user and the date. These representations can help us learn about the main ideas communicated via written language. However, our method not only looks at both the content of text and how it links to other key information, but also factors the relationship between words in our text as they appear in sequence and overlap as they appear across different bodies of text. We believe that date and user embeddings can especially shed light on event detection literature.

Keywords: Graph embeddings · Extracting time embeddings · Event detection · Transfer learning

1 Introduction

Time is a significant aspect in social network analysis. This is because it can not only help document individuals and their changing preferences and points of view but also the progress of events to a much larger scale. Information diffusion refers to the progress of events over time and it may manifest itself in different patterns. Attention given to an event may emerge in the beginning, peak and then fade away or it may continue to oscillate indefinitely. Time also has different

© Springer Nature Switzerland AG 2019
C. Ordonez et al. (Eds.): DaWaK 2019, LNCS 11708, pp. 266–278, 2019.
https://doi.org/10.1007/978-3-030-27520-4_19

levels of granularity ranging from seconds to months to years. Proposed models for information diffusion often consider time in units of days.

Recent years have seen an emergence in studies devoted to deep learning and language models that extract dense feature vectors called embeddings or representations. Graph representation learning has proven to be extremely useful for graph-based analysis and prediction. There has also been a rise in approaches that automatically encode graph structure (graph, subgraph, or node embeddings) into low and fixed dimensional embeddings.

In this study we represent social media data using a graph. The proposed graph embeds text information and key entities, such as date and user information corresponding to the tweets, and the relationships between them. Also we transferred a text-based news corpus and hierarchy of the entities to the training phase of the graph embedding extraction to extract the embeddings more accurately. The model that we used to extract vector representations from our graph is Node2Vec. The contributions of the paper are as follows:

1. **Gaining key information:** We can use the proposed model to extract date embeddings. To our knowledge, no graph embedding model currently exists for date embedding extraction. Although we test our method only for dates, the proposed model can be used for any other named entity.
2. **Understanding varying concepts:** The proposed method can extract varying levels of granularity for entities such as date, month, and year embeddings if we clearly define the hierarchy of these concepts. Extracting these date embeddings is the novel part of our study. It is also possible to extract village, city, and country embeddings.
3. **Graphs & Node2Vec:** The proposed model sheds light as to how graph data and the Node2Vec model can be used to model these concepts and extract embeddings.
4. **Enrichment of graph embeddings:** The time complexity is very high for graph mining algorithms. Thus we proposed to transfer non-graph data with the intention of making the model learn the embeddings more accurately. To our knowledge, transferring data to the graph domain from another domain to extract graph embeddings has never been done before. We believe this will pave the way for more advanced transfer learning methods to extract vector representations from graphs.
5. Although we did not propose a method specifically for event detection, we extracted related key concepts. We believe that these embeddings will shed light on event detection literature.

1.1 Literature Analysis

Event detection is identifying an event trigger (usually a single verb or noun [4]) from a body of text to determine what event(s) might be associated with it. An early study in event detection [8] clusters tweets through similarity measures and selects the most widely shared tweet to represent all other similar tweets. Each event trigger is introduced in [14] as a triple structure $(arg_s, verb, arg_o)$

containing a verb phrase and two noun phrases representing the subject and object with respect to the action. The model filters the frame elements for noise via a probabilistic model.

The results from [9] show that convolutional neural networks alone can be very helpful for sentence classification. An LSTM model in [10] uses mean pooling to combine user embeddings, community embeddings, and word embeddings to determine whether a given post will initiate a conflict. [3] uses a bidirectional gated recurrent unit to generate the most likely hashtag that will appear in a given tweet while [15] detects the sentiment and similarity of tweets using a CNN-LSTM model. The model proposed by [1] looks at subword information on words belonging to morphologically rich languages such as Turkish and Finnish. [16] mentions the graph neural network (GNN) as a framework that computes embeddings of a node by recursively aggregating and transforming the embeddings of each node's neighbors. These neighbors have no natural ordering unlike the elements within a lattice used for image processing [7].

Retrieving node embeddings from our proposed SnakeGraph to find correlations between written text and publication date is the focus of our study. We accomplish this with the help of an effective model that extracts embeddings for the nodes in our graph. Major kinds of embeddings identified in [2] are node embeddings, edge embeddings, subgraph embeddings, and graph embeddings. Node2Vec [6], Subgraph2Vec [12], DeepWalk [13], and Author2Vec [5] have successfully proposed models for extracting these node embeddings from already existing graphs. The differences in how these algorithms preserve graph properties affect how these algorithms will preserve distances between nodes in the embedding space.

Deepwalk learns node embeddings from random walks via a semi-supervised approach. It uses skipgram, which can essentially "skip" over words. Thus, skipgram models are unlike n-grams in that they are not necessarily consecutive. With little doubt, the skipgram method would enable us to capture the similarities between words that are not in sequence but still of relevance. Node2Vec is a variation of the DeepWalk approach combining breadth first search (BFS) and depth first search (DFS). Subgraph2Vec learns node embeddings and subgraph embeddings (generated through the embeddings of nodes and a small group of their neighboring nodes) with an unsupervised approach. Author2Vec uses a graph where a node represents each author and an edge represents each collaboration between two authors. The graph does not include text-based content. We have thus created a paradigm that builds connections between key information and text information which previous models such as doc2vec and word2vec cannot accomplish.

2 Our Approach

In this section we will discuss how we created our graph from a social media dataset and used the graph to extract node embeddings that will represent our key entities. We then introduce a transfer learning model to enrich the data.

2.1   Extraction of SnakeGraph

Here will discuss how we created the *SnakeGraph*. To show the relationship between all the words and the key entities from the dataset, we decided to create a graph.

We define a graph as $G = (V, E)$, where $v \in V$ is a node and $e \in E$ is an edge. The node mapping function of G is $f_v : V \to T^v$ and the edge mapping function of G is $f_e : E \to T^e$.

The set of node types and the set of edge types are given by T^v and T^e, respectively. Each node $v_i \in V$ belongs to one particular type, i.e., $f_v(v_i) \in T^v$. The same applies for each edge where for $e_{i,j} \in E, f_e(e_{i,j}) \in T^e$.

To incorporate the content of each tweet, we filtered the tweets through a preprocessing phase where we remove all stop words (punctuation marks) and any cluster of characters starting with a "@" or containing "http" or any other indication of a hyperlink. We then tokenized our preprocessed data and then added lowercase of these tokens to our graph as a node one by one, essentially in the form of a snake. Each word and the consecutive word after it is connected with an edge. Hence, we named the graph *SnakeGraph* for our proposed model. Each tweet from our dataset had a tweet id, given by a unique alphanumeric sequence to distinguish any tweet from all others that might have identical content. We added each tweet id to the graph as a node for every tweet we included in our graph. The node for each tweet id was attached to the node for first token or word of the corresponding tweet with an edge. Then the rest of the words were added in a sequence.

No two nodes in our graph are alike. The "snakes" or sequences of words overlap when the tweets have a mutual word. The words appearing in our graph are linked only to either the tweet id (if said token is the first in the tweet or sequence) or to one or two other token(s).

Figure 1 shows the appearance of two different short bodies of text as they would appear on our comprehensive *SnakeGraph* graph. The bodies of text in our figure are "besiktas ve fenerbahçe de vergi indiriminden" and "özel haber - fenerbahçe de sasirtan advocaat yasaklari". They translate to "Besiktas (football team) and fenerbahce (football team) granted tax deductions" and "Special news: Advocaat (manager) announces surprising prohibitions for fenerbahce."

2.2   Extracting Graph Embeddings

To determine the embeddings of entities such as date from our *SnakeGraph*, we use the Node2Vec model which creates node embeddings from graphs [6]. We provide the formal definitions of graph and node embeddings below.

Definition 1. Graph Embedding: For a given graph $G = (V, E)$, the graph embedding extraction converts G into a d-dimensional vector space where d $\ll$ |V|. The vector space preserves the graph attributes.

Definition 2. Node embedding: Node embedding provides an embedding vector (or feature vector) $ev(u)$ as a representation for each node u. These vectors appear in a low dimensional space and nodes that have similar characteristics have similar vector representations.

We may use the node2vec algorithm to generate the embeddings $e(w)$, $e(tw)$, $e(us)$, and $e(d)$ for the words w, tweet ids tw, user ids us, and/or dates d, respectively.

Each entity (word, tweet id, user id, and/or date) in our dataset is represented by a node in our graph. The node2vec algorithm enable us to generate the embeddings for each node appearing on our graph. These embeddings allow us to find similarities between words and other particular entities appearing on our *SnakeGraph*.

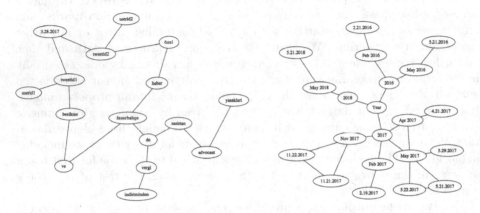

Fig. 1. Relationships between a given date on the graph and an id for a body of text, the words in it, and the author's userid.

Fig. 2. Relationships between the dates as they appear on our Snake-Graph.

We developed a biased random walk procedure to explore the neighborhoods of each node with a technique using both breath first search and depth first search. We created random walks in a similar way as given by the node2vec algorithm. For a node u we created random walks with a fixed length l. In a random walk, let c_i be the ith node starting with node $c_0 = u$. The following distribution is used to generate the random walks.

$$P(c_i = x | c_{i-1} = v) = \begin{cases} \frac{\pi_{vx}}{Z} & if \ (v, x) \in E \\ 0 & otherwise \end{cases}$$

where π_{vx} is the unnormalized transition probability between nodes v and x, and Z is the normalizing constant. Transition probability π_{vx} on an edge (v, x) is used to determine the next node x after node v in the walk. $\pi_{vx} = \alpha_{pq}(t, x).w_{vx}$ where

$$\alpha_{pq}(t, x).w_{tx} = \begin{cases} \frac{1}{p} & d_{tx} = 0 \\ 1 & d_{tx} = 1 \\ \frac{1}{q} & d_{tx} = 2 \end{cases}$$

and d_{tx} denotes the shortest path distance $\in \{0, 1, 2\}$ between nodes t and x. Parameters p and q determines the random walk using an interpolation of BFS and DFS.

We created a number of random walks $rw_u \in RW$ for each node u in the *SnakeGraph* with the methodology mentioned above where the list of random walks is given by RW.

To extract an embedding $ev(u)$ for each node u, we decided to use the Node2Vec methodology which is based on the skipgram method that extracts word embeddings in natural language processing (NLP). We seek to optimize the objective mentioned as mentioned below. The objective function maximizes the log-probability of neighbourhood $N(s)$ of a node u.

$$\max_f \sum_{n=1} log Pr(N(u)|f(u))$$

2.3   Enriching Graph Embeddings with Transfer Learning

Language models have limited performance when developing models for morphologically rich languages such as Turkish, Finnish, and Czech. As expected, the Node2Vec model has longer processing time as the graph grows. The time complexity of the Node2vec model depends on the number of random walks created.

Nonetheless, we need more data to learn word embeddings. Our text information embedded into our *SnakeGraph* will not be enough to learn text embeddings accurately enough. Thus, we decided to use a news text corpus C which includes sentences $s \in S$. Each sentence includes a number of words w. We transfer the text information in the skipgram model trained on C to the Node2Vec model trained on the random walks RW extracted from the graph information. This is known as transfer learning. After training the word2vec model with a text corpus, we used the parameters of the model and use them in the Node2Vec model. Then we fine-tuned the Node2Vec model with random walks RW extracted from the graph. In the softmax layer, the model uses the tokens in the last trained corpus. This means that the model extracts the embeddings from these tokens that are derived from the random walks. In order to construct sentences for event detection, we will need more words than the ones in the *SnakeGraph*. Thus we extended it to extract embeddings $ev(u)$ of both nodes in the last trained model and the embeddings $ev(w)$ of words w in the corpus C. We used a concatenation of the corpus C and random walks RW extracted from the graph. Then we give this data $C \cup RW$ to the embedding extraction step of the Node2Vec model and expect the model to learn the relationships between the tokens extracted from the graph and to learn the semantics of the corpus. Because we trained all of the nodes and words together, the Node2vec model will extract the relationships between them using the common words across both datasets. When we get the node embeddings $ev(n \in C)$, accuracy will also be higher for embeddings such as date embeddings.

To enrich the node embeddings we used a technique to embed the hierarchy of dates. We initialize the graph with the node "Year" in order to connect all the nodes representing each year on which the contents of our dataset were published. Then we add the nodes representing the months for each year, connecting these month nodes to their corresponding years. In this step, we also connect the nodes of each consecutive month. We then add the nodes representing each relevant date by connecting each date to its corresponding month. As we connected the nodes for each consecutive month we also connect the nodes for each consecutive date.

We created a node for each date and each date node is connected to a tweet node. To embed the date hierarchy of dates we added edges between consecutive dates. That way, date embeddings will also include the relationships between the dates. We provide a toy example in Fig. 2. In this paper we only focused on date embeddings, although it is possible to similarly enrich user embeddings and embeddings for other entities. It is also possible for us to add edges between two people who are mutual followers. This way the embeddings will also take into account the relationships between users as well.

2.4   Detecting Events Using the Embeddings

In this section we try to determine the events. For each date embedding $e(d)$, we get the set of k nearest words w in the embedding space. These words are likely to have close semantic relationships between the events that happened on that day. Because there is more than one event in a day, we clustered the embeddings of words $ev(w)$. Thus, each cluster EC which is called the **event cluster** will include the words $W_{EC} \in EC$ used to mention an event. To define what these words of events correspond to we used the cosine similarity metric to get the most related words to the words in the event cluster.

3   Experiments

For our experiments, we initially created a graph *SnakeGraph* which contains the words appearing in each body of text, the tweet ids, the user ids, and the dates. The dataset from which we extracted our graph nodes contained 23,801 tweets published between February 15, 2017 and May 31, 2017. The tweets are primarily published in Turkish, but there is some English content as many of the authors are either bilingual or use English words and phrases that are well known among Turkish speakers. Our dataset contains information about each tweet such as the date on which it was published, the unique user id of the author who published it, and the unique corresponding tweet id. We chose these particular dates because our total dataset had the largest density of tweets within this time period and because these months are among those during which events planned in advance, such as football championships, are most likely to take place. Our comprehensive graph contains all of this information because we wanted a framework for which we could find connections between different kinds of entities on our dataset, i.e. dates and topics. Our comprehensive graph G_s has 74,839 nodes and 254,073 edges. Every node on the graph is unique.

We extracted the embeddings from our graphs in two ways. The first method uses "non-enriched embeddings" and we refer to it as $SnakeGraph_{nonEE}$. The second method we refer to as having "enriched embeddings" because it involves using a text-based news corpus. We refer to it as $SnakeGraph_{EE}$. For the former method, we extracted embeddings using only the contents of the graph. We define the embedding of node u as $e(u)$. This algorithm walks only through the nodes of our graph using the Node2Vec method. It also includes a hierarchy of dates. The other method involves the extraction of embeddings using the contents of the graph including hierarchy of dates along with the text from a large news corpus entirely in the Turkish language. We essentially "transfer" the information from a Turkish text corpus onto our dataset containing tweets (most of which were in Turkish). This algorithm appends the walks from our graph to the contents of the news text corpus, the input being first the text corpus and then the walks from our graph. Thus, we extracted the embeddings from the data including both the graph contents and the text corpus. We define the embedding of node u for this method as $e_T(u)$, indicating that the method involves a transfer of data.

For our experiments, we use the gensim implementations of skipgram models. For the method involving non-enriched embeddings, we use a window of 5 and 10 and for the method involving enriched embeddings we use a window of 5 for a reasonable run time. For both methods, we set the size as 128. Our node2vec implementation is the one presented in [10].

To evaluate the extracted embeddings, we conducted several experiments.

Table 1. Major event for three selected dates within our dataset.

Date	Enriched Window = 5				Not Enriched Window = 5				Window = 10			
	Topic	Meaning	Similarity Measure	Event	Topic	Meaning	Similarity Measure	Event	Topic	Meaning	Similarity Measure	Event
2.19.2017	kalinic	Nikola Kalinic	0.5151	Football team transfer	soranın	Questioner's	0.5665	Football game	kalinic	Nikola Kalinic	0.6227	Football team transfer
	adıyamandayız	At Adıyaman	0.5032	Presidential visit	adıyamandayız	At Adıyaman	0.5595	Presidential visit	adıyamandayız	At Adıyaman	0.6011	Presidential visit
	teşekkürleradıyaman	Thank you Adıyaman	0.4845	Presidential visit	kalinic	Nikola Kalinic	0.5589	Football team transfer	teşekkürleradıyaman	Thank you Adıyaman	0.5646	Presidential visit
	tolunay	Tolunay Kafkas	0.4652	Football game	teşekkürlerkahta	Thank you Kahta	0.5353	Presidential visit	tolunay	Tolunay Kafkas	0.5319	Football game
	kararımızanet	Our decision is certain	0.4424	Referendum	öncememlekettablkievet	Nation first absolutely yes	0.5124	Referendum	poldilii	Lukas Podolski	0.5229	Football game
2.23.2017	türkvatanının	Turkish citizen's	0.6720	Referendum	türkvatanının	Turkish citizen's	0.7247	Referendum	türkvatanının	Turkish citizen's	0.7224	Referendum
	bekasiçnevet	Say yes for perpetuity	0.6356	Referendum	bekasiçnevet	Say yes for perpetuity	0.7013	Referendum	bekasiçnevet	Say yes for perpetuity	0.7002	Referendum
	nuno	Nuno Espirito Santo	0.6000	Football news update	nuno	Nuno Espirito Santo	0.6332	Football news update	nuno	Nuno Espirito Santo	0.6816	Football news update
	bundesbank	Bundesbank	0.5736	Financial statement	bundesbank	Bundesbank	0.6300	Financial statement	bundesbank	Bundesbank	0.6733	Financial statement
	türkbesd	TÜRKBESD Association	0.5700	Financial statement	haklisin	You are right	0.6207	Referendum	türkbesd	TÜRKBESD Association	0.6372	Financial statement
5.7.2017	emenikenin	Emmanuel Emenike	0.5677	Football game	emenikenin	Emmanuel Emenike	0.6194	Football game	emenikenin	Emmanuel Emenike	0.6559	Football game
	hayatlarının	To their lives	0.5411	Football game	kaptanımı	My captain	0.5953	Football game	takima	To the team	0.6206	Football game
	dabo	Basketball federation	0.5232	Outreach program	hayatlarının	To their lives	0.5914	Football game	hayatlarının	To their lives	0.5977	Football game
	takima	To the team	0.5048	Football game	dabo	Basketball federation	0.5709	Football game	yensen	If you win	0.5792	Football game
	yensen	If you win	0.4918	Football game	yensen	If you win	0.5562	Football game	dabo	Basketball federation	0.5763	Outreach program

3.1   Neighborhood Embeddings of Dates

We initialize the comprehensive $SnakeGraph_{nonEE}$ including dates by hierarchically organizing the relevant date information. The core node connects all other nodes representing each year. Each node representing a year is then connected to each month corresponding to that year and each month node is connected to each date corresponding to that month. The consecutive days, months, and years are connected to an edge. We construct a graph using the t-Distributed Stochastic Neighbor Embedding (t-SNE) technique that is used for representing date embeddings extracted from $SnakeGraph_{nonEE}$ defined above. The t-SNE is a technique for dimensionality reduction that is particularly used for the visualization of high-dimensional datasets [17]. The method through which we extracted the date embeddings on our t-SNE outputs non-enriched embeddings, has a window of ten, and sets the size to 128. Our t-SNE graph shows the neighborhood of the date embeddings spanning the months of June, July, and August of 2017. The dataset here contains 88 dates and 35,020 tweets from June 2017 to August 2017. We used a separate dataset than the one described above for event detection because we wanted our t-SNE graph to span three full months.

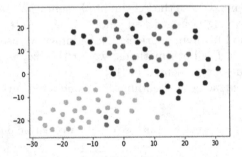

Fig. 3. The t-SNE graphic representing the dates appearing on our graph. The dates of June correspond to the indigo coordinates, the dates of July to the teal coordinates, and the dates of August to the yellow coordinates. (Color figure online)

Figure 3 shows the t-SNE graph of the dates appearing in the dataset between June 2017 and August 2017 and extracted from $SnakeGraph_{nonEE}$. Rather than showing three separate clusters for the three different months, the t-SNE graph shows clusters where each cluster corresponds to a group of days with similar characteristics. The overlap between some of the months, most notably between the months of June and July, most likely results from the repetition of events such as football matches.

To detect how the similarity between a specific day and its consecutive days changes we got 16 days and drew the similarity curves for each day. A similarity curve for a specific day is the cosine similarity of the day and the 33 consecutive days after it. The results of two days are given by Fig. 4(a). We then get the average of the similarity curves for 16 consecutive days. The average similarity curve is presented in Fig. 4(b). We observed that events have different trends. While some of the events can boost and diminish over time by having an oscillating behavior. This corresponds to the instant events. Some of the events can have slight oscillating behavior indefinitely. This may correspond to periodic events. For similar reasons the similarity curve may have an oscillating behavior. Football matches occurring repeatedly might cause the days of each of the match to have similar embeddings. As we see in Fig. 4(b) the trendline is decreasing which shows that, on the average, the events fade slightly.

3.2   Event Detection with Human Evaluation

Because we trained both the dates and the words appearing in our dataset on the same comprehensive graph, we map both the date and word embeddings onto the same embedding space. Thus, we find the cosine similarity between the date and word embeddings. For each date, we extract the N most similar entities (words, user ids, tweet ids, and/or dates) along with their respective cosine similarity measures, with N being the number of entities to extract. After getting the most similar entities, we filtered out all the entities that were not word tokens appearing in our dataset. We extracted the embeddings for each most similar word by using the graph data and, for the method involving enriched embeddings, the Turkish news text corpus. Our dataset for event detection used the dataset of tweets published between February 15, 2017 and May 31, 2017.

In our dataset, we selected 17 days and extracted the most similar words for each given date. The days that we selected were days on which one of Turkey's major football teams was competing as reported by the UEFA. Table 1 shows results for three different days on which some of the most similar words directly correlated to an event associated with that day or during that time period. Most of the events we found occurred on the same day as the day users mentioned or referenced them. A human evaluator helped to confirm the correlation between events and the similar words. All events in Table 1 occurred on the same date except for the referendum, which occurred on April 16, 2017. Nonetheless, the days associated with the referendum were days on which users mentioned the upcoming event.

Despite having a small dataset of only 23,801, our model made it possible for us to perform event detection by identifying crucial nouns or verbs appearing in our dataset based on the dates on which users mentioned them. These words were all associated with an event that occurred during or near the time on which they were mentioned. Although we had to look up each date on twitter in order to directly understand the correlation between these words and an event, a familiarity with the meanings and relevant topics of the words can give one an idea of what kind of event might have occurred that day. It appears that a large dataset will give us very strong results for detecting relevant events occurring on a given day.

In information retrieval the two most common evaluation techniques are precision, the percentage of extracted results which are relevant, and recall, the percentage of total relevant results correctly classified by the algorithm. We only

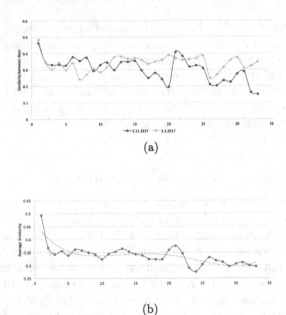

Fig. 4. (a) The similarities between a day and its consecutive 33 days for two specific days. (b) The average similarities of all days and their consecutive 33 days.

calculated the precision because of difficulties meeting the hardware requirements of processing all the tweets in our designated period of time. From the
dates within the span of our dataset, we selected the same 17 days as mentioned
earlier and collected the top ten relevant words for each date. We calculated the
percentage of words which correlated to an event either occurring on that day
or a very publicized upcoming event. Our calculated precision measure is given
by the percentage of our top ten words for each of 17 dates correlated to an
event. A human evaluator determined whether or not each of the 170 words was
relevant to any event. We calculated the precisions for the model with enriched
embeddings with a window of 5, the model without enriched embeddings with a
window of 5, and the model without enriched embeddings with a window of 10.
Because we could not train our model with enriched embeddings for a window of
10 within a reasonable amount of time, we did not calculate the corresponding
precision results. We observed that the average precision value is 72.9% for the
model with enriched embeddings and window of 5, 71.2% for the model without
enriched embeddings and window of 5, and 71.2% for the model without enriched
embeddings and window of 10. Table 2 shows our results.

With a larger dataset we can more automatically find the events by also
looking for the closest words to our potential events. For example, if a given date
is most closely related to the word "basketball" then that would be a potential
event. The most similar words to "basketball" would give better insight on what
kind of event occurred on that given date or in recent days or what kind of event
is projected to occur.

Table 2. Precision measures of top ten tokens for 17 selected dates in our dataset.

Evaluation results		
Enriched embeddings	Window size: 5	Window size: 10
With	72.9% (124/170)	-
Without	71.2% (121/170)	71.2% (121/170)

4 Conclusion

We introduced a new paradigm for extracting key information from a text-based dataset of social media posts. Our contribution is a novel way of creating a graph to map the relationships between the entities within our text-based dataset. The embeddings for these graphs, along with a new way of enriching our embeddings, have helped us successfully identify correlations between dates and major indicators of events. Our model works on even a small dataset, and we demonstrate that graphical models are powerful for building correlations between bodies of text, events, and dates. We hope that our research can pave the way for further studies in event detection.

Acknowledgment. This project (No. 117E566) is funded by the Scientific and Technological Research Council of Turkey (TUBITAK).

References

1. Bojanowski, P., et al.: Enriching word vectors with subword information. Trans. Assoc. Comput. Linguist. **5**, 135–146 (2017)
2. Cai, H., Zheng, V.W., Chang, K.C.-C.: A comprehensive survey of graph embedding: problems, techniques, and applications. IEEE Trans. Knowl. Data Eng. **30**(9), 1616–1637 (2018)
3. Dhingra, B., et al.: Tweet2Vec: character-based distributed representations for social media. In: The 54th Annual Meeting of the Association for Computational Linguistics (2016)
4. Feng, X., et al.: A language-independent neural network for event detection. In: Proceedings of the 54th Annual Meeting of the Association for Computational Linguistics (Volume 2: Short Papers), vol. 2 (2016)
5. Ganguly, S., et al.: Author2Vec: learning author representations by combining content and link information. In: Proceedings of the 25th International Conference Companion on World Wide Web. International World Wide Web Conferences Steering Committee (2016)
6. Grover, A., Leskovec, J.: Node2Vec: scalable feature learning for networks. In: Proceedings of the 22nd ACM SIGKDD International Conference on Knowledge Discovery and Data Mining. ACM (2016)
7. Hamilton, W., Ying, Z., Leskovec, J.: Inductive representation learning on large graphs. In: Advances in Neural Information Processing Systems (2017)
8. Ifrim, G., Shi, B., Brigadir, I.: Event detection in Twitter using aggressive filtering and hierarchical tweet clustering. In: SNOW-DC@ WWW (2014)

9. Kim, Y.: Convolutional neural networks for sentence classification. In: Proceedings of the 2014 Conference on Empirical Methods in Natural Language Processing (EMNLP) (2014)

10. Kumar, S., et al.: Community interaction and conflict on the web. In: Proceedings of The Web Conference (WWW) (2018)

11. Liu, Y., et al.: Topical word embeddings. In: Twenty-Ninth AAAI Conference on Artificial Intelligence (2015)

12. Narayanan, A., et al.: Subgraph2Vec: learning distributed representations of rooted sub-graphs from large graphs. arXiv preprint arXiv:1606.08928 (2016)

13. Perozzi, B., Al-Rfou, R., Skiena, S.: DeepWalk: online learning of social representations. In: Proceedings of the 20th ACM SIGKDD International Conference on Knowledge Discovery and Data Mining. ACM (2014)

14. Qin, Y., et al.: Frame-based representation for event detection on Twitter. IEICE Trans. Inf. Syst. **101**(4), 1180–1188 (2018)

15. Vosoughi, S., Vijayaraghavan, P., Roy, D.: Tweet2Vec: learning tweet embeddings using character-level CNN-LSTM encoder-decoder. In: Proceedings of the 39th International ACM SIGIR conference on Research and Development in Information Retrieval. ACM (2016)

16. Xu, K., et al.: How powerful are graph neural networks? arXiv preprint arXiv:1810.00826 (2018)

17. van der Maaten, L., Hinton, G.: Visualizing data using t-SNE. J. Mach. Learn. Res. **9**, 2579–2605 (2008)

Efficiently Computing Homomorphic Matches of Hybrid Pattern Queries on Large Graphs

Xiaoying Wu[1], Dimitri Theodoratos[2(✉)], Dimitrios Skoutas[3], and Michael Lan[2]

[1] Wuhan University, Wuhan, China
xiaoying.wu@whu.edu.cn
[2] New Jersey Institute of Technology, Newark, USA
{dth,ml122}@njit.edu
[3] R.C. Athena, Athens, Greece
dskoutas@imis.athena-innovation.gr

Abstract. In this paper, we address the problem of efficiently finding homomorphic matches for hybrid patterns over large data graphs. Finding matches for patterns in data graphs is of fundamental importance for graph analytics. In hybrid patterns, each edge may correspond either to an edge or a path in the data graph, thus allowing for higher expressiveness and flexibility in query formulation. We introduce the concept of answer graph to compactly represent the query results and exploit computation sharing. We design a holistic bottom-up algorithm called GPM, which greatly reduces the number of intermediate results, leading to significant performance gains. GPM directly processes child constraints in the given query instead of resorting to a post-processing procedure. An extensive experimental evaluation using both real and synthetic datasets shows that our methods evaluate hybrid patterns up to several orders of magnitude faster than existing algorithms and exhibit much better scalability.

1 Introduction

Graphs model complex relationships among objects in a variety of applications, including the Web, biological networks, etc. A fundamental operation for graph analytics is the graph matching operation, which identifies the matches of a query pattern in a large network. Due to its practical importance, pattern matching on graphs has received a lot of attention over the years [1,3,4,7,8,10,16].

Existing approaches can be characterized by: (a) the *type of edges* the patterns have, and (b) the *type of morphism* used to map the patterns to the data structure. An edge in a pattern can be either a *child* edge, which is mapped to an *edge* in the data graph, or a *descendant* edge, which is mapped to a *path* in the data graph. The morphism determines how a pattern is mapped to the data

---

The research of the first author was supported by the National Natural Science Foundation of China under Grant No. 61872276.

C. Ordonez et al. (Eds.): DaWaK 2019, LNCS 11708, pp. 279–295, 2019.
https://doi.org/10.1007/978-3-030-27520-4_20

graph. Over the years, research has advanced from considering *isomorphisms* for matching *patterns with child edges* [12,15] towards considering *homomorphisms* for matching *patterns with descendant edges* [2–4,9,10]. By allowing *edge-to-path* mappings, homomorphisms can extract matches "hidden" deep within large graphs which might be missed by isomorphisms. This is an important feature for numerous applications, such as search result classification, Web anomaly detection, plagiarism detection, and spam detection [7].

In this paper, we focus on finding homomorphic matches of tree patterns on large data graphs. Tree patterns are the basic building blocks of general graph patterns. Given a graph pattern Q, a common approach taken by several graph pattern matching methods [5,14] is to decompose Q into a set of tree patterns and use them as the basic unit for graph access. Thus, efficiently finding tree pattern matches on graphs is crucial in graph processing.

Existing algorithms for homomorphically matching tree patterns to graphs face two problems. First, they do not display satisfactory performance in terms of execution time and memory footprint when the size of the graph increases. Second, they cannot efficiently handle patterns that allow both descendant and child edges (*hybrid patterns*). In many applications, a parent-child relationship in the data graph is important. For instance, when child edges reflect causal effects, the patterns with child edges extract specific information that can be missed by patterns with descendant edges. *Hybrid patterns* generalize the other two types of patterns as they allow finding not only all the information extracted by them but also more detailed information which cannot be extracted by either one of them.

Contribution. In this paper we address the problem of efficiently evaluating hybrid tree patterns using homomorphisms over a large graph. The main contributions can be summarized as follows:

- We introduce the concept of *answer graph* to encode all the possible homomorphisms from a pattern to the graph. By losslessly summarizing the matches of a given pattern, the answer graph represents results more succinctly. Similar to factorized representations of query results studied in the context of classical and probabilistic databases [11], the answer graph exploits computation sharing to reduce redundancy in the representation and computation of query results.
- We design a holistic bottom-up algorithm called *GPM* (for '*Graph Pattern Matching*'). *GPM* adopts a holistic approach which greatly reduces the generation of redundant intermediate results, a problem typically occurring in decomposition-based algorithms [4,16]. As a consequence, it can efficiently process a whole spectrum of patterns including patterns with child edges and patterns with descendant edges and also hybrid patterns.
- Unlike most of the existing graph matching algorithms that allow edges to map to paths [4,10,16], *GPM* is not tied to a specific reachability indexing scheme. This not only makes *GPM* more generic and flexible, but also allows the algorithm to process child constraints in a given hybrid pattern directly instead of resorting to a postprocessing procedure. In our implementation,

we use the recent effective reachability scheme *Bloom Filter Labeling* (BFL), which greatly outperforms most existing approaches [13].

- We also present tuning strategies to further improve the performance of our algorithm.
- We run extensive experiments to evaluate the performance and scalability of our algorithm on real and synthetic datasets. We also compare with the best known previous algorithm *TwigStackD* [3,19], which follows a top-down approach. The experimental results show that our holistic bottom-up algorithm not only greatly outperforms *TwigStackD* with its original reachability index but also when it is modified to use BFL.

2 Related Work

We next discuss the existing homomorphism-based graph pattern evaluation methods in the literature.

Wang et al. [16] presented a decomposition-based graph matching algorithm called HGJoin. The method decomposes a pattern into a set of complete bipartite graphs, which are then evaluated according to a query plan. A hash-based structural join algorithm was designed to find matches for reachability relationships presented in the query. Cheng et al. [4] proposed another decomposition-based algorithm called R-Join. R-Join first decomposes the query into a set of binary (reachability) relationships between pairs of nodes; the query can then be matched by matching each binary relationship against the data graph and stitching together these basic matches. Similar to HGJoin [16], R-Join has to generate a huge amount of intermediate results and performs merge-join operations over the intermediate results. As a consequence its performance can degrade rapidly when the data graph becomes larger [10]. Follow-up work [10,18] has shown the limited scalability of the decomposition-based approaches, such as R-Join and HGJoin; hence, we do not consider them for comparison in our study.

Zeng et al. [10,18] proposed a holistic graph matching algorithm called TPQ-3Hop. Unlike R-Join and HGJoin, TPQ-3Hop tries to match the query against the input data as a whole. As such, it typically generates a smaller number of redundant intermediate results than the decomposition-based approach. However, like R-Join and HGJoin, TPQ-3Hop is closely coupled with a specific reachability indexing scheme supports descendant edge-only query patterns. As this approach is tied to a specific reachability index and suggests only expensive post-processing for handling possible child edges, it cannot be effectively applied towards hybrid pattern query evaluation.

The best-known algorithm for evaluating pattern queries on graphs is Twig-StackD [3,19]. Unlike other existing approaches, TwigStackD can be extended to directly handle hybrid query pattern evaluation, and it is not coupled with a specific reachability indexing scheme. Hence, we use TwigStackD as the baseline for comparison in our experimental study in Sect. 6. TwigStackD takes a holistic approach to evaluate a query pattern in a top-down manner. That is, a query node match is identified before its corresponding descendant matches are generated. To verify that one data node is in a match of a (sub)pattern, Twig-StackD has to scan, in the worst case, the whole graph, and repeats this process recursively for each possible descendant match of that node. Another limiting factor of this approach is on the pattern match generation. TwigStackD first generates path matches for each root-to-leaf path of the given query, and then produces the pattern matches by merge-joining the path solutions. This method can generate a large number of redundant results which do not participate in the final answer.

Our approach does not suffer from these issues. It works in a bottom-up way, where descendant matches are generated before their ancestor matches are identified. To verify whether a data node is in a match of a (sub)pattern of the query, our algorithm only scans previously generated child matches of that node, as there are significantly fewer of these than the nodes of the input graph. We further propose a data structure called *answer graph* to compactly encode all possible homomorphisms of a query in the graph. Using the answer graph, pattern occurrences can be enumerated without generating any redundant intermediate results. As we show in our experimental study, TwigStackD does not scale to larger graphs which can be efficiently processed with our approach.

In [17], we employed an initial version of the algorithm presented in this paper. However, the focus of [17] is on the efficient evaluation of pattern queries on large graphs using bitmap materialized views, which is orthogonal to what we study in this paper.

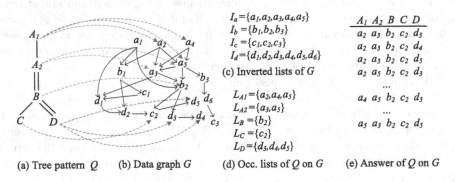

(a) Tree pattern Q (b) Data graph G (c) Inverted lists of G (d) Occ. lists of Q on G (e) Answer of Q on G

Fig. 1. A data graph G, a query Q, its inverted lists, occurrence lists, and answer on G.

3   Problem Definition

Data Graphs. A data graph is a directed node-labeled graph $G = (V, E)$ where V denotes the set of nodes and E denotes the set of edges (ordered pairs of nodes) of G. Let $\mathcal{L}$ be a finite set of node labels. Each node v in V has a label $label(v) \in \mathcal{L}$ associated with it. Given a label x in $\mathcal{L}$, the inverted list I_x is the list of nodes on G whose label is x.

A node u is said to *reach* node v, denoted $u \prec v$, if there exists a path from u to v in G. Clearly, if $(u, v) \in E$, $u \prec v$. Abusing tree notation, we refer to v as a *child* of u (or u as a *parent* of v) if $(u, v) \in E$, and v as a *descendant* of u (or u as an *ancestor* of v) if $u \prec v$.

Figure 1(b) shows a data graph G. Subscripts are used in the labels of nodes in G to distinguish between nodes with the same label (e.g., nodes a_1 and a_2 whose label is a). Figure 1(c) shows the inverted lists of G's labels.

Pattern Query. We focus on queries which are tree-patterns. Every node x in a pattern Q has a label $label(x)$ from $\mathcal{L}$. There can be two types of edges in Q. A *child* (resp. *descendant*) edge denotes a child (resp. descendant) structural relationship between the respective two nodes. Figure 1(a) shows a tree pattern Q. Single line edges denote child edges while double line edges represent descendant edges.

Query Answer. Given a tree pattern Q and a data graph G, a *homomorphism* from Q to G is a function m mapping the nodes of Q to nodes of G, such that: (1) for any node $x \in Q$, $label(x) = label(m(x))$; and (2) for any edge $(x, y) \in Q$, if (x, y) is a child edge, $(m(x), m(y))$ is an edge of G, while if (x, y) is a descendant edge, $m(x) \prec m(y)$ in G. Figure 1 shows all the possible mappings from the nodes of query Q to the nodes of G under different homomorphisms from Q to G.

The *occurrence* of a tree pattern Q on a data graph G is a tuple indexed by the nodes of Q whose values are the images of the nodes in Q under a homomorphism from Q to G. The *answer* of Q on G is a relation whose schema is the set of nodes of Q, and whose instance is the set of occurrences of Q under all possible homomorphisms from Q to G. Figure 1(e) shows the answer (relation of occurrences) of Q on G. One can see that there are 15 occurrences in this relation.

If x is a node in Q labeled by label a, the *occurrence list of x on G* is a sublist L_x of the inverted list I_a containing only those nodes that occur in the answer of Q on G for x (that is, nodes that occur in the column x of the answer). Figure 1(d) shows the occurrence lists of Q on G.

Let (q_i, q_j) be a query edge in Q, and v_i and v_j be two nodes in G, such that $label(q_i) = label(v_i)$ and $label(q_j) = label(v_j)$. The pair (v_i, v_j) is called an *occurrence* of the query edge (q_i, q_j) if: (a) (q_i, q_j) is a child edge in Q and (v_i, v_j) is an edge in G, or (b) (q, q_i) is a descendant edge in Q and $v_i \prec v_j$ in G.

Problem Statement. Given a large directed graph G and a tree pattern query Q, our goal is to efficiently find the answer of Q on G.

4   Query Answer Summarization

We introduce *answer graphs* to compactly encode all possible homomorphisms of a query in a graph.

Definition 1 (Answer Graph). *The answer graph G_A of a pattern query Q is a k-partite graph. Graph G_A has an independent node set for every node $q \in Q$ which is equal to the occurrence list L_q of q. There is an edge (v_x, v_y) in G_A between a node $v_x \in L_x$ and a node $v_y \in L_y$ if and only if there is an edge (x, y) in Q and a homomorphism from Q to G which maps x to v_x and y to v_y.*

The answer graph G_A losslessly summarizes all the occurrences of Q on G. It exploits computation sharing to reduce redundancy in the representation and computation of query results. The concept has analogies to the factorized representation of query results studied in the context of classical databases and probabilistic databases [11]. A useful property of G_A is that through a top-down traversal, the answer of Q on G can be obtained in time linear to the total number of occurrences of Q on G. Also, the cardinality of the query answer can be calculated without explicitly enumerating the occurrences of Q on G. Other benefits of succinct representations of query results include the possibility of speeding up subsequent data analysis [11].

Our answer graph generalizes the concept of *maximal matching graph* (MMG) introduced in [10,18]. While MMG is designed to encode query homomorphisms, it was defined to be a subgraph of the graph. MMG cannot represent pattern results that contain multiple appearances of the same data node in one pattern occurrence. This can happen when patterns involve nodes with the same label on a path.

A different way of compacting graph pattern query results uses a concept called *result graph* [6]. Unlike the answer graph which is a k-partite graph that encodes query homomorphisms, a result graph is a subgraph of the data graph which represents pattern matchings defined in terms of (an extension of) graph simulation. As such, it is not capable of compactly encoding homomorphic matches.

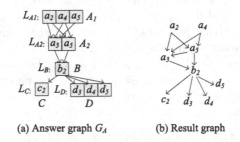

(a) Answer graph G_A (b) Result graph

Fig. 2. The answer graph of Q on G (Fig. 1) and the corresponding result graph.

Figure 2(a) shows the answer graph G_A of pattern query Q on the data graph G of Fig. 1. As a comparison, Fig. 2(b) shows the result graph of Q on G as defined in [6] (the maximal matching graph in [18] is similar). The answer graph representation of the query answer is much more compact than the tuple representation (Fig. 1(e)). The query answer in Fig. 1(e) has 90 values (18 tuples of 5 values each), while the corresponding answer graph in Fig. 2(a) only has 10 values (node labels).

5 A Bottom-Up Query Matching Algorithm

In this section, we present a bottom-up dynamic programming algorithm which incrementally builds the answer graph for the pattern query under consideration from the answer graphs of the sub-patterns rooted at the children of the pattern root.

5.1 The Algorithm

High-Level Structure. Figure 3 presents the algorithm, called GPM, which takes as input a data graph G, a BFL reachability index bfl [13], and a pattern query Q. Algorithm GPM builds the answer graph G_A of Q on G by performing a postorder traversal on Q; both bfl and G are used in the graph matching process. A *candidate* occurrence list CL_q of a query node q in Q is the occurrence list of the root of the subquery of Q rooted at q. Clearly, when q is a leaf node in Q, $CL_q = I_{label(q)}$. In general, we have $CL_q \subseteq I_{label(q)}$ and $L_q \subseteq CL_q$. Algorithm GPM has two phases:

In the first phase, GPM generates the candidate occurrence lists for the query nodes in Q and links their nodes with edges: a data node $v \in I_{label(q)}$ is put in the candidate occurrence list CL_q of a query node $q \in Q$ if there are data nodes $v_1, \ldots, v_k$ for the child query nodes $q_1, \ldots q_k$ of q such that: (a) for every $i \in [1, k]$, $v_i \in CL_{q_i}$, and (b) for every $i \in [1, k]$, (v, v_i) is an occurrence in G of the edge $(q, q_i) \in Q$. For every $i \in [1, k]$, an edge is added in G_A from a node $v \in CL_q$ to a node $v_i \in CL_{q_i}$ if and only if (v, v_i) is an occurrence in G of the edge $(q, q_i) \in Q$. Due to the bottom up traversal of Q, the candidate occurrence lists of the child nodes q_i of a q are available from the previous iteration of the algorithm.

In the second phase, GPM eliminates nodes in the CL_qs (and their incident edges) which are not in L_q, by doing a top-down traversal of the answer graph G_A under construction.

Detailed Description. The bottom-up processing phase in GPM is realized by procedure *traverse*. Let q be the current query node under consideration. For each node v_q in I_q, procedure *traverse* evokes procedure *expand* to potentially expand G_A by putting v_q into the candidate occurrence list CL_q and by adding incident edges to G_A (lines 6–7 in *traverse*).

Initially, CL_q is empty. Line 1 of Procedure *expand* puts v_q temporarily to CL_q. Then, line 2 iterates over every child q_i of q. For each node v_{q_i} in the candidate occurrence list CL_{q_i} of q_i, line 4 determines whether (v_q, v_{q_i}) is an occurrence of the query edge (q, q_i). If it is such a case, we add the edge (v_q, v_{q_i}) to G_A. The following two cases are considered: (a) (q, q_i) is a descendant edge, and (b) (q, q_i) is a child edge.

For case (a), we use the input BFL reachability index bfl to check whether $v_q \prec v_{q_i}$. For case (b), we use the adjacency lists of the graph G to check the

Input: Data graph G, a BFL reachability index bfl, and pattern query Q
Output: Answer graph G_A of Q on G

1. Initialize G_A to be a k-partite graph without edges having one data node set CL_q for every node $q \in Q$;
2. Initialize CL_q to be $\emptyset$ for every node $q \in Q$;
3. traverse($root(Q)$);
4. **for** (every $q \in Q$, $q \neq root(Q)$, in a top-down manner) **do**
5.   Remove the data nodes of CL_q which do not have an incoming edge;

Procedure traverse(q)

1. **if** ($isLeaf(q)$) **then**
2. $CL_q := I_q$;
3.   **return**
4. **for** ($q_i \in children(q)$) **do**
5.   traverse(q_i);
6. **for** ($v_q \in I_q$) **do**
7.   expand(q, v_q);

Procedure expand(q, v_q)

1. Append v_q to CL_q;
2. **for** ($q_i \in children(q)$) **do**
3.   **for** ($v_{q_i} \in CL_{q_i}$) **do**
4.   **if** ((v_q, v_{q_i}) is an occurrence of the query edge (q, q_i)) **then**
5.   Add the edge (v_q, v_{q_i}) to G_A;
6.   **if** (no match to (q, q_i) is found) **then**
7.   remove v_q from CL_q;
8.   **return**

Fig. 3. Algorithm *GMP*.

child/parent relationship between v_q and v_{q_i}. When the adjacency list is sorted (by node id), a binary search can be used to check if v_{q_i} is in the adjacency list of v_q. Finding all such v_{q_i}s in CL_{q_i} requires, for every element in CL_{q_i}, a repeated launching of a binary search against the adjacency list of v_q. Since a binary search incurs random memory access, the above computation can be costly when the cardinality of CL_{q_i} is large. We will later present a more efficient way to check child constraints.

After the last element of CL_{q_i} is accessed, if for some (q, q_i) no match is found, v_q is removed from CL_q, and the procedure terminates (lines 6–8 of Procedure *expand*).

When Procedure *traverse* terminates after processing the root of Q, $CL_{root(Q)} = L_{root(Q)}$. The candidate occurrence lists CL_q for other nodes q

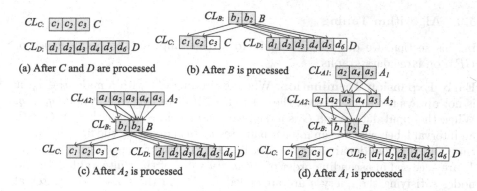

Fig. 4. Snapshots of the answer graph construction while running GMP on the query Q and data graph G of Fig. 1.

of Q might contain data nodes that are not in L_q. To eliminate these nodes, a breadth first traversal of Q is performed. For every node q of Q encountered (other than the root node), all the data nodes which do not have an incoming edge are removed from G_A along with their incident outgoing edges (lines 6–7 in the main procedure). The resulting graph is the answer graph G_A of Q.

An Example. Figure 4 shows the different snapshots of the answer graph construction when running GPM on the query Q and the data graph G of Fig. 1. The final answer graph G_A is shown in Fig. 2(a). In each snapshot, we mark in red those nodes which are not in the answer of the corresponding subquery of Q. A top-down traversal of the answer graph under construction recursively eliminates these marked nodes.

Correctness. In order to show the correctness of Algorithm GPM, we first provide the following lemma.

Lemma 1. *Algorithm* GPM *puts a data node* $v_q \in I_q$ *in the list under construction* CL_q *if and only if it is in the candidate occurrence list of the root of the subquery of Q rooted at q.*

Based on Lemma 1, we can show the following theorem.

Theorem 1. *Algorithm* GPM *correctly constructs the answer graph G_A for query Q on data graph G.*

Complexity. The time complexity of GPM is determined by the total number of pattern edge matches checked, which is $|Q| \times |I_{max}|^2$, where $|Q|$ is the number of pattern nodes and $|I_{max}|$ is the size of the largest inverted list. The time R for checking if a pair of data nodes is an edge occurrence is bounded by the time for checking reachability for a pair of nodes in the data graph. Therefore, the time complexity of GPM is $O(|Q| \times |I_{max}|^2 \times R)$. The memory consumption is determined by the size of the answer graph which is bound by $|Q| \times |I_{max}|^2$.

5.2   Algorithm Tuning

In this section, we discuss techniques to further improve the performance of *GPM* on large data graphs.

Early Expansion Termination. When expanding G_A with a node $v_q \in I_q$, it is not always necessary to scan the entire list CL_{q_i} for every child node q_i of q. When the input data graph G is a dag, we can associate each node u in G with an interval label, which is an integer pair (*begin, end*) denoting the first discovery time and the final departure time of u in a depth-first traversal of G. Nodes in I_q are accessed in ascending order of the *begin* value of their interval labels, and nodes satisfying the query q are appended to CL_q in the same order. Interval labelling, guarantees that node v_q does not reach node v_{q_i} if $v_q.end < v_{q_i}.begin$. Therefore, once such a node v_{q_i} is encountered, the expansion over CL_{q_i} can be safely terminated since all the subsequent nodes have a *begin* value which is larger than $v_{q_i}.begin$. Our experimental results show that the early expansion termination technique improves the performance by 30%.

Efficient Child Relationship Checking. As aforementioned, in order to find all the elements v_{q_i} in CL_{q_i} having a parent/child relationship with v_q, we can use binary search to check if v_{q_i} is in the sorted adjacency list of v_q. However, a binary search incurs random memory access, and repeatedly launching a binary search for every element in a large list can be costly. Below we present a more efficient method which can find all the v_{q_i}s in CL_{q_i} having a parent/child relationship with v_q with one step.

The Bloom Filter-based reachability scheme [13] converts node reachability checking into set containment testing implemented using bitwise operations. We likewise convert the parent/child relationship checking into set containment testing. More concretely, let A_{v_q} denote the adjacency list of v_q; every element v_{q_i} in the intersection $A_{v_q} \cap CL_{q_i}$ of A_{v_q} and CL_{q_i} is a child of v_q in CL_{q_i}. We store A_{v_q} and CL_{q_i} as bit vectors, and implement the intersection using a bitwise AND operation. Our method is able to obtain all the nodes in CL_{q_i} satisfying a child relationship with v_q in one batch, and avoids the random memory access problem. As verified by the experimental results in Sect. 6, this method outperforms, in many cases, the binary search-based method by orders of magnitude.

Cardinality-Based Expansion. Recall that in Procedure *expand*, the expansions of a node $v_q \in I_q$ with nodes from candidate occurrence lists CL_{q_i} of the child query nodes q_i of q are produced independently for every CL_{q_i}. Usually, it is beneficial to process the candidate occurrence lists in order of increasing cardinality. This way, a node added to the answer graph G_A that has to be removed later from G_A will be detected earlier on, before it is expanded with multiple incident edges. Our experimental results in Sect. 6 demonstrate that this heuristic can improve the performance of *GPM* by orders of magnitude when the cardinality difference among the lists is large.

Node Pre-filtering. As suggested by the complexity analysis, the performance of *GPM* can be improved by pruning nodes from the inverted lists. In order to prune nodes we employ the node pruning technique proposed in [3,19]. This pre-filtering technique filters out nodes not participating in the query answer before the query evaluation starts. In order to do so, it conducts two graph traversals and maintains data structures that record, for each data node, whether it has ancestors or descendants matching a particular query node. Our experimental results in Sect. 6 show that the pre-filter technique can largely improve the time performance of graph matching algorithms. However, it also incurs overhead costs.

6 Experimental Evaluation

We compare our pattern evaluation algorithm *GPM* with the well-known graph pattern matching algorithm *TwigStackD* [3,19] (abbreviated as *TSD*) in terms of time performance, memory usage and scalability.

6.1 Experimental Settings

Algorithms in Comparison. In our experimental evaluation, we used a number of different implementations of *TSD* and *GPM* for comparison purposes: (1) *GPM-bas*, a basic version of *GPM*. (2) *GPM-opt*, a version of *GPM-bas* enabled with two optimizations discussed in Sect. 5.2: Efficient Child Relationship Checking and Cardinality-based Expansion. (3) *GPM-flt*, the *GPM-opt* version enhanced with the node pre-filtering technique (Sect. 5.2). (4) *TSD-post*, a version of *TSD* that evaluates hybrid patterns in a postprocessing step. This method first finds all the solutions for the input pattern regarded as a descendant-only pattern, and then filters out solutions violating the child edge constraints. *TSD-post* returns query solution tuples (pattern occurrences) and it does so using merge-join operations over query path solutions. (5) *TSD-ag*, an extension of *TSD* which evaluates hybrid patterns directly, and, unlike *TSD-post*, produces answer graphs. (6) *TSD-flt*, an extension of *TSD-ag* with the node pre-filtering optimization. All the above implementations use *Bloom Filter Labeling* (BFL) [13] as their underlying reachability index scheme. (7) *TSD-sspi-flt*, a version of *TSD-flt* that uses the original *Surrogate Surplus Predecessor Index* (SSPI) [3] index scheme.

Our implementation was coded in Java. All the experiments reported here were performed on a workstation having an Intel Xeon CPU 1240V5@3.50 GHz processor with 32 GB memory.

Datasets. We ran experiments on two graph datasets with different structural properties. Their main characteristics are summarized in Table 1.

XMark[1] is a synthetic XML benchmark dataset modeling an auction website. We generated five XMark datasets using scaling factors, 5, 10, 15, 20, and 40,

---

[1] xml-benchmark.org.

Table 1. Dataset statistics. $|V|$, $|E|$ and $|L|$ are the number of nodes, edges and distinct labels, respectively. $maxout$ and $maxin$ are the maximum out-degree and in-degree of the graph. d_{avg} denotes the average degree of a graph.

dataset	$|V|$	$|E|$	$|L|$	maxout	maxin	d_{avg}
xm5	832911	960941	92	12750	168	2.31
xm10	1666315	1922640	92	25500	174	2.31
xm15	2502484	2887357	92	38250	180	2.31
xm20	3337649	3851680	92	51000	190	2.31
xm40	6688638	7718781	92	102000	243	2.31
eite-lb10000	6540401	15011260	8343	181247	203695	4.59
eite-lb7000	6540401	15011260	5662	181247	203695	4.59
eite-lb5000	6540401	15011260	3970	181247	203695	4.59
eite-lb3000	6540401	15011260	2434	181247	203695	4.59
eite-lb1000	6540401	15011260	885	181247	203695	4.59

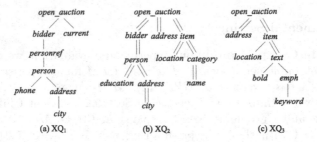

(a) XQ₁   (b) XQ₂   (c) XQ₃

Fig. 5. Pattern templates on XMark graphs, where edges with a single (resp. double) line denote child (resp. descendant) relationships.

named xm5, xm10, xm15, xm20, and xm40, respectively. For each dataset, we generated a graph by treating internal links (parent-child) and ID/IDREF links as edges. As in [10,18], we randomly classified *person* and *item* elements of the XMark graphs into ten groups and assigned a distinct label to each group.

citeseerx[2] represents a directed graph consisting of 6.3M publications (nodes) and 14.3M citations between them (edges). The original graph does not have labels. We wrote a label assignment program which randomly adds a specified number of distinct labels to graph nodes, following a Gaussian distribution. Using this program, we generated ten labeled citeseerx graphs whose number of labels ranges from 1,000 to 10,000. Each graph is named as cite-lb$x$, where x is the number of labels in the graph. Table 1 shows statistics for five of these datasets.

Since most of the reachability indexing schemes, including BFL and SSPI, work only with dags, in the experiments, we converted directed graphs with

---

[2] citeseerx.ist.psu.edu.

Table 2. Query set statistics. '//' denotes the descendant pattern edge.

query set	#queries	avg.$|V|$	avg. height	% of '//'	maxout	avg. #solutions
XQ_1	10	8	5	0.00	2	6037.6
XQ_2	10	11	4	100.00	3	7913.5
XQ_3	10	8	4	50.00	2	6734.3
cite-lb10000.qry	10	3.5	1.4	65.71	1.9	219.9
cite-lb7000.qry	10	3.1	1.1	67.74	2	1859.1
cite-lb5000.qry	10	3.2	1.1	65.63	2	582.9
cite-lb3000.qry	10	3.4	1.4	67.64	2	17543.3
cite-lb1000.qry	10	3.3	1.2	63.64	2	40064.1

Table 3. Parameters for query generation.

Parameters	Range	Description
Q	300 to 1800	Number of queries
D	6 to 16	Maximum depth of queries
DS	0 to 1	Probability of setting an edge to be a descendant edge ('//')
NP	1 to 3	Number of branches per query node

cycles into dags, by removing back edges. All the statistics shown in Table 1 are for dags.

Queries. For the XMark dataset, we used three different query templates XQ_1, XQ_2, and XQ_3, shown in Fig. 5, which denote child-only, descendant-only, and hybrid patterns, respectively. For each query template, we generated 10 queries by randomly choosing the labels on *person* and *item* nodes. We used randomly generated queries for the experiments on the citeseerx graphs. We implemented a query generator that creates a set of tree pattern queries based on the parameters listed in Table 3. Random queries are generated according to the input data graph and these parameters. For each data graph, we first generated a number of queries (in the range of 300 to 1800) using different value combinations of the parameters listed in Table 3, and then we formed a query set by randomly selecting 10 to 100 among them. Table 2 summarizes the statistics of the query sets.

6.2 Scalability of Graph Matching Algorithms

Varying Input Sizes. We measured the scalability of the different versions of *TSD* and *GPM* for evaluating instances of the query templates XQ_1–XQ_3 of Fig. 5 over XMark graphs. In Fig. 6, we report the execution time and memory usage of these algorithms over XMark graphs varying the scaling factor from 5 to 40. We next present our findings.

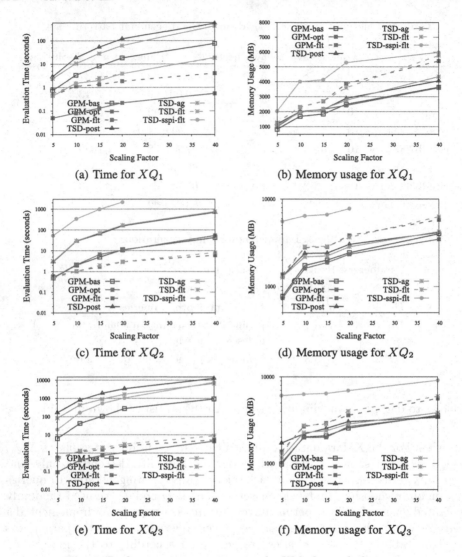

(a) Time for XQ_1

(b) Memory usage for XQ_1

(c) Time for XQ_2

(d) Memory usage for XQ_2

(e) Time for XQ_3

(f) Memory usage for XQ_3

Fig. 6. Performance comparison on the XMark-graph dataset.

The performance of *GPM* variants greatly outperforms *TSD* variants. In particular, *GPM-opt* has the overall best performance among all the algorithms in comparison. The max speedup over both *TSD-post* and *TSD-ag* is 3000×, 1000×, and 14× for hybrid (XQ_3), child-only (XQ_1), and descendant-only (XQ_2) queries, respectively. Also, *GPM-opt* consumes up to 78% less memory than *TSD-ag* and *TSD-post*. While *GPM-bas* has similar performance with *GPM-opt* on XQ_2, it is almost 155× and 100× slower on average than *GPM-opt* on XQ_1 and XQ_3. This demonstrates the effectiveness of the cardinality-based expansion and efficient child relationship checking optimization techniques. While *TSD-post* has a similar performance with *TSD-ag* for the descendant-only query template XQ_2, it is up to 1× slower than *TSD-ag* for the hybrid and child-only query templates XQ_3 and XQ_1, giving it the worst time performance among all the algorithms. This is due to the postprocessing strategy in *TSD-post* for handling the child constraints in hybrid (and child-only) queries, which results in a large amount of intermediate results that not participating in the final query answer.

We observe that the pre-filtering optimization can significantly improve the time performance of graph matching algorithms. *TSD-flt* has a max speedup of 22×, 43× and 737× over *TSD-ag* for XQ_1, XQ_2, and XQ_3 on the largest tested XMark graph (xm40), respectively. The improvement is less prominent for *GPM* though. The max speedup of over *GPM-opt* is around 8× by *GPM-flt* on XQ_2. On XQ_1 and XQ_3, *GPM-flt*, in fact, increases the average evaluation time of *GPM-opt* by nearly 8× and 2×, respectively. This is explained by the overhead cost associated with the node filtering, which negates the potential reduction in intermediate result generation. The pre-filtering process increases the memory usage of the algorithms as well. The average increase for *GPM-flt* over *GPM-opt* is 28%, 65% and 45% on XQ_1, XQ_2, and XQ_3, respectively. For *TSD-flt*, the increase is 21%, 32% and 36%, respectively.

TSD-sspi-flt exhibits the worst time and memory performance for evaluating XQ_2 (even with pre-filtering) mainly due to the bad performance of SSPI for node reachability checking. *TSD-sspi-flt* was unable to finish within 24 h on xm40. It, however, runs 27× and 1× faster than *TSD-post* for XQ_1 and XQ_3, respectively. Unlike *TSD-post*, *TSD-sspi-flt* evaluates hybrid (and child-only) queries directly. This fact makes *TSD-sspi-flt* generate a smaller amount of intermediate results than *TSD-post* on input data with reduced sizes.

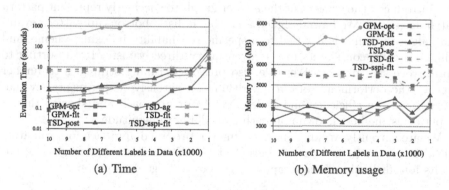

| (a) Time | (b) Memory usage |

Fig. 7. Scalability comparison on the citeseer graph.

Varying Data Labels. In this experiment, we examine the impact of the total number of distinct graph labels on the performance of the four algorithms. We used the aforementioned ten labeled citeseerx graphs *cite-lbx* (Table 1), where the number of labels x increases from 1000 to 10000. For each *cite-lbx* graph, we generated a query set *cite-lbx.qry* with 10 distinct queries (Table 2). Figure 7 reports the execution time and memory usage for the six algorithms. As in the previous experiment, *GPM-opt* has the best performance overall. We didn't show *GPM-bas* here as its performance is similar to *GPM-opt*. *TSD-sspi-flt* exhibits the worst time and memory performance.

As can be seen in Fig. 7(a), the execution time of the algorithms tends to increase while decreasing the total number of graph labels. This is reasonable since the average cardinality $(=|V|/|L|)$ of the input inverted lists per node label in a graph increases when the number of labels in the graph decreases. This observation confirms that the complexity results show dependency of the execution time on the input size.

In contrast to the results on XMark graphs, in citeseerx we observe a degenerate time performance of the algorithms when the pre-filtering technique is applied to graphs with a large (>1000) number of labels. The reason is that when the number of graph labels is large, the cardinality of the inverted lists is relatively small. In this case, the potential benefit of reducing intermediate results during graph matching can be offset by the overhead of node pre-filtering. The memory performance of the algorithms in citeseerx is consistent with the results on the XMark graphs.

7 Conclusion

In this paper, we have addressed the problem of evaluating hybrid tree pattern queries over a large data graph. Hybrid patterns allow the extraction of interesting information about the dataset which cannot be extracted with patterns that involve only child or descendant structural relationships between nodes. We have introduced the concept of the answer graph to succinctly represent pattern matching results. We have designed a flexible holistic bottom-up pattern evaluation algorithm that is not tied to a specific reachability indexing scheme, and it handles child constraints in the given query directly instead of resorting to a postprocessing process. We have also presented tuning strategies to further improve the performance of our algorithm. An extensive experimental evaluation verified the efficiency and scalability of our approach and showed that it outperforms the best-known approach by orders of magnitude.

We are currently working on extending the proposed approach for evaluating hybrid graph patterns over large graph data. We are also exploring using our results for efficiently mining graph patterns from large data graphs.

References

1. Barceló, P., Libkin, L., Reutter, J.L.: Querying graph patterns. In: PODS (2011)
2. Bruno, N., Koudas, N., Srivastava, D.: Holistic twig joins: optimal XML pattern matching. In: SIGMOD (2002)
3. Chen, L., Gupta, A., Kurul, M.E.: Stack-based algorithms for pattern matching on DAGs. In: VLDB (2005)

4. Cheng, J., Yu, J.X., Yu, P.S.: Graph pattern matching: a join/semijoin approach. IEEE Trans. Knowl. Data Eng. **23**(7), 1006–1021 (2011)
5. Cheng, J., Zeng, X., Yu, J.X.: Top-k graph pattern matching over large graphs. In: ICDE, pp. 1033–1044 (2013)
6. Fan, W., Li, J., Ma, S., Tang, N., Wu, Y., Wu, Y.: Graph pattern matching: from intractable to polynomial time. PVLDB **3**(1), 264–275 (2010)
7. Fan, W., Li, J., Ma, S., Wang, H., Wu, Y.: Graph homomorphism revisited for graph matching. PVLDB **3**(1), 1161–1172 (2010)
8. Gallagher, B.: Matching structure and semantics: a survey on graph-based pattern matching. In: AAAI FS, vol. 6, pp. 45–53 (2006)
9. Grimsmo, N., Bjørklund, T.A., Hetland, M.L.: Fast optimal twig joins. PVLDB **3**(1), 894–905 (2010)
10. Liang, R., Zhuge, H., Jiang, X., Zeng, Q., He, X.: Scaling hop-based reachability indexing for fast graph pattern query processing. IEEE Trans. Knowl. Data Eng. **26**(11), 2803–2817 (2014)
11. Olteanu, D., Schleich, M.: Factorized databases. SIGMOD Rec. **45**(2), 5–16 (2016)
12. Shang, H., Zhang, Y., Lin, X., Yu, J.X.: Taming verification hardness: an efficient algorithm for testing subgraph isomorphism. PVLDB **1**(1), 364–375 (2008)
13. Su, J., Zhu, Q., Wei, H., Yu, J.X.: Reachability querying: can it be even faster? IEEE Trans. Knowl. Data Eng. **29**(3), 683–697 (2017)
14. Sun, Z., Wang, H., Wang, H., Shao, B., Li, J.: Efficient subgraph matching on billion node graphs. PVLDB **5**(9), 788–799 (2012)
15. Ullmann, J.R.: An algorithm for subgraph isomorphism. J. ACM **23**(1), 31–42 (1976)
16. Wang, H., Li, J., Luo, J., Gao, H.: Hash-base subgraph query processing method for graph-structured XML documents. PVLDB (2008)
17. Wu, X., Theodoratos, D., Skoutas, D., Lan, M.: Evaluating mixed patterns on large data graphs using bitmap views. In: Li, G., Yang, J., Gama, J., Natwichai, J., Tong, Y. (eds.) DASFAA 2019. LNCS, vol. 11446, pp. 553–570. Springer, Cham (2019). https://doi.org/10.1007/978-3-030-18576-3_33
18. Zeng, Q., Jiang, X., Zhuge, H.: Adding logical operators to tree pattern queries on graph-structureddata. PVLDB **5**(8), 728–739 (2012)
19. Zeng, Q., Zhuge, H.: Comments on "stack-based algorithms for pattern matching on dags". PVLDB **5**(7), 668–679 (2012)

Databases

From Conceptual to Logical ETL Design Using BPMN and Relational Algebra

Judith Awiti[1]([⊠]), Alejandro Vaisman[2], and Esteban Zimányi[1]

[1] Université Libre de Bruxelles, Brussels, Belgium
{judith.awiti,ezimanyi}@ulb.ac.be
[2] Instituto Tecnológico de Buenos Aires, Buenos Aires, Argentina
avaisman@itba.edu.ar

Abstract. Extraction, transformation, and loading (ETL) processes are used to extract data from internal and external sources of an organization, transform these data, and load them into a data warehouse. The Business Process Modeling Notation (BPMN) has been proposed for expressing ETL processes at a conceptual level. This paper extends relational algebra (RA) with update operations for specifying ETL processes at a logical level. In this approach, data tasks can be automatically translated into SQL queries to be executed over a DBMS. An extension of RA is presented, as well as a translation mechanism from BPMN to the RA specification. Throughout the paper, the TPC-DI benchmark is used for comparing both approaches. Experiments show the efficiency of the resulting ETL flow with respect to the Pentaho Data Integration tool.

Keywords: OLAP · ETL · BPMN · Data warehousing

1 Introduction

Extraction, transformation, and loading (ETL) processes extract data from internal and external sources of an organization, transform these data, and load them into a data warehouse (DW). Since ETL processes are complex and costly, it is important to reduce their development and maintenance costs. Modeling these processes at a conceptual level would contribute to achieve this goal. Since there is no agreed-upon conceptual model to specify such processes, existing ETL tools use their own specific language to define ETL workflows. Considering this, the paper discusses two methods for designing ETL processes. The first one, called BPMN4ETL, is based on the Business Process Modeling Notation (BPMN), a de-facto standard for specifying business processes, which provides a conceptual and implementation-independent specification of such processes, that can be then translated into executable specifications for ETL tools. The second is a logical model based on Relational Algebra (RA), a formal language that provides a solid basis to specify ETL processes for relational databases.

© Springer Nature Switzerland AG 2019
C. Ordonez et al. (Eds.): DaWaK 2019, LNCS 11708, pp. 299–309, 2019.
https://doi.org/10.1007/978-3-030-27520-4_21

Running Example. The TPC-DI benchmark [6] is used as running example throughout the paper, with focuses on the processes that update customers and their accounts. The benchmark has two phases: *Historical Load*, and *Incremental Updates*. In the former, destination tables are initially empty and then populated with new data. The OLTP database represents transactional information about securities market trading and the entities involved, e.g., customers, accounts, and so on. A CustomerMgmt.xml file represents actions resulting in new or updated customer and account information. For each action, only the properties involved in the update are given. For example, a 'NEW' action (an insertion of a new customer) will contain customer identifying information, many properties (e.g. name, address), and information about the customer's account. An 'UPDCUST' action updates a customer and all her current accounts. For that action, only the properties used to identify the updated properties of the customer are given. All actions have at least one related customer, and each account is associated with a single customer. There are also other related tables (e.g., Prospect list, Financial Newsware, and so on) that are used by the different processes.

Contributions. The paper discusses the modeling of Slowly Changing Dimensions with Dependencies, that is, the case when updating a SCD table impacts on associated SCD tables (Sect. 2). As a key contribution, an ETL development approach is proposed, which begins with a BPMN4ETL conceptual model (Sect. 3) translated into RA extended with update operations (Sect. 4) at the logical level. Common ETL tasks and their extended RA specifications are also shown. Although BPMN4ETL has been already proposed (see [1, 11]), the problem of modeling SCDs with dependencies using this technique is discussed here for the first time. Related work is covered in Sect. 7. Experiments over the TPC-DI benchmark are carried out and results are reported, suggesting that the above-mentioned approach results in more efficient processes than the ones produced by BPMN4ETL conceptual model translated into the Pentaho Data Integration (PDI) tool (Sect. 6). Conclusions are given in Sect. 8.

2   Slowly Changing Dimensions with Dependencies

Slowly Changing Dimensions (SCD) [3, 11] are used in a DW to keep the history of changes that occurs in data sources. Kimball [3] defined seven types of SCD. With SCDs of type 2, the history of changes is kept by augmenting the schema of the dimension table with two temporal attributes, called StartDate and EndDate. The former stores the time when the tuple was inserted into the dimension table. The latter stores the date when an update of the attribute was made in the dimension table. In general, a currently valid record has a NULL value or a date far off into the future as its EndDate. When a tuple is deleted from the source table, the EndDate attribute of its corresponding tuple in the dimension table is set to the current date. An additional current indicator attribute, IsCurrent, contains a Boolean value which is set to True to indicate that a tuple is the current record corresponding to the natural key.

$$DimA = <SkA,PkA,...,IsCurrent,StartDate,EndDate> \tag{1}$$
$$DimB = <SkB,SkA,PkB,...,IsCurrent,StartDate,EndDate> \tag{2}$$

Fig. 1. Schema of DimA and DimB

This paper (as well as the TPC-DI benchmark) tackles SCDs of type 2 with dependencies. Such a Slowly Changing Dimension table contains a surrogate key reference to another dimension table. If an update occurs to the referenced dimension table, the referencing table must be updated as well. To illustrate this, the schemas of two dimension tables are defined, namely DimA and DimB, in Equations 1 and 2. DimB references the surrogate key (SkA) of DimA. DimA and DimB have 'natural' keys PkA and PkB, respectively. When an update occurs in DimA, the (SkA) value of DimB must be replaced by the most current (SkA) value in DimA. Below, the general steps of an ETL to update a tuple in DimA are listed, implementing the dependency stated above.

1. Retrieve the current tuple (the one with IsCurrent = True) from DimA.
2. Rename SkA to SkAOld.
3. Retrieve the maximum SkA value from DimA and increase it by 1.
4. Retire the current tuple from DimA (set IsCurrent and EndDate attributes to False and the current date, respectively).
5. Insert a new tuple in DimA (set IsCurrent and EndDate attributes to True and NULL, respectively).
6. Retrieve the corresponding current tuples from DimB. These are the tuples with IsCurrent and SkA values of True and SkAOld, respectively.
7. Retire the current tuples in DimB (set IsCurrent and EndDate attributes to False and the current date, respectively).
8. Insert new tuples in DimB (set IsCurrent and EndDate to True and NULL, respectively).

3   Conceptual BPMN for Slowly Changing Dimensions

The BPMN4ETL conceptual model [1,11], represents ETL processes as a combination of control and data tasks. Control tasks orchestrate groups of tasks, and data tasks detail how input data are transformed and output data are produced. For example, populating a DW is a control task composed of multiple subtasks, while populating fact or dimension tables is a data task. This section shows how BPMN4ETL can be applied to handle SCD with dependencies described above in the TPC-DI benchmark. First, a description of the possible changes is given.

In the TPC-DI benchmark introduced in Sect. 1, changes that occurred in the data sources before the historical load, are stored in the `CustomerMmgt.xml` file. The type of changes can be: (a) NEW, where a new customer is inserted, always associated with a new account; (b) ADDACCT, where one or more new accounts are associated with an existing customer; (c) UPDACCT, which updates the information in one or more existing accounts; (d) UPDCUST, which updates existing customer's information; (e) CLOSEACCT, which closes one or more existing accounts; (f) INACT, which sets the status of an existing customer, and her associated active accounts, to "inactive". Note that for UPDCUST and INACT actions, no account fields are included in the file. Updates over the dimensions DimCustomer and DimAccount are not only present during the incremental load, but also in the historical load. The DimAccount table has a surrogate key that references the surrogate key (Sk_CustomerID) of the DimCustomer table. These two tables are modelled as type 2 SCDs with dependencies.

Figure 2 shows the conceptual design of the UPDCUST action for the historical load. Initially, an Add Column task updates the phone attributes of the sources. The next five tasks (Lookup, Update Column, Drop Column, Lookup and Add Column), implement the updates specified in the TPC-DI benchmark specifications. Note that Status is set to "active" in the Add Column task to indicate that an updated customer is still active. Since DimCustomer is a SCD-type 2 table, upon updating a customer record, a new Sk_CustomerID value must be inserted for the new current tuple. Thus, the current Sk_CustomerID value will not refer to the correct value in the DimAccount table anymore. Further, the only link to DimAccount is the Sk_CustomerID value of the current DimCustomer record. Thus, a lookup is performed using the customer identifier, followed by a Rename Column task that renames the current Sk_CustomerID attribute to Sk_CustomerIDOLD. In the next Add Column task, the maximum Sk_CustomerID value of DimCustomer is retrieved and incremented by one, becoming the Sk_CustomerID for the new tuple. This is followed by an Update Data task that sets the IsCurrent and EndDate values of the current tuple in DimCustomer to 'False' and 'ActionTS' (the action timestamp) respectively. Then the new tuple is inserted into DimCustomer, with IsCurrent and EndDate values to 'True' and '9999-12-31' respectively. Now, all the accounts of the updated customer must be updated, setting the new value of Sk_CustomerID. These accounts are found through a Lookup task with the Sk_CustomerIDOLD value that has been saved in the flow. This value is matched with the Sk_CustomerID value of DimAccount. After this, the current tuple of this account in DimAccount, for this customer, is logically deleted, by setting the IsCurrent and EndDate values to 'False' and 'ActionTS', respectively, using an Update Data task. Since a rule in the TPC-DI specification document requires that there must be at most one update to a 'natural' key record on any given day (even when the source data contains more than one update), a tuple is only deleted if the EffectiveDate value is not equal to the ActionTS value (which will become the EndDate of the deleted tuple). Otherwise, another Update Data task sets their Sk_CustomerID value to the new Sk_CustomerID value in the flow. Then, an Add Column task adds the Sk_AccountID column to the flow. The row number value of each tuple is added to the maximum Sk_AccountID value since more than one account could belong to the same customer being updated. Finally, an Insert Data task inserts the tuples in DimAccount.

4   An Extended Relational Algebra for ETL Processes

This section presents an extended RA that can be used for implementing ETL processes. RA can be used to automatically generate SQL queries to be executed in any Relational Database Management System (RDBMS). The typical

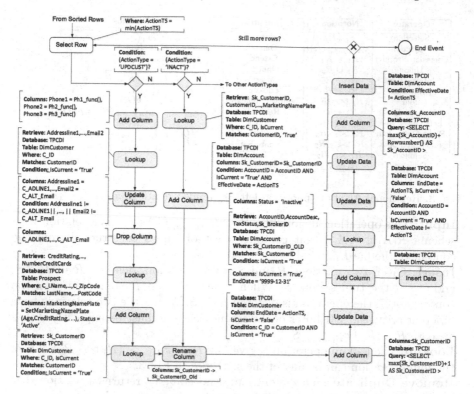

Fig. 2. BPMN4ETL design for updating a customer (SCD type 2 table).

RA operations are shown on the left-hand side of Fig. 3. A description of these operations can be found in classic database literature. In order to model different scenarios of ETL processes, these operations are extended with additional operators as indicated on the right-hand side of the figure, and detailed next.

- **Aggregate:** Let F be an aggregate function such as Count, Min, Max, Sum, or Avg. The aggregate operator $\mathcal{A}_{A_1,\ldots,A_m|C_1=F_1(B_1),\ldots,C_n=F_n(B_n)}(R)$ partitions the tuples of R in groups that have the same values of attributes A_i and computes for each group new attributes C_i by applying the aggregate function F_i to the values of attribute B_i in the group, or the cardinality of the group if $F_i(B_i)$ is Count(*). If no grouping attributes are given (i.e., $m = 0$), the aggregate functions are applied to the whole relation R. The schema of the resulting relation has the attributes $(A_1,\ldots,A_m,C_1,\ldots,C_n)$.
- **Delete:** The delete operation, denoted by $R \leftarrow R - \sigma_C(R)$, removes from relation R the tuples that satisfy the Boolean condition C.
- **Extend:** Given a relation R, the extension operation, denoted $\mathcal{E}_{A_1=Exp_1,\ldots,A_n=Exp_n}(R)$, returns a relation where each tuple in R is extended with new attributes A_i obtained by computing the expression Exp_i.

Operator	Notation	Operator	Notation
Selection	$\sigma_C(R)$	Aggregate	$\mathcal{A}_{A_1,..,A_m \mid C_1=F_1(B_1),..,C_n=F_n(B_n)}(R)$
Projection	$\pi_{A_1,..,A_n}(R)$	Delete	$R \leftarrow R - \sigma_C(R)$
Cartesian Product	$R_1 \times R_2$	Extend	$\mathcal{E}_{A_1=\text{Expr}_1,...,A_n=\text{Expr}_n}(R)$
Union	$R_1 \cup R_2$	Input	$R \leftarrow \mathcal{I}_{A_1,...,A_n}(F)$
Intersection	$R_1 \cap R_2$	Insert	$R \leftarrow R \cup S$ or $R \leftarrow S$
Difference	$R_1 - R_2$	Lookup	$R \leftarrow \pi_{A_1,...A_n}(R_1 \bowtie_C R_2)$
Join	$R_1 \bowtie_C R_2$	Remove duplicates	$\delta(R)$
Natural Join	$R_1 * R_2$	Rename	$\rho_{A_1 \leftarrow B_1,...,A_n \leftarrow B_n}(R)$ or $\rho_S(R)$
Left Outer Join	$R_1 ⟕_C R_2$	Sort	$\tau_A(R)$
Right Outer Join	$R_1 ⟖_C R_2$	Update	$\mathcal{U}_{A_1=\text{Expr}_1,...,A_n=\text{Expr}_n \mid C}(R)$
Full Outer Join	$R_1 ⟗_C R_2$	Update Set	$R \leftarrow \mathcal{U}(R)_{A_1=\text{Expr}_1,...,A_n=\text{Expr}_n \mid C}(S)$
Semijoin	$R_1 ⋉_C R_2$		
Division	$R_1 \div R_2$		

Fig. 3. Relational Algebra operators (left). Extended relational operators (right).

- **Input:** This operation, denoted by $R \leftarrow \mathcal{I}_{A_1,...,A_n}(F)$ returns a relation with schema $R(A_1,...,A_n)$ that contains a set of tuples constructed from the content of the file F.
- **Insert:** Given two relations R and S, this operation, denoted $R \leftarrow R \cup S$, adds to R the tuples from S. When a new relation R is created with the contents of S, the operation is denoted $R \leftarrow S$. Also if the two relations have different arity, the attributes of the second relation must be explicitly stated as, e.g., $R \leftarrow R \cup \pi_{B_1,...,B_n}S$.
- **Lookup:** The lookup operation is given by $R \leftarrow \pi_{A_1,...A_n}(R_1 \bowtie_C R_2)$, where the join operation can be any of the six types in Fig. 3.
- **Remove Duplicates:** This operation, denoted $\delta(R)$, returns a relation that contains the tuples of R without duplicates.
- **Rename:** This operation is applied over relation names or over attribute names. For the former, the operation is denoted by $\rho_S(R)$, where the input relation R is renamed to S. For attributes, $\rho_{A_1 \leftarrow B_1,...,A_n \leftarrow B_n}(R)$, returns a relation where the attributes A_i in R are renamed to B_i, respectively.
- **Sort:** This operation, denoted $\tau_A(R)$, sorts a relation that contains the tuples of R sorted by the attribute A .
- **Update Column:** This operation, denoted $\mathcal{U}_{A_1=Expr_1,...,A_n=Expr_n \mid C}(R)$, returns a relation where for each tuple in R that satisfies the Boolean condition C, the value of the attribute A_i is replaced, respectively, by the value of $Expr_i$.
- **Update Set:** Denoted $R \leftarrow \mathcal{U}(R)_{A_1=Expr_1,...,A_n=Expr_n \mid C}(S)$, this operation updates tuples in R that correspond to tuples in S that satisfy the Boolean condition C. The value of attribute A_i is replaced by the value of the expression $Expr_i$. Unlike the Update Column operation, the condition of the Update Set operation includes matching the tuples of two relations.

Figure 4 shows how tasks in the BPMN4ETL methodology can be translated into RA. For example, the Aggregate Data task is translated as an Aggregate operation. The Drop Column operation is specified in RA as a projection of all columns except the removed ones (in this case, A_n and A_{n-1}). The Update Data task is translated as an Update Set operation with matching attributes explicitly stated in the condition.

BPMN Data task	Relational Algebra Expression
Add Column	$\mathcal{E}_{A_1=\text{Expr}_1,\ldots,A_n=\text{Expr}_n}(R)$
Aggregate	$\mathcal{A}_{A_1,\ldots,A_m|C_1=F_1(B_1),\ldots,C_n=F_n(B_n)}(R)$
Delete Data	$R \leftarrow R - \sigma_C(R)$
Drop Column	$R \leftarrow \pi_{A_1,\ldots,A_{n-2}} R$
Input Data	$R \leftarrow \mathcal{I}_{A_1,\ldots,A_n}(F)$
Union	$R \leftarrow R_1 \cup R_2$
Insert Data	$R \leftarrow R \cup \pi_{B_1,\ldots,B_n} S$
Join	$R_1 \bowtie_C R_2$
Lookup	$R \leftarrow \pi_{A_1,\ldots,A_n}(R_1 \bowtie_C R_2)$
Sort	$\tau_A(R)$
Rename	$\rho_{A_1 \leftarrow B_1,\ldots,A_n \leftarrow B_n}(R)$
Update Column	$\mathcal{U}_{A_1=\text{Expr}_1,\ldots,A_n=\text{Expr}_n|C}(R)$
Update Data	$R \leftarrow \mathcal{U}(R)_{A_1=\text{Expr}_1,\ldots,A_n=\text{Expr}_n|R.A_1=S.B_1\wedge,\ldots,}(S)$

Fig. 4. Translation of BPMN4ETL tasks to RA operations.

5   Relational Algebra Specification for Type 2 SCDs with Dependencies

The implementation in RA of the process described in the running example, is shown in Fig. 5. Variable Temp0 holds all tuples except for the ones with ActionType = 'NEW', and Temp1 holds the tuple pointed to by a cursor from Temp0 at any particular time. For the sake of space, only the part concerning SCDs will be explained (Eqs. 13 through 24). Equations 13 and 14 obtain the Sk_CustomerID of the current customer tuple in DimCustomer, and rename it to Sk_CustomerIDOLD, to keep the current surrogate key of DimCustomer in the flow, for the reasons already explained. Equations 15–16 add Sk_CustomerID to the flow (the new surrogate key value is computed by adding 1 to the maximum Sk_CustomerID value in DimCustomer). The corresponding current tuple in Dim-Customer is then "deleted" (Eq. 17). Then, the remaining columns needed are added (Eq. 18), and the tuple is inserted into DimCustomer (Eq. 19). After this, all current accounts of the customer are obtained (Eq. 20) and "deleted" (by setting, e.g., IsCurrent as 'False', in Eq. 21). Again, only one tuple is inserted, for accounts such that EffectiveDate $\neq$ ActionTS. For accounts with EffectiveDate = ActionTS, only their Sk_CustomerID values are updated (Eq. 22). Finally, Eq. 23 adds Sk_AccountID to the flow. This is the maximum Sk_AccountID in DimAccount plus the rownumber() value of each current account tuple. Finally, the tuples are inserted into DimAccount (Eq. 24).

6   Performance Evaluation

This section briefly describes and reports results of the experimental evaluation on the TPC-DI benchmark described in Sect. 1. The benchmark contains one historical load and two identical incremental loads. Data sources are of different formats (xml, csv, txt, and so on). The experiments implement the historical and incremental loads in two ways: (a) Using PDI[1], translating the BPMN4ETL

---

[1] https://github.com/pentaho/pentaho-kettle.

$$Temp1 \leftarrow \sigma_{FetchCursorRow()}(Temp0) \tag{3}$$

$$UCTemp1 \leftarrow \sigma_{ActionType = 'UPDCUST'}(Temp1) \tag{4}$$

$$UCTemp2 \leftarrow \mathcal{E}_{Phone1 = GetPhone1(),\ldots, Phone3 = GetPhone3()}(UCTemp1) \tag{5}$$

$$UCTemp3 \leftarrow \pi_{\ldots,Addressline1,\ldots,Email2}(UCTemp2 \bowtie_{C\_ID = CusomerID \wedge} \tag{6}$$
$$_{IsCurrnet = 'True'} DimCustomer)$$

$$UCTemp4 \leftarrow \sigma_{Addressline1 \neq C\_ADLINE1 \vee, \ldots, \vee Email2 \neq C\_ALT\_EMAIL}(UCTemp3) \tag{7}$$

$$UCTemp5 \leftarrow \sigma_{Addressline1 = C\_ADLINE1 \vee, \ldots, \vee Email2 = C\_ALT\_EMAIL}(UCTemp4) \tag{8}$$

$$UCTemp6 \leftarrow \mathcal{U}_{Addressline1 = C\_ADLINE1, \ldots, Email2 = C\_ALT\_EMAIL}(UCTemp4) \tag{9}$$

$$UCTemp7 \leftarrow UCTemp5 \cup UCTemp6 \tag{10}$$

$$UCTemp8 \leftarrow \pi_{\ldots,AgencyID,Age,\ldots}(UCTemp7 \bowtie_{C\_LName, \ldots, C\_ZipCode} \tag{11}$$
$$_{= LastName, \ldots PostCode} Prospect)$$

$$UCTemp9 \leftarrow \mathcal{E}_{MarketingNamePlate = SetMarketingNamePlate(Age,CreditRating,\ldots)}(UCTemp8) \tag{12}$$

$$UCTemp10 \leftarrow \pi_{\ldots,Sk\_CustomerID}(UCTemp9 \bowtie_{C\_ID=CustomerID \wedge IsCurrent = 'True'} DimCustomer) \tag{13}$$

$$UCTemp11 \leftarrow \rho_{\ldots,Sk\_CustomerID \leftarrow Sk\_CustomerIDOLD}(UCTemp10) \tag{14}$$

$$UCTemp12 \leftarrow \sigma_{Sk\_CustomerID = max(Sk\_CustomerID) + 1}(DimCustomer) \tag{15}$$

$$UCTemp13 \leftarrow \pi_{\ldots,Sk\_CustomerID}(UCTemp11 \times UCTemp12) \tag{16}$$

$$DimCustomer \leftarrow \mathcal{U}(DimCustomer)_{EndDate = ActionTS, IsCurrent = False|CustomerID=C\_ID} \tag{17}$$
$$_{\wedge IsCurrent = 'True'}(UCTemp13)$$

$$UCTemp14 \leftarrow \mathcal{E}_{Status = 'Active', IsCurrent = 'True', EffectiveDate = ActionTS,} \tag{18}$$
$$_{EndDate = '9999-12-30'}(UCTemp14)$$

$$DimCustomer \leftarrow DimCustomer \cup (\pi_{C\_ID,\ldots,EndDate}(UCTemp14)) \tag{19}$$

$$UCTemp15 \leftarrow \pi_{AccountID, AccountDesc, TaxStatus}(UCTemp14 \bowtie_{Sk\_CustomerIDOLD = Sk\_CustomerID} \tag{20}$$
$$_{\wedge IsCurrent = 'True'} DimAccount)$$

$$DimAccount \leftarrow \mathcal{U}(DimAccount)_{EndDate = ActionTS, IsCurrent = False|AccountID=AccountID} \tag{21}$$
$$_{\wedge IsCurrent = 'True' \wedge EffectiveDate \neq ActionTS}(UCTemp15)$$

$$DimAccount \leftarrow \mathcal{U}(DimAccount)_{Sk\_CustomerID = Sk\_CustomerID |AccountID=AccountID} \tag{22}$$
$$_{\wedge IsCurrent = 'True' \wedge EffectiveDate=ActionTS}(UCTemp15)$$

$$UCTemp16 \leftarrow \sigma_{Sk\_AccountID = max(Sk\_AccountID) + rownumber()}(DimAccount) \tag{23}$$

$$DimAccount \leftarrow DimAccount \cup (\pi_{AccountID, SK\_BrokerID, Status,\ldots,EndDate}(\sigma_{EffectiveDate \neq ActionTS} UCTemp16)) \tag{24}$$

Fig. 5. RA expressions to model the historical load for an updated customer

specification directly into PDI; (b) Translating the BPMN4ETL specification into RA, and then implementing the RA operations using Postgres PLSQL.

To optimize the performance of the PDI implementation, the PDI performance tuning tips were applied[2]. The PDI memory limit was increased from 2G to 4G in order to avoid java out of memory exceptions and improve performance. Both tests were run over an Intel i7 computer, with a RAM of 16 GB, running the Windows 10 Enterprise operating system, using the PostgresSQL database as the DW storage. The total execution times of the processes, for different scales factors, are reported in Table 1. For scale factor 3, the benchmark processes 4.5 million records. For scale factors 5 and 10, the benchmark processes 7.8 and 16.1 million records, respectively. It can be noticed that for the historical load, PLSQL implementation is orders of magnitude faster, for all scale factors. Differences are also relevant for incremental loads.

---

[2] https://help.pentaho.com/Documentation/7.1/0P0/100/040/010.

Table 1. Results of implementing TPCDI with relational Algebra (PLSQL) and PDI in Hours.Minutes.Seconds

		Historical	Incremental 1	Incremental 2
SF-3	PLSQL	00:12:50	00:00:09	00:00:07
	PDI	11:23:52	00:01:32	00:01:40
SF-5	PLSQL	00:22:31	00:00:15	00:00:14
	PDI	20:25:32	00:03:03	00:03:11
SF-10	PLSQL	02:11:15	00:00:39	00:00:36
	PDI	25:08:13	00:11:35	00:12:38

These results are partially explained by the poor performance of PDI when it comes to implementing loops, needed for updates of the SCDs with dependencies. Except for the 'NEW' action type, which is loaded in batch form, all other action types are loaded using loops, one row at a time. PDI handles loops with the Copy Rows to Result step, that stores the rows in memory and retrieves them one row at a time. It takes PDI about 15 h out of the total running time for the historical load of scale factor 5, to finish running the DimCustomer and DimAccount dimension tables due to this loop. The same applies to the DimCompany and DimSecurity tables. Also, in spite of the tips applied to improve performance mentioned above, certain steps that cause slow execution could not be avoided. For example Merge Join Steps required input data to be sorted in advance in PDI. This slows the ETL flow since the complete results of the sorted input data had to be obtained prior the execution of the update process.

To conclude, the results reported here suggest that the alternative of implementing ETL SQL-based processes based on a translation from a RA specification is plausible and competitive.

7 Related Work

Several different strategies have been proposed to model ETL processes. The conceptual model proposed in [12] analyzes the structure and data of the data sources and their mapping to a target DW. The conceptual model proposed in [10], uses UML to design ETL processes, where each ETL process is represented by a stereotyped class. This model is refined in [4]. The work in [1,2] proposes a vendor-independent conceptual metamodel for designing ETL processes based on BPMN, which combines two perspectives, a control process view, and a data process view. Using BPMN to specify ETL processes makes this model simple and easy to understand at the cost of expressiveness: it is not possible to visualize the transformation of attributes and attribute constraints at any point in the workflow. Relational algebra was first used to model ETL processes in [7–9]. ETL processes for slowly changing dimensions specifications [7], data quality enforcement tasks [8], and ETL conciliation tasks [9] were modelled with relational algebra and applied to a real-world ETL scenario. Finally, relevant to the work presented here, a research reported in [5] presents a framework for ETL development based on writing software code, instead of specifying the process using commercial tools like PDI, Integration Services, etc., and discusses the advantages of this approach.

8   Conclusion

This paper proposed RA as a language to specify ETL processes at the logical level. To illustrate the proposal, and show the plausibility of the approach in real-world scenarios, the TPC-DI benchmark was used in two ways: one, using RA to translate the BPMN4ETL specification of the benchmark into SQL. The other one implements the benchmark using BPMN4ETL, and translates this directly into the PDI tool. The experiments showed that the SQL implementation runs orders of magnitude faster than that of PDI. This work did not consider structural changes of the data sources, which will be addressed in future research.

Acknowledgments. Alejandro Vaisman was partially supported by PICT-2017 Project 1054 from the Argentinian Scientific Agency.

References

1. El Akkaoui, Z., Zimányi, E.: Defining ETL worfklows using BPMN and BPEL. In: Proceedings DOLAP, pp. 41–48. ACM (2009)
2. El Akkaoui, Z., Zimányi, E., Mazón, J.N., Trujillo, J.: A BPMN-based design and maintenance framework for ETL processes. Int. J. Data Warehouse. Min. (IJDWM) **9**(3), 46–72 (2013)
3. Kimball, R., Caserta, J.: The Data Warehouse ETL Toolkit: Practical Techniques for Extracting, Cleaning, Conforming, and Delivering Data. Wiley, New York (2011)
4. Muñoz, L., Mazón, J.-N., Pardillo, J., Trujillo, J.: Modelling ETL processes of data warehouses with UML activity diagrams. In: Meersman, R., Tari, Z., Herrero, P. (eds.) OTM 2008. LNCS, vol. 5333, pp. 44–53. Springer, Heidelberg (2008). https://doi.org/10.1007/978-3-540-88875-8_21
5. Pedersen, T.B.: Programmatic ETL. Business Intelligence and Big Data: 7th European Summer School, eBISS 2017, Bruxelles, Belgium, July 2–7, 2017, Tutorial Lectures 324, 21 (2018)
6. Poess, M., Rabl, T., Jacobsen, H.A., Caufield, B.: TPC-DI: the first industry benchmark for data integration. Proc. VLDB Endowment **7**(13), 1367–1378 (2014)
7. Santos, V., Belo, O.: Slowly changing dimensions specification a relational algebra approach. Int. J. Inf. Technol. **1**(3), 63–68 (2011)
8. Santos, V., Belo, O.: Modeling ETL data quality enforcement tasks using relational algebra operators. Procedia Technol. **9**, 442–450 (2013)
9. Santos, V., Belo, O.: Modelling ETL conciliation tasks using relational algebra operators. In: Proceedings of the 2014 European Modelling Symposium, pp. 275–280. IEEE, Pisa (2014)

10. Trujillo, J., Luján-Mora, S.: A UML based approach for modeling ETL processes in data warehouses. In: Song, I.-Y., Liddle, S.W., Ling, T.-W., Scheuermann, P. (eds.) ER 2003. LNCS, vol. 2813, pp. 307–320. Springer, Heidelberg (2003). https://doi.org/10.1007/978-3-540-39648-2_25
11. Vaisman, A., Zimányi, E.: Data Warehouse Systems. DSA. Springer, Heidelberg (2014). https://doi.org/10.1007/978-3-642-54655-6
12. Vassiliadis, P., Simitsis, A., Skiadopoulos, S.: Conceptual modeling for ETL processes. In: Proceedings of DOLAP, pp. 14–21. ACM (2002)

Accurate Aggregation Query-Result Estimation and Its Efficient Processing on Distributed Key-Value Store

Kosuke Yuki[1], Atsushi Keyaki[1], Jun Miyazaki[1(✉)], and Masahide Nakamura[2]

[1] School of Computing, Tokyo Institute of Technology, Tokyo, Japan
{zhang,keyaki}@lsc.cs.titech.ac.jp, miyazaki@cs.titech.ac.jp
[2] Kobe University, Hyogo, Japan
masa-n@cs.kobe-u.ac.jp

Abstract. We propose four methods for improving the accuracy of aggregation query-result estimation using histograms and/or kernel density estimation and the efficiency of query processing on a distributed key-value store (D-KVS). Recently, aggregation queries have played a key role in analyzing a large amount of multidimensional data generated from sensors, Internet-of-Things devices, etc. A D-KVS is a platform to manage and process such large-scale multidimensional data. However, querying large-scale multidimensional data on a D-KVS sometimes requires a costly data scan owing to its insufficient support for indexes. Since aggregation-query results do not always need to be accurate, our four methods are not only for estimating accurate query results rather than obtaining accurate results by scanning all data, but also improving query-processing performance. We first propose two kernel density estimation-based methods. To further improve query-result estimation accuracy, we combined each of these two methods with a histogram-based scheme so that we can dynamically select an optimal estimation method based on the relationship between a query and the data distribution. We evaluated the efficiency and accuracy of the proposed methods by comparing them with a current method and showed that the proposed methods perform better.

Keywords: Query-result estimation · Data summarization · Histogram · kernel density estimation · Distributed key-value store

1 Introduction

Due to the spread of the Internet and smartphone, large-scale multidimensional data, called big data have recently been collected and it is becoming increasingly important to analyze and use big data. One of the most useful operations that enable such analysis is an aggregation query; however, there are various challenges in computing aggregation queries for large-scale multidimensional data.

© Springer Nature Switzerland AG 2019
C. Ordonez et al. (Eds.): DaWaK 2019, LNCS 11708, pp. 310–320, 2019.
https://doi.org/10.1007/978-3-030-27520-4_22

A key-value store (KVS) [10] is suitable for managing large-scale data. A KVS is a simplified table-type database in which a tuple consists of two attributes: key and value. Because of its simple structure, it is easy to decentralize data over several servers by horizontal partitioning, which is also called a distributed KVS (D-KVS). However, in many cases, D-KVSs support an index only on a key. Therefore, it is difficult to execute flexible and complex queries because of the costs incurred in carrying out a data scan over a large amount of data. There have been several studies on efficiently processing aggregate operation for large-scale multidimensional data on D-KVSs [3,12,18–20]. Watari et al. [19] proposed a method for mutually using a relational database (RDB) [4] and D-KVS. With their method, the data space is split into several hyper-rectangles, which are called *grids*. A partial aggregation value for each grid is computed and stored in the D-KVS. Given a query, scans of the data in grids that are completely included in the query range are omitted because the aggregation values of such grids have already been computed. This optimization reduces the amount of data to be scanned. However, for grids that partially overlap the query range, it will still execute data scans for these grids, which may result in a large number of data scans in some cases.

To omit these costly data scans, we propose four methods for estimating query results, rather than obtaining accurate results, using a histogram and kernel density estimation (KDE). Histograms [7–9,14,15] are known as a lightweight and less accurate estimation scheme, whereas KDE [17] is more accurate and costly. With these estimation schemes, we construct a histogram based on the data in each grid and/or carry out KDE for each bucket of the histogram. At query processing, the value of the query result is estimated using only a pre-computed histogram and KDE results. We also aim at further improvement in estimation accuracy and query throughput with the advantages of this histogram and KDE.

Besides histograms and KDE, there have been many approaches for approximate query processing, such as wavelets [2], cardinality estimation algorithms [6], deterministic algorithms [5]. However, only a few studies refer to the distributed multidimensional approximate query processing.

We evaluated the efficiency of the proposed methods by comparing their query throughput performance with Watari et al.'s method [19] on a cluster machine and showed that the proposed methods perform better. We also compared the query-result estimation accuracy of each of the proposed methods.

2   Related Work

2.1   Partial Pre-Aggregation

Partial pre-aggregation [13,19] is a method of improving aggregation-processing efficiency by reducing the number of data scans. This method splits the database into several blocks and pre-computes aggregation results for each block. When processing aggregate operations, the pre-computed results are reused as much as possible. Some aggregation queries can be efficiently evaluated with partial pre-aggregation values.

For example, consider an aggregate operation for a relation B to obtain the sum of the *heights* of records satisfying $age < 20$, where B consists of *age* and *height*. Assume that B is split into two blocks B_1 and B_2, each of which has its own partial sum of the *heights*. B_1 and B_2 have only the records that satisfy $age < 15$ and $age \geq 15$, respectively. According to this information, it is not necessary to scan the data in B_1, and only B_2 needs to be scanned.

Watari et al. [19] proposed a method of efficiently processing aggregation queries for large-scale multidimensional data using partial pre-aggregation with a cluster combining an RDB and a D-KVS. With their method, the data space is split into several hyper-rectangles (*grids*) by designating the number of data entries (*grid size*) in each grid. After the split, partial pre-aggregation values for each grid are pre-computed, and the results are stored. At query processing, high efficiency is achieved using the pre-computed partial pre-aggregation values.

Given a query range QR, the aggregation query of the data within QR with their method is processed as follows.

Step 1. Find all grids that intersect a QR by using the metadata of grids in the RDB. Let Gs be a set of the obtained grids. Check if each grid range is completely included in QR.

Step 2. Combine the partial pre-aggregation results of the grids in the Gs that are completely included in the QR. These partial pre-aggregation values can be obtained quickly because they are pre-computed and stored in the D-KVS.

Step 3. Scan all data in the grids in the Gs that partially overlap the QR (hereafter, referred to as *surrounding grids*) and aggregate the values within the QR.

Step 4. Combine the results obtained in Steps 2 and 3.

However, for grids that partially overlap the query range, it will still perform data scans with their method (hereafter, we call this method the *All-Scan* method), which could reduce query throughput.

2.2 Query Processing with Histograms

Poosala et al. proposed a technique called MHIST of dividing the multidimensional data distribution into a histogram [16]. The essential idea of MHIST is to choose the top $p - 1$ attributes $A_1 \ldots A_{p-1}$, which belong to the distribution D, whose marginal distributions in D are the most in need of partitioning, and split D into p buckets along $A_1 \ldots A_{p-1}$. This procedure is repeated until the number of buckets reaches the predetermined upper limit.

The meaning of "the most in need of partitioning" varies depending on which histogram is adopted. From the experiment by Poosala et al., the best estimation accuracy was obtained by MHIST adopting the max-diff histogram [15] with 2 splits (hereafter, this method is abbreviated as *MHIST*). The aim of a max-diff histogram is to avoid putting attribute values with vastly different frequencies into the same bucket.

Muralikrishna et al. [11] proposed a scheme for using histograms to estimate the query results called *Uniform Scheme*. In this scheme, the data distribution in each bucket of a histogram is assumed a uniform distribution. Therefore, the estimated result for a given query Q is calculated as the following expression

$$\sum_{B_i \in Q} F_{B_i} \cdot \frac{Volume(B_i \cap Q)}{Volume(B_i)}$$

where F_{B_i} is the frequency of bucket B_i. The validity of this assumption will enhance as the volume of each bucket becomes smaller, regardless of the actual data distribution.

3   Proposed Methods

3.1   Naïve Methods

In addition to pre-aggregation methods [13,19], we introduce our two naïve data summarization methods using histograms and KDE. We store the summarized result (hereafter, *statistical data*) and estimate the query result by using these statistical data to avoid full data scans at query processing. The processes of both data summarization as preprocessing and query-result estimation using statistical data are described as follows.

Step1. Divide the data space into several grids using the All-Scan method and store the aggregated values for each grid. This step in Fig. 1 shows that the entire data space is divided into six grids.

Step2. Construct a histogram for each grid using MHIST and save these results. This step in Fig. 1 means a histogram was constructed based on the data that grid 000 contains.

Step3. In case of use KDE, which is used based on the data in each bucket of the histogram constructed at Step 2 and save the results. This step in Fig. 1 is for the result of KDE based on the data in one of the buckets of the histogram from grid 000.

Given a query Q, we use pre-aggregated values for the grids that are completely included in the query range. This part follows the All-Scan method. On the other hand, for the surrounding grids, we use statistical data obtained from MHIST and KDE to estimate aggregation results (Fig. 2).

We now discuss our two naïve methods for estimating aggregation results using statistical data and compare them with a previous method.

1. Previous method: MHIST
   First, we consider a previous study that only used MHIST and Uniform Scheme in Sect. 2.2 for estimation. We call this method *MHIST*.

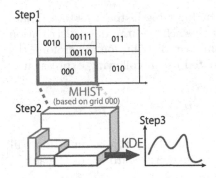

Fig. 1. Process of data summarization   **Fig. 2.** Process of query-result estimation

2. Proposed naïve method 1: All-KDE

We considered obtaining more detail summarized data by using KDE to achieve more accurate estimation than MHIST since KDE can be used to understand the data distribution in more detail. We call this method *All-KDE*.

After histograms are constructed using MHIST, KDE is carried out for data distribution in each bucket and store several points from estimated distribution. Therefore, $Sum(B_i, Q)$ is calculated as the following expression, where $K_{ij} \in Q$ means that the j-th estimated point in B_i is included in Q.

$$Sum(B_i, Q) = \sum_{K_{ij} \in Q} K_{ij}$$

3. Proposed naïve method 2: Part-KDE

Though All-KDE can be used to understand the distribution in more detail, the amount of the data obtained with KDE becomes large and this could cause a decrease in query throughput. As a result of the histogram constructed with MHIST, in some cases, there are several buckets that do not have a domain range for a certain dimension (the minimum and maximum values of the attribute's domain are the same). We refer to such a bucket as a *dimension diminished bucket*.

We assume that the data distribution in the bucket can be regarded as a uniform distribution for a dimension diminished bucket; thus, KDE is not carried out. For normal buckets, KDE is carried out in the same manner as with All-KDE. Given a Q, if the dimension diminished bucket satisfies Q, the estimated value is calculated using Uniform Scheme in the same manner as with MHIST, and for normal buckets that satisfy Q, precomputed estimated values obtained with KDE are used.

Therefore, $Sum(B_i, Q)$ is expressed as follows.

$$Sum(B_i, Q) = \begin{cases} F_{B_i} \cdot \frac{Volume(B_i \cap Q)}{Volume(B_i)} & \text{(if } B_i \text{ is a dimension diminished bucket)} \\ \sum_{K_{ij} \in Q} K_{ij} & \text{(otherwise)} \end{cases}$$

By modifying All-KDE to carry out KDE only for normal buckets, the data size of the statistical data reduces. Therefore, it is possible to achieve higher query throughput than with All-KDE and prevent deterioration in estimation accuracy. We call this method *Part-KDE*.

By combining histogram and KDE, it is expected that All-KDE and Part-KDE can improve the estimation accuracy.

3.2   Hybrid Methods

In this section, we focus on the relationship between the overlap rate OR_{ij} of query Q_i to a surrounding grid G_{ij} and the estimation error in each surrounding grid. We aim to further improve estimation accuracy by dynamically selecting the estimation method based on OR_{ij}.

Before explaining our further methods, we introduce the overlap rate and the error index. The overlap rate OR_{ij} between Q_i and one of the surrounding grids of Q_i, G_{ij}, is defined as follows.

$$OR_{ij} = \frac{Volume(G_{ij} \cap Q_i)}{Volume(G_{ij})}$$

On the other hand, the error index EI_{ij} for grid G_{ij} is defined as

$$EI_{ij} = \frac{Error\_of\_G_{ij}}{TrueValue\_of\_Q_i}$$

This error index means how much the error of the individual grid affects that of the entire query result compared with its true value. The error index of a grid is higher as its error to the entire query result becomes greater.

We introduce our two hybrid methods which dynamically select an estimation method applied to each grid to improve estimation accuracy and query throughput. This idea is based on the assumption that the best estimation method changes according to the overlap rate.

By combining MHIST and All-KDE (hereafter, *MA*), it is expected that both estimation accuracy and query throughput would improve because MHIST is the lightest among the naïve estimation methods. We also consider combining MHIST and Part-KDE (hereafter, *MP*) to enhance throughput, because Part-KDE has less overhead than All-KDE.

4   Experimental Evaluations

The experimental evaluations were conducted on a cluster with 13 PCs running HBase 1.2.0 (D-KVS) and PostgreSQL 9.6.1 (RDB) with the following two datasets.

Extended Indoor Sensor Data: Based on 2 million real data entries collected
from indoor environmental sensors, each of which data entry has 16 attributes,
we generated pseudo dataset by replicating the data, resulting in 100 million
data entries. We call such data *extended indoor sensor data (EIS Data)*. Four-
dimensional range queries are carried out for EIS data.

San Francisco Bay Area Data: We generated 22 million points of moving
objects in the San Francisco Bay Area using a network-based generator [1].
Each data entry has two attributes, latitude and longitude. In addition, we
assumed each data entry as a running car and gave it car speed as the third
attribute. We call such data *San Francisco bay area data (SFB Data)*. Two-
dimensional range queries are carried out for SFB data.

Some variables were configured for the experiments as follows.

- Maximum number of data entries in a grid (grid size): 1000
- Upper limit of the number of buckets with MHIST: 25
- Kernel function for KDE: Gaussian

4.1 Preliminary Experiment

We measured OR_{ij} and EI_{ij} for the datasets and randomly generated aggregation range queries, and plotted it. For these plots, we drew trendlines to clarify
the relationship between the overlap rate and the error index for each estimation
method.

The results are shown in Figs. 3 and 4. From these results, we observed that
there are transition points at which the error index was inverted. For range-sum
query, the transition point for MHIST and All-KDE was 1.8% of the overlap rate,
and 3.7% for MHIST and Part-KDE. For range-count, that for MHIST and All-
KDE was 49.4%, and 85.6% for MHIST and Part-KDE. When the overlap rate is
less than these values, the error index of MHIST is the best. In other words, it is
possible to optimize the estimation accuracy and query throughput by switching
estimation methods based on the overlap rate. Although not shown here, SFB
Data case also has the similar transition points.

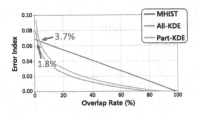
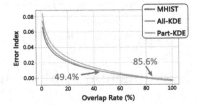

Fig. 3. EIS Data (Sum)   **Fig. 4.** EIS Data (Count)

4.2   Evaluation of Estimation Accuracy

We compared the estimation accuracy of the hybrid methods using the transition points obtained by the preliminary experiment to that of the naïve methods. We conducted randomly generated multidimensional range-sum and range-count queries, so that their selectivity can be set as 0.001% and 10%. We measured the error rate of each query for both the entire range and only the surrounding grids.

Figures 5 and 6 show the average error rates of each method when changing each selectivity for EIS data. Figure 5 indicates that MA achieved higher estimation accuracy than the naïve methods at all selectivities. The accuracy of the surrounding grids at 0.001% selectivity markedly improved because there tend to be many grids with low OR_{ij}. However, Fig. 6 shows that MA did not always lead to the best estimation. We concluded that this result was caused by the high/low relationship of the trendlines of the count queries (Fig. 4). The high/low relationship of the trendlines of the count queries was not as clear as the trendlines of the sum queries (Fig. 3). Therefore, MHIST could be inferior to the other two naïve methods even at OR_{ij} of less than 49.4%. Though MA is not always the best, its accuracy degradation is insignificant from the viewpoint of the error rate of the entire query result. On the other hand, MP was inferior to MA in many cases, but its accuracy did not deteriorate much compared with that of MA.

Also, we observed the same tendency for SFB data (Figs. 7 and 8). For both of the sum and count queries at all selectivities, MA achieved the best estimation accuracy. As for MP, we confirmed the improvement of estimation accuracy compared with Part-KDE.

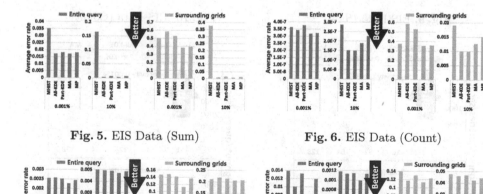

Fig. 5. EIS Data (Sum)   **Fig. 6.** EIS Data (Count)

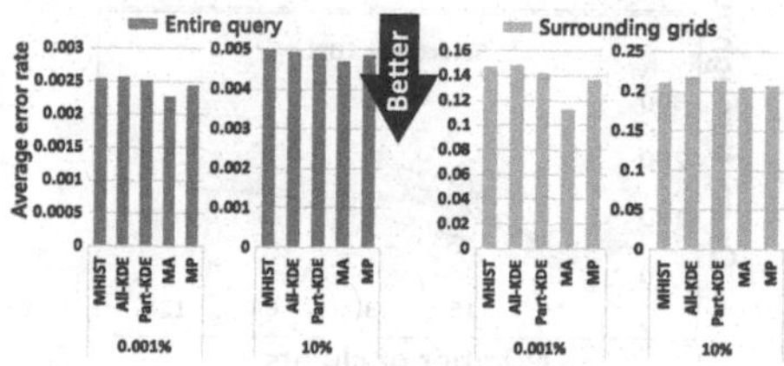

Fig. 7. SFB Data (Sum) 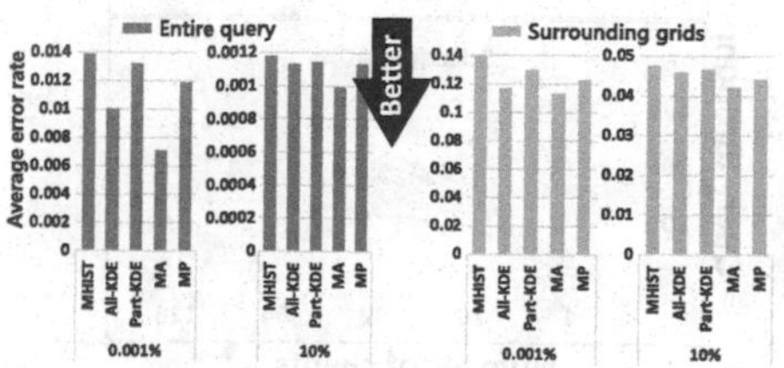   **Fig. 8.** SFB Data (Count)

4.3 Evaluation of Query Throughput

We also evaluated the query throughput performance for the four proposed methods and others (All-Scan and MHIST). We ran the same queries under the same
conditions as mentioned in Sect. 4.2. The queries were issued from 1, 16, 32, 64,
and 128 clients simultaneously while varying selectivity.

Figures 9, 10, 11, 12 depict the results of query throughput performance for
each dataset and query. As an overall view, these results indicate that the hybrid
methods achieved higher throughput than All-KDE and Part-KDE. In addition,
Figs. 11 and 12 show that MA can achieve higher throughputs than Part-KDE
when the number of clients increased, and MP can achieve high throughputs,

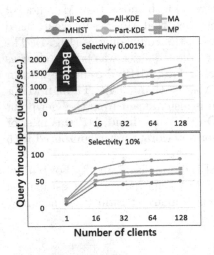

Fig. 9. EIS Data (Sum)

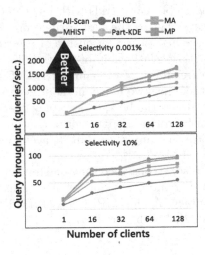

Fig. 10. EIS Data (Count)

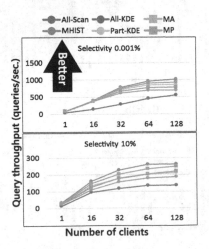

Fig. 11. SFB Data (Sum)

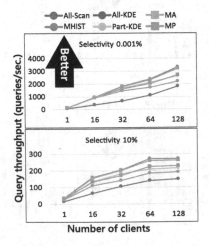

Fig. 12. SFB Data (Count)

which is comparable to MHIST. This is because MHIST is adapted at almost half or higher OR_{ij}.

In summary, MA can achieve highly accurate and efficient estimation. In contrast, MP can provide much higher query throughput than MA with small accuracy deterioration.

5   Conclusion

We proposed four methods for highly efficient and accurate estimation of aggregation queries on large-scale multidimensional data on a cluster of an RDB and a D-KVS. In particular, MA and MP dynamically select estimation methods among MHIST, All-KDE, and Part-KDE to adapt the most accurate estimation method based on the overlap rate. MA and MP can also achieve higher query throughput through the partial use of MHIST.

For future work, we will consider a method for automatically finding transition points to switch estimation methods for arbitrary data.

Acknowledgments. This work was partly supported by JSPS KAKENHI Grant Numbers 18H03242, 18H03342, and 19H01138.

References

1. Brinkhoff, T.: A framework for generating network-based moving objects. GeoInform. **6**(2), 153–180 (2002)
2. Chakrabarti, K., Garofalakis, M.N., Rastogi, R., Shim, K.: Approximate query processing using wavelets. Proc. VLDB **2000**, 111–122 (2000)
3. Eldawy, A., Mokbel, M.F.: Spatialhadoop: a mapreduce framework for spatial data. Proc. of IEEE ICDE **2015**, 1352–1363 (2015)
4. Garcia-Molina, H., Ullman, J.D., Widom, J.: Database Systems: The Complete Book. Prentice Hall, New Jersey (2002)
5. Han, X., Wang, B., Li, J., Gao, H.: Efficiently processing deterministic approximate aggregation query on massive data. Knowl. Inf. Syst. **57**(2), 437–473 (2018)
6. Heule, S., Nunkesser, M., Hall, A.: Hyperloglog in practice: algorithmic engineering of a state of the art cardinality estimation algorithm. In: Proceedings of EDBT 2013. pp. 683–692. ACM (2013)
7. Ioannidis, Y.: The history of histograms (abridged). Proc. VLDB **2003**, 19–30 (2003)
8. Jagadish, H.V., Koudas, N., Muthukrishnan, S., Poosala, V., Sevcik, K.C., Suel, T.: Optimal histograms with quality guarantees. In: Proceedings of VLDB 1998, pp. 275–286 (1998)
9. Kooi, R.P.: The Optimization of Queries in Relational Databases. Ph.D. thesis (1980)
10. Lakshman, A., Malik, P.: Cassandra: a decentralized structured storage system. ACM SIGOPS Operating Syst. Rev. **44**(2), 35–40 (2010)
11. Muralikrishna, M., DeWitt, D.J.: Equi-depth multidimensional histograms. In: Proceedings of ACM SIGMOD 1988. pp. 28–36 (1988)

12. Nishimura, S., Agrawal, S.D.D., Abbadi, A.E.: MD-HBase: design and implementation of an elastic data infrastructure for cloud-scale location services. Distrib. Parallel Databases **31**(2), 289–319 (2013)
13. Papadias, D., Kalnis, P., Zhang, J., Tao, Y.: Efficient OLAP operations in spatial data warehouses. In: Jensen, C.S., Schneider, M., Seeger, B., Tsotras, V.J. (eds.) SSTD 2001. LNCS, vol. 2121, pp. 443–459. Springer, Heidelberg (2001). https://doi.org/10.1007/3-540-47724-1_23
14. Piatetsky-Shapiro, G., Connell, C.: Accurate estimation of the number of tuples satisfying a condition. In: Proceedings of ACM SIGMOD 1984, pp. 256–276 (1984)
15. Poosala, V., Haas, P.J., Ioannidis, Y.E., Shekita, E.J.: Improved histograms for selectivity estimation of range predicates. In: Proceedings of ACM SIGMOD 1996, pp. 294–305 (1996)
16. Poosala, V., Ioannidis, Y.E.: Selectivity estimation without the attribute value independence assumption. In: Proceedings of VLDB 1997, pp. 486–495 (1997)
17. Silverman, B.W.: Density Estimation for Statistics and Data Analysis. No. 26 in Monographs on Statistics and Applied Probability. CRC Press (1986)
18. Wang, J., Wu, S., Gao, H., Li, J., Ooi, B.C.: Indexing multi-dimensional data in a cloud system. Proc. ACM SIGMOD **2010**, 591–602 (2010)
19. Watari, Y., Keyaki, A., Miyazaki, J., Nakamura, M.: Efficient Aggregation Query Processing for Large-Scale Multidimensional Data by Combining RDB and KVS. In: Hartmann, S., Ma, H., Hameurlain, A., Pernul, G., Wagner, R.R. (eds.) DEXA 2018. LNCS, vol. 11029, pp. 134–149. Springer, Cham (2018). https://doi.org/10.1007/978-3-319-98809-2_9
20. Zhang, X., Ai, J., Wang, Z., Lu, J., Meng, X.: An efficient multi-dimensional index for cloud data management. In: Proceedings of CloudDB 2009, pp. 17–24. ACM (2009)

Author Index

Printed in the United States
By Bookmasters